# ANATOMY AND PHYSIOLOGY

## The Easy Way

I. Edward Alcamo, Ph.D.
Professor of Biology
State University of New York
Farmingdale, New York

BARRON'S

I am pleased to dedicate this book to my colleagues and friends at the State University of New York at Farmingdale.

*All inquiries should be addressed to:*
Barron's Educational Series, Inc.
250 Wireless Boulevard
Hauppauge, New York 11788

Library of Congress Catalog Card No. 95-45605

International Standard Book No. 0-8120-9134-5

**Library of Congress Cataloging-in-Publication Data**
Alcamo, I. Edward.
  Anatomy & physiology the easy way / I. Edward Alcamo.
    p.   cm. — (Easy way)
  Includes index.
  ISBN 0-8120-9134-5
  1. Human anatomy—Outlines, syllabi, etc. 2. Human anatomy—
Examinations, questions, etc. 3. Human physiology—Outlines, syllabi, etc.
4. Human physiology—Examinations, questions, etc. I. Title.  II. Series.
QM31.A43   1996
612'.0076—dc20
                                                              95-45605
                                                              CIP

PRINTED IN THE UNITED STATES OF AMERICA
19 18 17 16

# CONTENTS

# PREFACE

Learning anatomy and physiology can be difficult. There are new terms to memorize, new processes to grasp, and new structures to study.

We've tried to make your job easier by presenting a no-frills approach to anatomy and physiology. *Anatomy and Physiology the Easy Way* contains the foundation material for the standard anatomy and physiology course taken in the health science curriculum. Students preparing for careers in nursing, physical therapy, dental hygiene, medical technology, pharmacy, or other allied health sciences will find the book useful. Premedical and medical students will find it a valuable adjunct to their professional courses. And advanced high school students of biology will benefit from the book's direct approach to the human body.

The content material of *Anatomy and Physiology the Easy Way* conforms to the standard textbooks and is presented in clear, concise detail that can be read easily and understood immediately. The paragraphs are brief and the essential vocabulary has been included to make you confident in using the concepts of anatomy and physiology. We have avoided technical language wherever possible, and we have minimized the jargon in favor of direct, no-nonsense science. A complete glossary is included for clarification and review purposes. And to help you assess your progress, each chapter contains 100 questions of the short-answer type most often asked on examinations.

This book can also be used to prepare for cumulative, comprehensive, and board exams since it contains the basics of anatomy and physiology in a highly readable form. We've planned the chapter contents carefully to help you feel comfortable in the sea of new terms you will be encountering. And please write your notes all over the pages—this book is meant to be a welcome friend where you can express yourself freely.

I would like to take a moment to thank the following individuals who contributed their talents to this book: Grace Freedson supervised the project; Anna Damaskos oversaw the production process with a deft and experienced hand; Maryellen Lo Bosco copyedited the manuscript to ensure that every *i* was dotted and every *t* was crossed; Samuel Collins and his staff rendered the illustrations accurately; and Barbara Dunleavy typed the original manuscript and lent her special publishing expertise to the project.

I also want to say thanks to the lights of my life, my children Michael, Elizabeth, and Patricia. Their love and support have encouraged me to spend the endless hours of research and preparation required for a book such as this.

Now it's your turn. I hope you enjoy learning from this book, and I will be very pleased if you achieve an A in anatomy and physiology. Please write and let me know how well the book works for you and how we can improve it. Your comments will be extremely valuable in the preparation of any future editions.

I can be reached at the Biology Department, State University of New York, Farmingdale, NY 11735. If you wish to call, I'm at (516) 420-2423. And if you subscribe to e-mail, you can try me at alcamoie@snyfarva.cc.farmingdale.edu.

In closing, I would like to wish you well in your studies. Should things become overwhelming, take a deep breath and remember this little saying I heard some years ago: "Life is hard, yard by yard. Inch by inch, it's a cinch."

E. Alcamo

# CHAPTER 1

# INTRODUCTION TO ANATOMY AND PHYSIOLOGY

The study of anatomy and physiology is essential to understanding the human body. These disciplines concern the body's structural framework and how it works. What the body is able to do depends intimately on how it is constructed, and the body's construction gives a strong indication to what it does. For example, the lungs are composed of millions of air sacs with extremely thin walls. This construction permits them to serve as a site for exchanging oxygen and carbon dioxide gases.

The topic of anatomy has many subdivisions. For example, **gross anatomy** concerns body structures examined by observation without the use of a microscope. **Histologic anatomy** is the study of cells, tissues, and organs as they are observed with a microscope. **Developmental anatomy** deals with the development of the individual from the fertilized egg to the adult form.

Within the science of physiology, there are also many subdivisions. For instance, **cytology** is the study of cells and how they function; **neurophysiology** is the study of nerve function; and **renal physiology** deals with the excretory system and its activities. **Reproductive physiology** is the study of reproductive organs and the methods for reproduction.

## LEVELS OF STRUCTURE

The human body has several levels of structural organization. At the simplest level, the human body is composed of **atoms**, the ultramicroscopic building blocks of matter. Atoms are typified by oxygen, carbon, nitrogen, and sodium. Atoms combine with one another to form **molecules**. Molecules are typified by water, sodium chloride, proteins, carbohydrates, and lipids (Figure 1.1).

The association of molecules with one another yields the next level of organization, the **cell**. The cell is the fundamental unit of living things. It has levels of subcellular structures such as the nucleus, mitochondria, ribosomes, and lysosomes (Chapter 3). Among the different cells in the body are nerve cells, muscle cells, and blood cells.

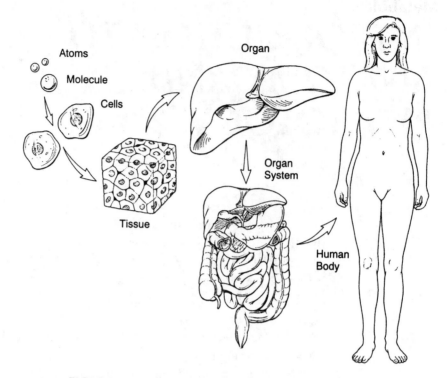

**FIGURE 1.1**  *The levels of structure in the human body.*

The next level of structural organization, the **tissue**, is a group of cells of similar structure performing the same function (Chapter 4). The body has four basic types of tissues: epithelial tissue (such as in the organ linings), connective tissue (such as blood and bone), muscle tissue, and nervous tissue. Each tissue type plays a unique role in the body.

The next level of organization is the **organ**. An organ is composed of two or more different kinds of tissues. The stomach, for example, is an organ composed of epithelial tissue, muscle tissue, nerve tissue, and connective tissue. An organ functions as a specialized physiological center for a particular activity.

The final level of structural organization is the **organ system**, composed of several organs with related functions. Organ systems in the body include the digestive, respiratory, nervous, and circulatory systems (Table 1.1). The systems operating together form the **organism**, which is the highest level of organization.

## CHARACTERISTICS OF THE LIVING BODY

The human body, like other living things, has certain characteristics distinguishing it from nonliving things. These characteristics enable the cells of the body to carry on the activities necessary for growth and survival.

# Metabolism

An important characteristic of living things is **metabolism**, which is the sum total of all chemical processes occurring in the body. Metabolism is divided into the subcategories of catabolism and anabolism. **Catabolism** is the breakdown of organic matter, usually with the release of energy. **Anabolism** is the buildup of organic matter, usually requiring an input of energy. Such life processes as digestion, respiration, circulation, and excretion are adapted for supplying the building blocks for metabolism and removing the waste products of metabolism.

**TABLE 1.1** *Major Organ Systems in Humans*

| Organ System | Physiological Role | Components |
|---|---|---|
| Integumentary | Covers the body and protects it | Skin, hair, nails, and sweat glands |
| Skeletal | Protects the body and provides support for locomotion and movement | Bones, cartilage, and ligaments |
| Nervous | Receives stimuli, integrates information, and directs the body | Brain, spinal cord, nerves, and sense organs |
| Endocrine | Coordinates and integrates the activities of the body | Pituitary, adrenal, thyroid, and other ductless glands |
| Muscular | Produces body movement | Skeletal muscle, smooth muscle, and cardiac muscle |
| Digestive | Absorbs soluble nutrients from ingested food | Teeth, salivary glands, esophagus, stomach, intestines, liver, and pancreas |
| Respiratory | Collects oxygen and exchanges it for carbon dioxide | Lungs, pharynx, trachea, and other air passageways |
| Circulatory | Transports cells and materials throughout the body | Heart, blood vessels, blood, and lymph structures |
| Immune | Removes foreign chemicals and microorganisms from the bloodstream | T-lymphocytes, B-lymphocytes, and macrophages; lymph structures |
| Urinary | Removes metabolic wastes from the bloodstream | Kidney, bladder, and associated ducts |
| Reproductive | Produces sex cells for the next generation of organism | Testes, ovaries, and associated reproductive structures |

# Movement and Other Characteristics

Another important characteristic is **movement**, the result of contracting muscle cells. Movement can be voluntary, such as occurs in the muscles of the skeleton, or it may be involuntary, such as occurs in the lining of internal organs. The bones of the skeletal system assist movement by providing attachment sites for the muscles.

Another characteristic, **growth**, refers to an increase in the size of body cells or the body itself. Growth is the process in which an organism obtains materials from the environment and forms more of itself. An increase in the number of cells, the size of existing cells, or the substance surrounding the cells constitutes growth.

A fourth characteristic is **conductivity**. Conductivity refers to the ability of cells to receive stimuli and carry them from one body part to another. This characteristic is associated with nerve cells and muscle cells.

Still another important characteristic of the living body is **reproduction**, the ability of the body to replicate itself. Reproduction can refer to the formation of new cells for growth, repair, or replacement, or the production of an entirely new individual. Reproduction in humans involves the production of sperm and egg cells and their union to form a fertilized egg cell, which develops into a new individual. This form of reproduction is known as **sexual reproduction**. It compares to **asexual reproduction**—the duplication of a single cell, which produces two identical daughter cells.

Other characteristics of living things include **irritability**, the response of the body to an internal or external stimulus; and **excretion,** the process of removing waste products from the body.

# Homeostasis

**Homeostasis** refers to the steady-state equilibrium existing in the body and the maintenance of this state. It is associated with the relative constancy of the chemical and physical environment in the cells and in the organism itself. Such things as water, nutrients, and oxygen are part of the chemical requirements to maintain homeostasis; a constant temperature and atmospheric pressure are part of the physical requirements for homeostasis.

The body is said to be in homeostasis when the needs of its cells are met and its activities are occurring smoothly. All organ systems play a role in homeostasis, and the composition of fluids within the body is maintained precisely at all times. Stress, such as heat, pain, or lack of oxygen, creates an imbalance in the internal environment and disturbs the homeostasis of the body.

Because internal conditions vary constantly, the body is protected against extremes by self-regulating systems known as **feedback systems**. With feedback systems the body sends information back into the system to induce a response. The **setpoint** of a feedback system is the normal value of a variable, such as temperature. A sensor or **receptor** detects any deviation from the setpoint, and a **control center** receives information from various receptors to integrate and determine the response to return to the setpoint. **Effectors** then implement the response to return the body to homeostasis.

A feedback system is a **negative feedback system** when the information decreases the system's output to bring the system back to its setpoint. For example, the level of glucose rises in the body after a meal, and the glucose stimulates the release of insulin from the pancreas. Insulin facilitates the passage of glucose into cells and reduces the glucose level. The lowered glucose level then influences insulin-secreting cells to decrease their output of insulin to maintain homeostasis.

Homeostasis may also be obtained by a **positive feedback system**. In this situation, the information returned to the system increases the deviation from the setpoint. For example, stimulating a nerve cell membrane causes more sodium ions to flow across the membrane into the cell, and the sodium flow increases the membrane's passageways to encourage still more sodium ions to flow inward. The result is a nerve impulse.

## DIRECTIONAL TERMS

Directional terms are used in relation to one another to denote where body parts are located. The point of reference for the body is the **anatomical position**. In this position, the body is erect with eyes forward, feet together, arms at the sides, and palms up with the thumbs pointing away from the body (see Figure 1.2).

In the anatomical position, the **anterior** aspect of the body is toward the front of the body on the belly side. Anterior is often used interchangeably with the term **ventral** (even though ventral refers to the belly side of a four-legged animal such as a dog). The **posterior** aspect refers to the back side of the human. This term is often used interchangeably with the term **dorsal**.

In the human body, the term **superior** refers to the aspect toward the head or upper part of the body. For example, the nose is superior to the mouth. The terms **cephalic** and **cranial** are sometimes used instead of superior. The **inferior** aspect of the body refers to a direction away from the head or toward the lower part of the body. The term **caudal** is an alternative expression. In this context, the abdomen is inferior to the thorax.

The term **medial** refers to a direction closer to the midline of the body or to one of its structures—the nose is medial to the eyes. The term **lateral** refers to a location off to the side and away from the midline—the eyes are lateral to the nose. The terms **ipsilateral** and

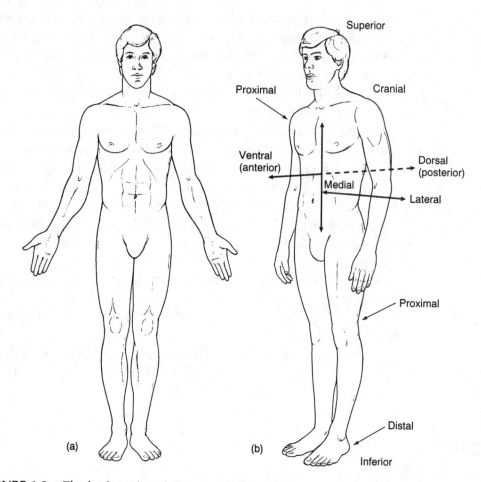

**FIGURE 1.2** *The body and its directions. (a) The body in the anatomical position standing with the feet together, arms at side, and palms up (b) Several directional terms associated with anatomical structures of the body.*

**contralateral** refer to structures on the same side of the body or opposite sides of the body, respectively (Table 1.2). For instance, the ascending colon and gall bladder are ipsilateral, while the ascending colon and descending colon are contralateral.

In the human body the term **proximal** refers to a direction closer to the attachment point of an extremity to the trunk; thus, the femur is proximal to the body trunk as compared to the ankle. In comparison, **distal** refers to a region farther from the attachment of a limb to the trunk. The ankle is distal to body trunk relative to the femur. The terms **superficial** and **deep** refer to a location closer to the body surface or removed from it; the skin is superficial to the muscles; the heart lies deep to the muscles.

## Planes

The structural plan of the human body has various imaginary flat surfaces called **planes**. Planes pass through the body and provide reference points for the organs of the body. A **sagittal plane**, for example, is a vertical plane dividing the body into right and left sides. Such a plane may be **midsagittal** if it divides the body into equal right and left halves, or **parasagittal** if it divides the body into unequal right and left halves.

**TABLE 1.2**  *A Summary of Directional Terms*

| Term | Definition | Example |
|---|---|---|
| Anterior (ventral) | Nearer to or at the front of the body | Sternum is anterior to the heart |
| Posterior (dorsal) | Nearer to or at the back of the body | Esophagus is posterior to the trachea |
| Superior (cephalic or cranial) | Toward the head or the upper part of a structure; generally refers to structures in the trunk | Heart is superior to the liver |
| Inferior (caudal) | Away from the head or toward the lower part of the structure; generally refers to structures in the trunk | Stomach is inferior to the lungs |
| Medial | Nearer to the midline of the body or a structure | Ulna is on the medial side of the forearm |
| Lateral | Away from the midline of the body | Lungs are lateral to the heart |
| Ipsilateral | On the same side of the body | Gall bladder and ascending colon of the large intestine are ipsilateral |
| Contralateral | On the opposite side of the body | Ascending and descending colons of the large intestine are contralateral |
| Proximal | Nearer to the attachment of an extremity to the trunk or structure | Femur is proximal to the tibia |
| Distal | Farther from the attachment of an extremity to the trunk of a structure | Phalanges are distal to the carpals (wrist bones) |
| Superficial | Toward the surface of the body | Muscles of the thoracic wall are superior to the viscera in the thoracic cavity |
| Deep | Away from the surface of the body | Ribs are deep to the skin of the chest |

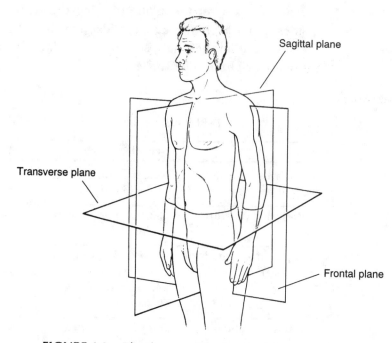

**FIGURE 1.3**  *The three major planes of the human body.*

The second important plane is the **frontal**, or **coronal, plane**. Like the sagittal plane, the frontal plane runs longitudinally, but it divides the body into anterior and posterior portions. The frontal plane lies at a right angle to a sagittal plane (Figure 1.3).

The third important plane is the **transverse plane**, also known as the **horizontal plane**. This plane divides the body into superior and inferior portions. Organs sectioned across the transverse plane for study are referred to as **cross-sections**.

## Body Cavities and Regions

The body cavities are areas within the body containing its internal organs. The two principal cavities are the **dorsal body cavity** and the **ventral body cavity**. The dorsal body cavity is located along the posterior (dorsal) surface of the body, where it is subdivided into the **cranial cavity** housing the brain and the **spinal cavity**, formed by the vertebral column and housing the spinal cord.

The second body cavity is the **ventral body cavity**, located on the anterior (ventral) aspect of the body. Its two major subdivisions are the **thoracic cavity** and the **abdominopelvic cavity** (Figure 1.4). The thoracic cavity is surrounded by ribs and muscles of the chest and is further subdivided into the two **pleural cavities**, each having a lung. In addition, there is a third cavity called the **pericardial cavity**, medial to the pleural cavities.

The pericardial cavity houses the heart and is located in a region called the **mediastinum**. The mediastinum includes all the contents

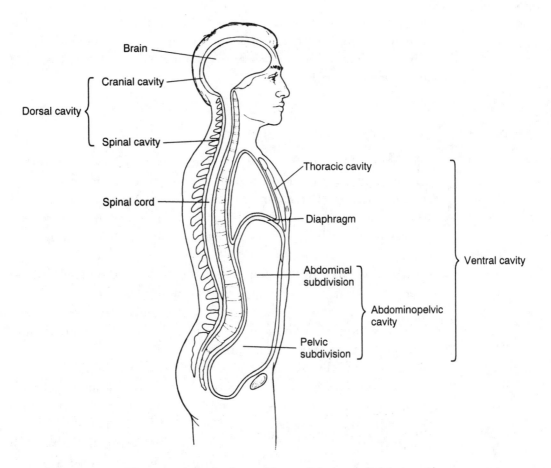

**FIGURE 1.4**   *The two major body cavities and their subdivisions and components.*

of the thoracic cavity except the lungs. Within the mediastinum are the heart, thymus, esophagus, trachea, bronchi, and many blood and lymphatic vessels. The pericardial cavity is actually a small space between the visceral pericardium and parietal pericardium, which are the membranes covering the heart.

The abdominopelvic cavity is separated from the thoracic cavity by the large, dome-shaped diaphragm muscle. This cavity is also known as the **peritoneal cavity**. It contains the visceral organs of the abdomen and pelvis. Located within the **abdominal subdivision** are the stomach, intestines, spleen, liver, and other organs. The inferior portion, called the **pelvic subdivision**, contains the bladder, certain female reproductive organs, and rectum.

Additional divisions of the abdominopelvic cavity yield nine designations for various regions. The **umbilical region** is at the center of the abdomen, while the **epigastric region** is immediately superior to the umbilical region, and the **hypogastric region** is immediately inferior. Lateral to the epigastric region are the right and left **hypochondriac regions**, and lateral to the umbilical region are the right and left **lumbar regions**. Lateral to the hypogastric region are the right and left **inguinal (iliac) regions**. In addition, four abdominopelvic designations are used clinically from imaginary horizontal and vertical lines intersecting at the center. These regions are the **right and left upper quadrants** and the **right and left lower quadrants** (Figure 1.5).

## Membranes

The walls of the ventral body cavity and its organs are covered with a thin, double-layered membrane called the **serous membrane**, so-named because it contain a lubricating fluid called **serous fluid** secreted by both membranes. The fluid permits organs to slide easily across cavity walls and pass by one another without causing friction. Serous membranes lie very close to one another.

The body has three major serous membranes: the **pleura**, which lines the pleural cavities; the **pericardium**, which lines the heart; and the **peritoneum**, which surrounds some organs and covers others. The covering is peritoneal if the organs are surrounded, and retroperitoneal, if the organs are covered.

Each of the three serous membranes has a **parietal portion** and a **visceral portion**. The parietal portion lines a cavity, while the visceral portion covers an organ. For example, the parietal peritoneum lines the abdominal pelvic cavity, while the visceral peritoneum covers the different organs within this cavity. The area between the parietal peritoneum and the visceral peritoneum is the peritoneal cavity. The pleural cavity is the space between the parietal pleura and the visceral pleura.

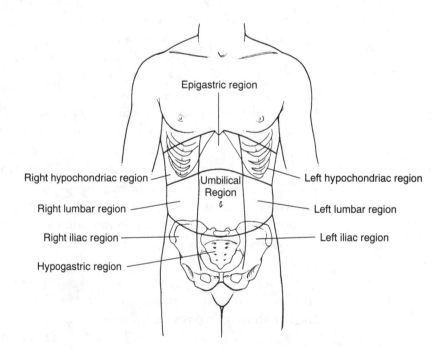

**FIGURE 1.5** *The important anatomical regions of the abdominopelvic cavity.*

Other small cavities of the body are found in the head. They include the oral cavity, the nasal cavity, the middle ear cavities, and the orbital cavities of the eyes.

# REVIEW QUESTIONS

**PART A—Completion: Add the word or words that correctly complete each of the following statements.**

1. The study of the body structures without the use of a microscope is known as _____ .

2. For histologic anatomy, it is important that one employ a _____ .

3. One of the branches of physiology is cytology, which is the study of _____ .

4. The function of the excretory system is a major topic of a branch of physiology known as _____ .

5. At its most simple level of structure the body is composed of _____ .

6. Sodium chloride, proteins, lipids, and water typify the level of structure of the body in which the main components are _____ .

7. The fundamental unit of all living things, including the human body is _____ .

8. A group of cells working together to perform the same function represents a _____ .

9. A type of tissue represented by the blood and bone is _____ .

10. The organs of the body are lined with a type of tissue known as _____ .

11. Various types of tissue work together in the human body to compose a _____ .

12. The sum total of all chemical processes occurring in the body is _____ .

13. When organic matter is built up from smaller molecules, and the process usually requires an input of energy, the overall process is called _____ .

14. The metabolic process in which organic matter is broken down, usually with the release of energy is _____ .

15. Two general types of movement in the body are voluntary and _____ .

16. To assist the process of movement in the body, the muscles are usually attached to _____ .

17. The body obtains materials from the environment and forms more of itself in the process of _____ .

18. The characteristic of conductivity is associated in the body with muscle cells and _____ .

19. In addition to producing an entirely new individual, new cells are formed in the body for the three purposes of replacement, growth, and _____ .

20. The form of reproduction in which a fertilized egg cell forms is called _____ .

21. In the duplication of a single cell to form two identical daughter cells, the reproduction is known as _____ .

22. The steady state equilibrium existing in the body is known as _____ .

23. The cells of an organism and the organism itself remains relatively constant in its chemical environment and in its _____ .

24. Part of the chemical requirements of the human body to maintain homeostasis include water, nutrients, and _____ .

25. Among the systems that contribute to the maintenance of homeostasis are the nervous system and the _____ .

26. An imbalance in the internal environment and disturbance of homeostasis are both created by _____ .

27. A system in which information decreases the system's output and brings the system back to its setpoint is a _____ .

28. The system in which information is returned in order to increase the deviation from the original setpoint is a _____ .

29. The body is erect with eyes forward, feet together, arms at the side, and palms up in the _____ .

30. In speaking of a direction toward the front of the body on the belly side, one uses the term _____ .

31. Although the term dorsal is sometimes used, the preferred term when referring to the back side of a human is _____ .

32. In anatomical nomenclature, the term superior refers to an aspect of the body toward the _____ .

33. In anatomical terms, the abdomen is said to be inferior to the _____ .

34. The anatomical term referring to a side away from the midline is _____ .

35. The term proximal refers to a direction closer to the attachment point of an extremity to the body _____ .

36. In the anatomical literature, the hand would be considered distal to the _____ .

37. Two structures on the same side of the body such as the left arm and left leg are said to be _____ .

38. A vertical plane dividing the body into right and left sides represents a _____ .

39. A longitudinal plane dividing the body into anterior and posterior portions is a frontal plane, also known as a _____ .

40. A horizontal plane divides the body into superior and inferior parts and is also known as a _____ .

41. A midsagittal plane divides the body into equal right and left halves, but if the halves are unequal the plane is said to be _____ .

42. The dorsal body cavity is subdivided into the spinal cavity and the _____ .

43. Two major subdivisions of the ventral body cavity are the abdominopelvic cavity and the _____ .

44. The heart, esophagus, trachea, and bronchi are all located in a portion of the body called the _____ .

45. The abdominal subdivision and pelvic subdivision are portions of the abdominopelvic cavity, which is also known as the _____ .

46. The large dome-shaped muscle separating the abdominopelvic cavity from the thoracic cavity is the _____ .

47. Immediately superior to the umbilical region is a region of the abdominopelvic cavity known as the _____ .

48. Lateral to the hypogastric region of the abdominopelvic cavity is the iliac region, also called the _____ .

49. Among the three major serous membranes of the body are the peritoneum, the pleura, and the _____ .

50. The three serous membranes of the body have both visceral portions and _____ .

**PART B—Multiple Choice: Circle the letter of the item that correctly completes each of the following statements.**

1. The discipline of histologic anatomy is concerned primarily with
   (A) the development of an individual
   (B) the study of the kidney
   (C) the microscopic observations of cells and tissues
   (D) the study of the brain

2. The fundamental unit of living things is the
   (A) tissue
   (B) cell
   (C) organ
   (D) organ system

3. All the following are basic types of human tissues except
   (A) connective tissue
   (B) nervous tissue
   (C) epithelial tissue
   (D) squamous tissue

4. Several organs having related functions and working together constitute
   (A) a tissue
   (B) an organ system
   (C) an organism
   (D) a cell

5. The metabolic process of catabolism involves the
   (A) buildup of organic matter
   (B) utilization of energy
   (C) breakdown of organic matter
   (D) absorption of nutrients

6. The bones of the skeletal system assist the function of movement
   (A) by producing white blood cells
   (B) by storing minerals
   (C) by providing sites for the attachments of muscles
   (D) by conducting nerve impulses

7. All of the following are characteristics of growth except
   (A) an organism obtains materials from the environment
   (B) an organism uses its muscles to move
   (C) an organism forms more of itself
   (D) the size of the body's cells increases

8. When the needs of the body's cells are met and the activities of the body are occurring smoothly, then the body is said to be in
   (A) homeostasis
   (B) anabolic metabolism
   (C) growth
   (D) ipsilateral motion

9. A feedback system operating in the body may be
   (A) positive or negative
   (B) lateral or contralateral
   (C) regulatory or enabling
   (D) proximal or distal

10. All the following characteristics apply to the body in the anatomical position except
    (A) the eyes are forward
    (B) the arms are at the sides
    (C) the feet are together
    (D) the palms are down with the thumbs pointing to the body.

11. Which of the following structures would be considered on the anterior aspect of the body?
    (A) the back of the head
    (B) the navel
    (C) the ears
    (D) the elbows

12. In relationship to the stomach, the spinal cord is
    (A) lateral
    (B) posterior
    (C) anterior
    (D) ventral

13. The directional term referring to an area toward the lower part of the body and away from the head is
    (A) cephalic
    (B) cranial
    (C) superior
    (D) inferior

14. If structure A lies lateral to the body, and structure B is in the opposite direction, then structure B is in the direction referred to as
    (A) proximal
    (B) medial
    (C) ventral
    (D) anterior

15. Relative to each other, the right and left arms are
    (A) proximal
    (B) frontal
    (C) contralateral
    (D) caudal

16. In comparison to the knee joint, the hip joint of the body is said to be
    (A) proximal
    (B) superficial
    (C) inferior
    (D) dorsal

17. In comparison to the skin, the muscles are
    (A) superficial
    (B) deep
    (C) lateral
    (D) cephalic

18. Because the left arm and left leg are on the same side of the body, they are said to be
    (A) transverse
    (B) epigastric
    (C) ventral
    (D) ipsilateral

19. Compared to the upper arm, the fingers are
    (A) superficial
    (B) distal
    (C) parasagittal
    (D) horizontal

20. A sagittal plane divides the body into
    (A) right and left sides
    (B) anterior and posterior portions
    (C) superior and inferior portions
    (D) lateral and contralateral portions

21. A cross section of an organ is made in an organ when it is divided across the
    (A) sagittal plane
    (B) midsagittal plane
    (C) coronal plane
    (D) transverse plane

22. The cranial cavity and spinal cavity make up the
    (A) inferior body cavity
    (B) superior body cavity
    (C) dorsal body cavity
    (D) ventral body cavity

23. The mediastinum is located within the
    (A) pelvic portion of the abdominopelvic cavity
    (B) abdominal portion of the abdominopelvic cavity
    (C) spinal cavity
    (D) thoracic cavity

24. The hypogastric, iliac, and umbilical regions may all be located in the
    (A) abdominopelvic cavity
    (B) right lower quadrant of the thoracic cavity
    (C) upper left quadrant of the spinal cavity
    (D) dorsal cavity

25. The serous fluid secreted by serous membranes
    (A) is the site of red blood cell production
    (B) contains digestive enzymes
    (C) permits organs to slide easily across cavity walls
    (D) is the medium for transport of hormones in the body

*PART C—True/False:* **For each of the following statements, mark the letter "T" next to the statement if it is true. If the statement is false, change the <u>underlined</u> word to make the statement true.**

1. In the levels of structure in the body, molecules associate with one another to form a <u>tissue</u>.

2. Two or more different kinds of tissues associate to form a level of structure called an <u>organ</u>.

3. In the metabolic process of catabolism, organic matter is broken down, usually with the <u>input</u> of energy.

4. In the characteristic of <u>irritability</u>, cells receive stimuli and transport those stimuli from one cell part to another.

5. <u>Asexual</u> reproduction is that form of reproduction in which a single cell duplicates to yield two identical daughter cells.

6. Homeostasis is associated with the relative constancy of the physical and <u>chemical</u> environment in the cells of an organism and in the organism itself.

7. The self-regulating systems that function in the body to protect it against extremes are known as <u>control</u> systems.

8. A structure found on the belly side of the body is said to exist on the <u>posterior</u> aspect.

9. The <u>horizontal</u> position is the reference point for all the directional terms referring to the body.

10. A structure lying in the inferior aspect of the body may also be regarded as lying in the <u>caudal</u> aspect.

11. The eyes are said to be <u>medial</u> to the nose.

12. A structure lying closer to the body surface than a second structure is said to be <u>superficial</u>.

13. The term dorsal is similar to but not exactly the same as <u>anterior</u>.

14. Relative to the femur, the ankle is said to be <u>deep</u>.

15. A <u>sagittal</u> plane is a longitudinal plane that divides the body into left and right sides.

16. A coronal plane of the body is the same as a <u>transverse</u> plane.

17. A coronal plane lies at a right angle to a <u>sagittal</u> plane.

18. The two major subdivisions of the <u>dorsal</u> body cavity are the thoracic cavity and the abdominopelvic cavity.

19. The esophagus, trachea, and heart lie in a region of the thoracic cavity called the <u>pericardium</u>.

20. In the abdominopelvic cavity, the epigastric region lies immediately superior to the <u>hypogastric</u> region.

21. Lateral to the umbilical region of the abdominopelvic cavity are the right and left <u>inguinal</u> regions.

22. The walls of the <u>dorsal</u> body cavity and its organs are covered by the serous membranes.

23. The abdominal organs and many pelvic organs are covered by a serous membrane called the <u>pleura</u>.

24. Serous membranes generally lie very <u>distant</u> to one another.

25. The two main portions of the serous membranes are the visceral portion and the <u>parietal</u> portion.

**Answers**

## PART A—Completion

1. gross anatomy
2. microscope
3. cells
4. renal physiology
5. atoms
6. molecules
7. cell
8. tissue
9. connective tissue
10. epithelial tissue
11. organ
12. metabolism
13. anabolism
14. catabolism
15. involuntary movement
16. bones
17. growth
18. nerve cells
19. repair
20. sexual reproduction
21. asexual reproduction
22. homeostasis
23. physical environment
24. oxygen
25. endocrine system
26. stress
27. negative feedback system
28. positive feedback system
29. anatomical position
30. anterior
31. posterior
32. head
33. thorax
34. lateral
35. trunk
36. lower arm
37. ipsilateral
38. sagittal plane
39. coronal plane
40. transverse plane
41. parasagittal
42. cranial cavity
43. thoracic cavity
44. mediastinum
45. peritoneal cavity
46. diaphragm
47. epigastric region
48. inguinal region
49. pericardium
50. parietal portions

## PART B—Multiple Choice

| | | | | |
|---|---|---|---|---|
| 1. C | 6. C | 11. B | 16. A | 21. D |
| 2. B | 7. B | 12. B | 17. B | 22. C |
| 3. D | 8. A | 13. D | 18. D | 23. D |
| 4. B | 9. A | 14. B | 19. B | 24. A |
| 5. C | 10. D | 15. C | 20. A | 25. C |

## PART C—True/False

1. cell
2. true
3. release
4. conductivity
5. true
6. true
7. feedback
8. anterior
9. anatomical
10. true
11. lateral
12. true
13. posterior
14. distal
15. vertical
16. frontal
17. true
18. ventral
19. mediastinum
20. umbilical
21. lumbar
22. ventral
23. peritoneum
24. close
25. true

# CHAPTER 2

# THE CHEMICAL BASIS OF ANATOMY AND PHYSIOLOGY

During the 1800s, scientists first realized that the compounds found in living things could be synthesized. Friedrich Wohler's production of urea in 1828 was one of the first such syntheses. Since that time it has become apparent that a study of chemistry is intimately linked with the study of biology, especially the physiology of the body.

Chemical substances associated with living things are called organic compounds, while all other compounds are said to be inorganic. The four major organic substances occurring in all living things, including the body, are carbohydrates, lipids, proteins, and nucleic acids. They are the main subject matter of this chapter.

## CHEMICAL PRINCIPLES

The formation of organic substances depends on the binding together of atoms to form molecules. Atoms are units of which elements are composed, while molecules are the units of compounds.

## Elements

All matter in the universe is composed of one or more fundamental substances known as **elements**. Ninety-two elements are known to exist naturally, and certain others have been synthesized by scientists. An element cannot be decomposed to a more basic substance by chemical means. Such things as oxygen, iron, calcium, sodium, hydrogen, carbon, and nitrogen are elements.

Elements are designated by symbols often derived from Latin (Table 2.1). For example, sodium (from the Latin word *natrium*) is abbreviated to Na; potassium (from *kalium*) is expressed as K; and iron (from *ferrum*) is expressed as Fe. Other symbols are derived from English names: H stands for hydrogen; O for oxygen; N for nitrogen; and C for carbon. Carbon, oxygen, hydrogen, and nitrogen make up over 90 percent of the weight of the human body.

**TABLE 2.1** *Some Elements of Living Things and Their Symbols and Atomic Weights*

| Element | Atomic Symbol | Atomic Weight |
|---|---|---|
| Carbon | C | 12 |
| Hydrogen | H | 1 |
| Oxygen | O | 16 |
| Nitrogen | N | 14 |
| Sulfur | S | 32 |
| Phosphorus | P | 31 |
| Potassium | K | 39 |
| Calcium | Ca | 40 |
| Iron | Fe | 56 |
| Magnesium | Mg | 24 |
| Copper | Cu | 64 |
| Boron | B | 11 |
| Zinc | Zn | 65 |
| Chlorine | Cl | 35 |
| Sodium | Na | 23 |
| Manganese | Mn | 55 |
| Cobalt | Co | 59 |
| Iodine | I | 127 |

## Atoms

Elements are composed of **atoms**, the smallest part of an element entering into combinations with atoms of other elements. An atom cannot be broken down further without losing the properties of the element.

Atoms consist of positively charged particles called **protons** surrounded by negatively charged particles called **electrons** (Figure 2.1). A proton is about 1835 times the weight of an electron. A third particle called the **neutron** has no electrical charge; it has the same weight as a proton. Protons and neutrons adhere tightly to form the dense, positively charged nucleus of the atom. Electrons spin around the nucleus. The **atomic number** is the number of protons found in an atom, while the **mass number** is the total of protons and neutrons in an atom.

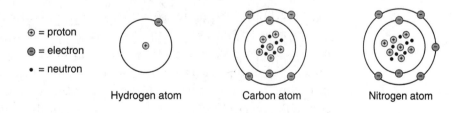

⊕ = proton
⊚ = electron
• = neutron

Hydrogen atom          Carbon atom          Nitrogen atom

**FIGURE 2.1** *The structure of three atoms showing the essential particles of the atom.*

The arrangement of electrons in an atom is important in the chemistry of the atom. Atoms are most stable when their outer shell of electrons has a full quota. This quota may be two electrons or eight electrons. An atom tends to gain or lose electrons until its outer shell is full and the atom becomes stable. An element with atoms that have full outer shells is an **inert element** because its atoms do not enter into reactions with other atoms.

The gain or loss of electrons contributes to the chemical reactions in which an atom participates. When a reaction results in a loss of electrons, it is called an **oxidation**. When a reaction results in a gain of electrons, it is called a **reduction**. Reactions as these often occur together and are called oxidation-reduction reactions.

Atoms are uncharged when they contain the same number of protons and electrons. When they lose or gain electrons, however, they acquire a charge and become **ions**. An ion may have a positive charge if it has an extra proton, or a negative charge if it possesses an extra electron. Sodium ions, calcium ions, potassium ions, and numerous other types of ions are important in body physiology.

Although the number of protons is the same for all atoms of an element, the number of neutrons may vary. Variants such as these are called **isotopes**. Isotopes have the same atomic numbers, but different mass numbers.

## Molecules

**Molecules** are precise arrangements of atoms derived from different elements. An accumulation of molecules is a compound, and a molecule may also be defined as the smallest part of the compound that retains the properties of the compound. For example, water is a compound composed of water molecules ($H_2O$), while glucose is composed of carbon, oxygen, and hydrogen ($C_6H_{12}O_6$) . In these situations, there are different kinds of atoms in the molecule. In other situations—for example, hydrogen gas ($H_2$) or oxygen gas ($O_2$)—the compound is composed of a single type of atom.

The arrangements of the atoms in a molecule accounts for the properties of a compound. The **molecular weight** is equal to the atomic weights of the atoms in the molecule. For example, the molecular weight of water is 18. Molecular weights are expressed in **daltons**; (a dalton is the weight of one hydrogen atom; a compound with a weight of 18 is therefore 18 times heavier than a hydrogen atom) and give a relative idea of the sizes of a molecule.

Atoms are linked to one another in molecules by associations called chemical bonds. In order for a bond to be created, the atoms must come close enough for their electron shells to overlap. Then, an electron exchange or sharing will occur.

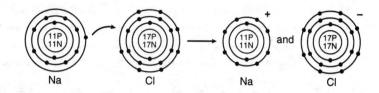

**FIGURE 2.2**    *The formation of an ionic bond using sodium (Na) and chlorine (Cl) atoms. An electron moves from the Na to the Cl atom thereby creating ions whose electrical attractions form the ionic bond.*

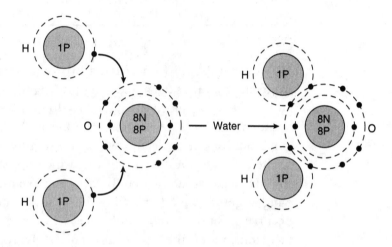

**FIGURE 2.3**    *The formation of a covalent bond in a water molecule. The oxygen atom shares its electrons with two hydrogen atoms, thereby completing the outer electron shells of each.*

An **ionic bond** forms when the electrons of one atom transfer to a second atom. This transfer results in electrically charged atoms, or ions (Figure 2.2). The electrical charges are opposite to one another (that is, positive and negative), and the oppositely charged ions attract one another. The attraction results in the ionic bond. Sodium chloride is formed from sodium and chloride ions drawn together by ionic bonding.

The second type of linkage is called a **covalent bond**. This bond forms when two atoms share one or more electrons (Figure 2.3). For example, carbon shares its electrons with four hydrogen atoms in methane molecules ($CH_4$), and oxygen and hydrogen atoms share electrons in water molecules ($H_2O$). When a single pair of electrons is shared, the bond is a **single bond**; when two pairs are shared, then the bond is a **double bond**.

Another type of linkage is the **hydrogen bond**. This is a weak bond. It forms between protons and free pairs of electrons, and is named because it exists in water molecules. The hydrogen bond is also known as an electrostatic bond. It is found between the components of nucleic acids and helps hold the strands of DNA together in the double helix (Table 2.2).

**TABLE 2.2** *Three Types of Chemical Bonds in Organic Molecules*

| Type | Chemical Basis | Strength | Example |
|------|----------------|----------|---------|
| Ionic | Attraction between oppositely charged ions | Strong | Sodium chloride |
| Covalent | Sharing of electron pairs between atoms | Strong | Carbon to carbon bonds |
| Hydrogen | Attraction of a hydrogen nucleus (a proton) to negatively charged atoms in neighboring molecules | Weak | Cohesiveness of water |

Hydrogen bonds form when a hydrogen atom linked to an electronegative atom, such as oxygen or nitrogen, is attracted to another atom that is also electronegative. Hydrogen bonds are often found between dipoles, such as those in water molecules, since the hydrogen atoms of one molecule, which are slightly positive, are attracted to the atoms of another molecule, which are slightly negative. The slightly negative atom may be an oxygen atom. Water molecules cling together and form films because of the dipole attractions occurring between water molecules. This clinging increases the surface tension of the water molecules. Hydrogen bonds are also important for the three-dimensional shape of many organic molecules, such as proteins and nucleic acids.

Carbon is renowned for its ability to enter into myriad covalent bonds, because it has only four electrons in its outer shell. Thus, it can combine with four other atoms or groups. So diverse are the possible carbon compounds that the chemistry of the body and the discipline of organic chemistry are essentially the chemistry of carbon.

When molecules interact to form new bonds, the process is called a **chemical reaction**. The **reactants** in a chemical reaction may form various **products**. For example, a molecule may be the reactant separated, or there may be a switch of parts among reactant molecules, or water may be introduced in a reaction known as a hydrolysis, or an oxidation-reduction reaction involving an exchange of electrons may occur.

**Water** is an important component of many chemical reactions either as something added to the reactants or as the molecule resulting from the reaction. Water is the universal solvent in the human body, and virtually all the chemical reactions of physiology occur in water. Over 75 percent of the body is water. It participates in most functions.

## Acids and Bases

An acid is a chemical compound that releases hydrogen atoms when placed in water. When hydrochloric acid is placed in water, for example, it releases hydrogen atoms (protons). An acid can be strong (such as hydrochloric, sulfuric, and nitric acids) if it releases the maximum amount of hydrogen ions, or it can be weak (for example, carbonic acid) if it releases few hydrogen ions. Acids have a sour taste and react with many metals. The concentration of protons released by an acid determines the acidity of the solution. When the molecular formula for an acid is written, the hydrogen is the first atom expressed.

Certain chemical compounds attract hydrogen atoms when they are placed in water. These substances are called **bases**. Typical bases are sodium hydroxide ($NaOH$) and potassium hydroxide ($KOH$). When these compounds are placed in water, they attract hydrogen ions from the water molecules, leaving behind the hydroxyl (-OH) ions. A basic (or alkaline) solution results. Both $NaOH$ and $KOH$ are strong bases, while substances such as guanine and adenine are weak bases. Bases have a bitter taste and a slippery feel. Ammonia, which is a waste product of protein digestion, forms the base ammonium hydroxide when it reacts with water, or it forms the basic ammonium ion when it takes on a proton.

The measure of acidity or alkalinity of a substance is the **pH**. The term pH refers to the hydrogen ion concentration of a substance. When the number of hydrogen ions and hydroxyl ions is equal, the pH of the substance is 7.0. (Pure water has a pH of 7.0) Decreasing numbers represent more acidic substances, and the most acidic substance has a pH of 1.0 (Table 2.3). Alkaline substances have pH numbers higher than 7, and the most alkaline substance has a pH of 14.0.

As we have already noted, when the pH of a solution is 7, the number of hydrogen and hydroxyl ions is equal. A solution having a pH of 6 has 10 times the number of hydrogen ions as a neutral solution, and a solution with a pH of 5 has 100 times the number. A solution having a pH of 8 has $\frac{1}{10}$ the number of hydrogen ions of a neutral solution, and a solution of pH 9 has $\frac{1}{100}$ the number of hydrogen ions of a neutral solution. Conversely, the smaller the concentration of hydrogen ions, the greater the concentration of hydroxyl ions. The pH is the negative log of the ionization constant of a solution.

**TABLE 2.3**  *The pH Values of Some Body Fluids*

| Material | pH | Material | pH |
|---|---|---|---|
| Gastric juice | 1.4 | Tears | 7.2 |
| Urine | 6.0 | Blood | 7.4 |
| Saliva | 6.8 | Intestinal juice | 7.8 |
| Milk | 7.1 | Pancreatic juice | 8.0 |

# ORGANIC COMPOUNDS

Among the numerous types of organic compounds are four major categories found in all living things including the human body. The four categories are carbohydrates, lipids, proteins, and nucleic acids.

## Carbohydrates

Carbohydrates serve as structural materials and energy sources for the human body. They are composed of carbon, hydrogen, and oxygen; the ratio of hydrogen atoms to oxygen atoms is 2:1. Often the carbohydrates are called **saccharides**.

Simple carbohydrates are commonly referred to as sugars. Sugars are known as **monosaccharides** if they are composed of single molecules. The most widely encountered monosaccharide in the human body is **glucose** $(C_6H_{12}O_6)$. Glucose is the basic form of fuel in the body. It is soluble and is transported by body fluids to all its cells, where it is metabolized to release its energy. Other monosaccharides are **fructose** and **galactose** (Figure 2.4). All have the same molecular formula $(C_6H_{12}O_6)$, but the atoms are arranged differently. Such molecules are called **isomers**.

**Disaccharides** are composed of two monosaccharide molecules covalently bonded. Three important disaccharides are associated with the human body: **maltose**, a combination of two glucose units resulting from the breakdown of starch in the mouth and intestine; **sucrose**, the table sugar formed by linking glucose to fructose; and **lactose**, composed of glucose and galactose molecules (Figure 2.5). Lactose is the principal carbohydrate in milk.

Complex carbohydrates are known as **polysaccharides**. Polysaccharides are formed by combining innumerable numbers of mono-

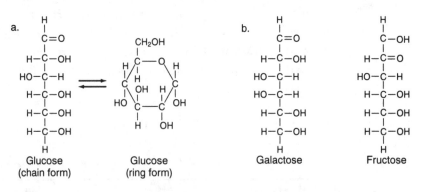

**FIGURE 2.4**   *A display of monosaccharides. (a) Glucose is the most frequently encountered monosaccharide in anatomy and physiology and metabolism. It exists both in a chain form and a ring form, as shown. (b) Two other monosaccharides, galactose and fructose, have the same number and type of atoms, but the arrangements are different.*

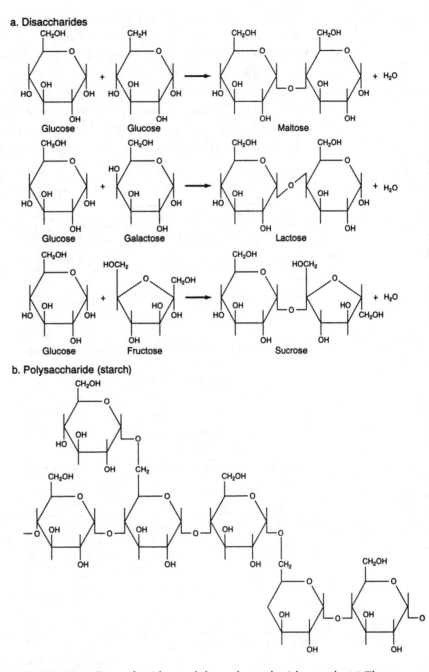

a. Disaccharides

b. Polysaccharide (starch)

**FIGURE 2.5**  *Disaccharides and the polysaccharide starch. (a) The disaccharides maltose, lactose, and sucrose are formed from their constituent monosaccharides. (b) Starch is formed from numerous glucose units. Note that the carbohydrates are shown in their ring forms. Water molecules are formed in the synthesis of disaccharides and polysaccharides.*

saccharides. Among the most important polysaccharides are **starches**, which are composed of thousands of glucose units. Starches serve as a storage form for carbohydrates; much of the world's population satisfies its energy needs with the starches of rice, wheat, corn, and potatoes.

Another important polysaccharide is **glycogen**. Glycogen is also composed of thousands of glucose units, but the units are bonded in a different pattern than in starch. Glycogen is the important form in which glucose is stored in the human liver.

Still another important polysaccharide is **cellulose**. Cellulose is also composed of glucose units, but the covalent linkages cannot be broken except by a few species of bacteria. Cellulose is found in the cell walls of plants and is used as roughage in the body (Table 2.4).

**TABLE 2.4** *Major Macromolecules of Living Things*

| Macromolecule | Building Block | Major Functions | Examples |
|---|---|---|---|
| Carbohydrate | Monosaccharides | Energy storage; physical structure | Starch, glycogen, cellulose |
| Lipid: | | | |
| Fats | Fatty acids and glycerol | Energy storage; thermal insulation; shock absorption | Fat; oil |
| Phospholipids | Fatty acids, glycerol, phosphate, and an R group* | Foundation for membranes | Plasma (cell) membranes |
| Waxes | Fatty acids and long-chain alcohols | Waterproofing; protection against elements | Cutin; suberin; ear wax; beeswax |
| Steroids | 4-ringed structure | Membrane stability; hormones | Cholesterol; testosterone; estrogen |
| Protein | Amino acids | Catalysts for metabolic reaction; hormones; oxygen transport; physical structure | Hormones (insulin, human growth hormone); hemoglobin, keratin; collagen, and enzymes |
| Nucleic acid | Nucleotides | Inheritance; instructions for protein synthesis | DNA, RNA |

* R group = a variable portion of a molecule

# Lipids

Lipids are organic molecules composed of carbon, hydrogen, and oxygen atoms. The ratio of hydrogen atoms to oxygen atoms is much higher in lipids than in carbohydrates. Lipids include steroids (the material of which many hormones are composed), waxes (such as earwax), and the fats. Other lipids include the phospholipids, which contain phosphorus and are found in the membranes of cells.

**Fat molecules** are composed of a glycerol molecule and one, two, or three molecules of fatty acid (thus forming mono-, di-, and triglycerides). A fatty acid is a long chain of carbon atoms with associated hydroxyl (-OH) groups and an organic acid (-COOH) group. The fatty acids in a fat may be alike or different. They are bound to the glycerol molecule in the process of **dehydration synthesis**, a process involving the removal of water components during covalent bond formation (Figure 2.6). The number of carbon atoms in a fatty acid may be as few as 4 or as many as 24.

Certain fatty acids have one or more double bonds in the molecule where hydrogen atoms are missing. Fats that include these molecules are called **unsaturated** fats. Other fatty acids have no double bonds and are called **saturated** fats. Unsaturated fats are preferred to saturated fats in certain health situations.

When fats are stored in cells, they are usually held in clear oil droplets called **globules**. Animals, including humans, store fats in large, clear globules in the cells of **adipose tissue**. Fats in adipose tissue contain much energy (Chapter 4), and they are extremely useful to the body. In digestion, the enzyme lipase breaks down fats to yield fatty acids and glycerol.

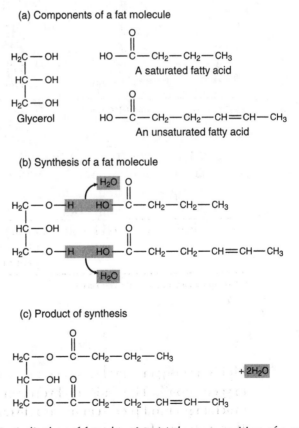

**FIGURE 2.6** *A display of fats showing (a) the composition of a molecule of fat; (b) the synthesis of a fat molecule; and (c) the new fat molecule.*

# Proteins

Proteins are immense in both size and complexity, but all proteins are composed of units called **amino acids**. Amino acids contain carbon, hydrogen, oxygen, and nitrogen atoms; sulfur or phosphorus atoms are sometimes present. There are 20 different kinds of amino acids, each having an amino (-NH$_2$) group and an organic acid (-COOH) group, and usually an attached radical (-R) group. Amino acids vary with the radical group attached. Differences in the arrangement, types, and number of atoms in the R group make each amino acid unique, and it expresses this uniqueness in its bonding behavior and its degree of acidity or alkalinity. Examples of amino acids are alanine, valine, glutamic acid, tryptophan, tyrosine, and histidine.

To form a protein, amino acids are linked to one another by removing the hydrogen atom from the amino group of one amino acid and the hydroxyl group from the acid group of the second amino acid. The amino and acid groups then link up (see Figure 2.7). Since the components of water are removed in the process, the reaction is a dehydration synthesis. The linkage forged between the amino acids is called a **peptide bond**. Small proteins are often called **peptides**.

The human body depends upon proteins for the construction of cellular parts (for example, muscle proteins) and for the synthesis of hormones and enzymes. **Enzymes** are chemical substances that catalyze most of the chemical reactions taking place within cells. Enzymes are not used up in the reaction, but rather they remain available to catalyze succeeding reactions. Without enzymes, the chemistry of the cell and organism could not take place, because enzymes provide a location where chemical substances can interact during a chemical reaction that either joins them or breaks them apart. Thus, synthesis and digestion reactions both depend heavily upon enzymes for their completion.

Proteins are also found outside of cells as supporting and strengthening materials in cells. Bone, cartilage, tendons, and ligaments all

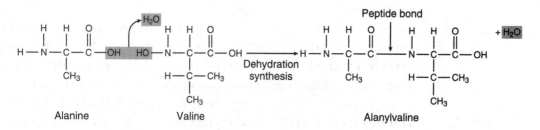

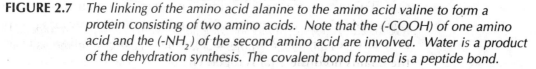

**FIGURE 2.7**  *The linking of the amino acid alanine to the amino acid valine to form a protein consisting of two amino acids. Note that the (-COOH) of one amino acid and the (-NH$_2$) of the second amino acid are involved. Water is a product of the dehydration synthesis. The covalent bond formed is a peptide bond.*

contain extracellular protein. Many hormones, such as insulin and human growth hormone, are composed of protein.

Various cells of the body manufacture proteins unique to themselves. The information for synthesizing these proteins is found in the nucleus of the cell, where a **genetic code** specifies the sequence of amino acids occurring in the final protein. The chromosomes of the cell contain the genetic code in functional units of activity called **genes**.

The amino acids of proteins also can serve as a reserve source of energy for the cell. When the need arises, the liver removes the amino group from an amino acid and uses the resulting compound for energy.

## Nucleic Acids

Like proteins, nucleic acids are very large molecules composed of building blocks. The units of nucleic acids are called **nucleotides**. Each nucleotide contains a carbohydrate molecule bonded to a phosphate group and to a nitrogen-containing molecule called a **nitrogenous base** because it has basic properties.

Two important kinds of nucleic acids are found in cells of the human body. One type is **deoxyribonucleic acid**, or **DNA**; the other is known as **ribonucleic acid**, or **RNA**. DNA is found primarily in the 46 chromosomes of the cell nucleus; it is the material of which the genes are composed. RNA is found in the nucleus, nucleolus, and cytoplasm of the cell. It participates with DNA in the synthesis of protein (Chapter 3).

DNA and RNA differ slightly in the components of their nucleotides. DNA contains the five-carbon carbohydrate **deoxyribose**, while RNA has **ribose**. Both DNA and RNA have **phosphate groups** derived from a molecule of phosphoric acid. The phosphate groups connect the deoxyribose or ribose molecules to one another in the nucleotide chain. Both compounds contain the nitrogenous bases **adenine**, **guanine**, and **cytosine**, but DNA contains the base **thymine**, while RNA has **uracil** (Table 2.5). Adenine and guanine are **purine** molecules, while cytosine, thymine, and uracil are **pyrimidine** molecules.

In 1953, the biochemists James D. Watson and Francis H. C. Crick proposed a model for the structure of DNA that is now almost universally accepted. In the Watson-Crick model, DNA consists of two long chains of nucleotides. Guanine and cytosine line up opposite one another, and adenine and thymine oppose each other. Adenine and thymine are said to be **complementary**, as are guanine and cytosine. This is the principle of complementary base pairing. The two nucleotide chains then twist to form a **double helix** resembling a spiral staircase (Figure 2.8). Weak hydrogen bonds existing between the chains hold the chains together.

**TABLE 2.5**   *A Comparison of DNA and RNA*

| DNA (Deoxyribonucleic Acid) | RNA (Ribonucleic Acid) |
|---|---|
| Found only in the nucleus | Found in the nucleolus of the nucleus and in the cytoplasm |
| Always associated with chromosomes (genes) | Found mainly in combinations with proteins in ribosomes in the cytoplasm, as messenger RNA and transfer RNA |
| Contains a pentose (5-carbon) sugar called deoxyribose | Contains a pentose (5-carbon) sugar called ribose |
| Contains the bases adenine, guanine, cytosine, thymine | Contains the bases adenine, guanine, cytosine, uracil |
| Contains phosphate groups that connect sugars to one another | Contains phosphate groups that connect sugars to one another |
| Functions in protein synthesis and as the molecule of inheritance | Functions in protein synthesis |

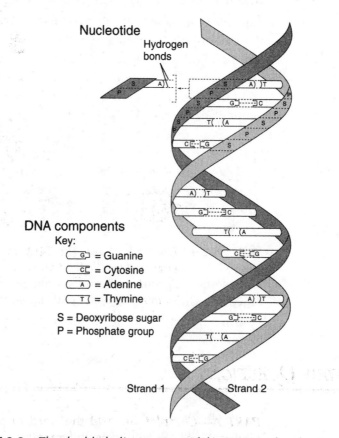

Nucleotide

Hydrogen bonds

DNA components
Key:

G = Guanine
C = Cytosine
A = Adenine
T = Thymine

S = Deoxyribose sugar
P = Phosphate group

Strand 1          Strand 2

**FIGURE 2.8**   *The double helix structure of the DNA molecule. A nucleotide contains the sugar deoxyribose (S), a phosphate group (P), and a nitrogenous base. Nucleotides then bind together to form the two DNA chains. Complementary base pairing (A–T) and (C–G) is involved in the binding that holds the chains together.*

The sequence of these bases is the essential element in the proper placement of amino acids in proteins. Indeed, how these bases are arranged is the essence of the genetic code. (Chapter 3 explores the significance and operation of the genetic code in cells.)

Before a cell divides, the DNA replicates itself. In human cells, this implies that 46 chromosomes (or 46 molecules of DNA) replicate to form 92 chromosomes. Later, the replicated chromosomes separate, and 46 pass into each new cell.

The process of **DNA replication** begins when specialized enzymes pull apart or "unzip" the DNA double helix. As the two strands separate, the purine and pyrimidine bases on each strand are exposed. The bases attract their complementary bases, causing them to stand opposite. Deoxyribose molecules and phosphate groups are brought into the molecule, and the enzyme DNA polymerase unites all the nucleotide components to form a long strand of nucleotides (Figure 2.9).

By this process, the old strand of DNA directs the synthesis of a new strand of DNA through complementary base pairing. The old strand then unites with the new strands to reform a double helix. This process is called **semiconservative replication**, because one of the old strands is conserved in the new DNA double helix.

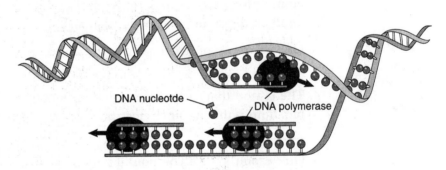

**FIGURE 2.9**   *DNA replication. The DNA strands unwind, and the enzyme DNA polymerase unites nucleotides in each chain according to bases present on the old chain. The new strand then unites with an old strand to form two new double helixes.*

# REVIEW QUESTIONS

**PART A—Completion: Add the word or words that correctly complete each of the following statements.**

1. Chemical substances associated with living things are known as _____.

2. All matter in the universe is composed of one or more _____ .

3. The letters Na represent the abbreviation for _____ .

4. Atoms consist of neutral particles called neutrons and positively charged particles called _____ .

5. The number of protons found in an atom is referred to as the _____ .

6. Atoms are most stable when their outer electron shell is filled with either two electrons or _____ .

7. A reaction that results in the gain of electrons by one of the participants is called a _____ .

8. An ion is an atom that has lost or gained one or more electrons and has acquired a _____ .

9. Precise arrangements of atoms derived from different elements constitute a _____ .

10. The compound glucose is composed of atoms of oxygen, hydrogen, and _____ .

11. The sum of the atomic weights of the atoms in a molecule is expressed as the _____ .

12. When the electrons of one atom are transfered to a second atom, two ions form, and when they are attracted to one another, the result is a bond called an _____ .

13. One example of a compound formed as a result of covalent bonding is _____ .

14. The reason that carbon enters into innumerable chemical combinations is that its outer electron shell has electrons numbering _____ .

15. One example of a compound resulting from ionic bonding is _____ .

16. When placed in water, an acid is a chemical compound that releases _____ .

17. Sodium hydroxide and potassium hydroxide are both examples of substances called _____ .

18. Carbohydrates serve the human body as structural material and as sources of _____ .

19. All carbohydrate molecules are composed of three different types of atoms, namely carbon, oxygen, and _____ .

20. The simplest carbohydrates are composed of single molecules and are _____ .

21. Those carbohydrates composed of two molecules covalently bonded to each another are _____ .

22. Starches, glycogen, and cellulose are examples of complex carbohydrates referred to as _____ .

23. Three examples of monosaccharides are fructose, galactose, and _____ .

24. Lactose is the principal carbohydrate found in _____ .

25. Cellulose is the carbohydrate that cannot be broken down in the body and is therefore used as _____ .

26. Fats, steroids, and waxes all belong to the class of organic compounds known as _____ .

27. The essential elements of a fat molecule are one, two, or three molecules of fatty acid bonded to a molecule of _____ .

28. Those fats in which hydrogen atoms are missing in the fatty acid molecules are said to be _____ .

29. The tissue in humans where fat is stored in large globules is called _____ .

30. A dehydration synthesis is a process in which fatty acids are linked to glycerol molecules with the removal of the components of _____ .

31. Every fatty acid contains an organic acid group and numerous -OH groups known as _____ .

32. During the digestive process, fats are broken down to yield fatty acids and glycerol by the enzyme _____ .

33. The four component elements of all amino acids are carbon, hydrogen, oxygen, and _____ .

34. The number of different amino acids that makes up virtually all the proteins in the body is _____ .

35. When amino acids link together to form a protein, the linkage that binds them is called a _____ .

36. Proteins are used for the construction of cellular parts and for the synthesis of chemical catalysts known as _____ .

37. Every amino acid molecule has an amino group and an organic acid group represented as _____ .

38. The linking of amino acids together to form a peptide occurs by the process of _____ .

39. Supporting and strengthening materials of the body that are composed in part of protein include bone, tendons, ligaments, and _____ .

40. Two hormones of the body that are composed exclusively of protein are human growth hormone and _____ .

41. When proteins are to be used as energy sources, the conversion to an energy compound is accomplished in the _____ .

42. The building blocks that compose all the body's nucleic acids are called _____ .

43. The material of which the genes are composed is _____ .

44. The nitrogenous base uracil is found only in the nucleic acid called _____ .

45. The nitrogenous bases adenine and guanine belong to a class of molecules called _____ .

46. In the double helix of DNA, adenine and thymine stand opposite one another, and the other two bases that stand opposite each other are _____ .

47. The double helix form that DNA takes resembles a spiral _____ .

48. Every human cell, with the exception of red blood cells and reproductive cells, contains a set of chromosomes that numbers _____ .

49. During DNA replication, one old strand unites with one new strand in the process called _____ .

50. The backbone of the DNA molecule consists of phosphate groups and molecules of the carbohydrate _____ .

## PART B—Multiple Choice: Circle the letter of the item that correctly completes each of the following statements.

1. All matter in the universe is composed of one or more fundamental substances known as
   (A) compounds
   (B) molecules
   (C) ions
   (D) elements

2. Organic compounds are those that
   (A) are composed solely of ions
   (B) are found in all living things
   (C) break down to yield acids and bases
   (D) react only with inorganic compounds

3. Which of the following describes an atom?
   (A) it cannot be broken down further without losing the properties of the element
   (B) it consists of negatively charged protons and positively charged electrons
   (C) the nucleus consists of protons and electrons
   (D) neutrons spin about the nucleus of the atom

4. An oxidation-reduction reaction is one in which
   (A) a double bond forms
   (B) an acid breaks down into hydrogen ions and hydroxyl ions
   (C) a gain or loss of electrons takes place
   (D) monosaccharides react with disaccharides

5. The molecular weight of a molecule consists of
   (A) the sum of the atomic weights of its atoms
   (B) the sum of the weights of its protons and electrons
   (C) the same number as its mass number
   (D) the sum of the number of bonds it forms

6. When the electrons of one atom transfer to a second atom, the bond that forms is known as
   (A) a covalent bond
   (B) a molecular bond
   (C) a hydrogen bond
   (D) an ionic bond

7. Atoms of carbon enter into numerous bonds with other atoms because the carbon atom
   (A) lacks a single electron in its outer shell
   (B) has four electrons in its outer shell
   (C) breaks down easily to form an acid
   (D) forms ionic bonds easily

8. When a base is placed in water
   (A) it attracts oxygen atoms from the water molecules
   (B) it releases numerous hydrogen atoms
   (C) it attracts hydrogen atoms and leaves an accumulation of hydroxyl ions
   (D) it reacts with weak acids only

9. Organic compounds composed exclusively of carbon, hydrogen, and oxygen include
   (A) proteins and fats
   (B) proteins and nucleic acids
   (C) nucleic acids and fats
   (D) carbohydrates and fats

10. All the following are monosaccharides except
    (A) fructose
    (B) sucrose
    (C) galactose
    (D) glucose

11. Glycogen and starch are similar because
    (A) both are composed of glucose units
    (B) both are used for structural materials in the cell
    (C) both are monosaccharides
    (D) both contain nitrogen

12. The polysaccharide used as roughage in the human body is
    (A) glycogen
    (B) lactose
    (C) cellulose
    (D) glucose

13. The breakdown of starch in the human mouth and intestine results in
    (A) the monosaccharide lactose
    (B) the polysaccharide glycogen
    (C) the polysaccharide cellulose
    (D) the disaccharide maltose

14. Glycogen may be found in large amounts in the
    (A) human bone marrow
    (B) human brain
    (C) human liver
    (D) human spleen

15. The two major components of fats are
    (A) glucose and amino acids
    (B) glycerol and fatty acids
    (C) nitrogen and sulfur
    (D) amino acids and organic acids

16. Fats are stored in large, clear globules in the cells of
    (A) connective tissue
    (B) nerve tissue
    (C) epithelial tissue
    (D) adipose tissue

17. The linkage forged between amino acids in a protein is known as a
    (A) ionic bond
    (B) peptide bond
    (C) hydrogen bond
    (D) amino bond

18. All the following apply to proteins except
    (A) they are used for the synthesis of enzymes
    (B) they are composed of amino acids linked together
    (C) they are found in tendons, ligaments, and cartilage
    (D) they are composed exclusively of carbon, nitrogen, and oxygen

19. The information for determining the sequence of amino acids in a protein is located in the
    (A) mitochondrion of the cell
    (B) endoplasmic reticulum in the cell's cytoplasm
    (C) cell membranes and their proteins
    (D) nucleus of the cell

20. All the following are amino acids except
    (A) valine
    (B) glutamic acid
    (C) tyrosine
    (D) guanine

21. RNA differs from DNA because RNA contains
    (A) adenine but no cytosine
    (B) cytosine but no adenine
    (C) adenine but no uracil
    (D) uracil but no thymine

22. The carbohydrate portion of a DNA molecule consists of
    (A) ribose
    (B) galactose
    (C) deoxyribose
    (D) cellulose

23. RNA may be found
    (A) in the nucleus and cytoplasm of the cell
    (B) only in the nucleus of the cell
    (C) only in the endoplasmic reticulum of the cell
    (D) in the lysosome and mitochondrion of the cell

24. During the replication process of DNA
    (A) two old strands reunite
    (B) two new strands of DNA form a double helix
    (C) a new and an old strand bind together
    (D) the old strand dissolves and only the new strands are left

25. The essential element of DNA that determines the placement of amino acids in proteins is
    (A) the placement of phosphate molecules
    (B) the existence of ribose molecules
    (C) the sequence of the nitrogenous bases
    (D) how the deoxyribose molecules are linked to the phosphate groups.

**PART C—True/False:** For each of the following statements, mark the letter "T" next to the statement if it is true. If the statement is false, change the <u>underlined</u> word to make the statement true.

1. Sodium chloride is an example of <u>an atom</u>.

2. Atoms react with one another in order to fill their outer shell with <u>protons</u>.

3. When atoms lose or gain an electron, they acquire a charge and become <u>an isotope</u>.

4. The molecular weight of a compound is expressed in units called <u>angstroms</u>.

5. When a single pair of electrons is shared in a covalent bond, the bond is described as a <u>single</u> bond.

6. An acid may be described as a chemical compound that releases <u>hydroxyl</u> atoms when it is placed in water.

7. Those chemical compounds found in all living things are known as <u>organic</u> compounds.

8. The most widely encountered monosaccharide in the physiology of the human body is <u>sucrose</u>.

9. Molecules having the same molecular formula but different arrangements of atoms are known as <u>isotopes</u>.

10. The polysaccharide starch is composed of thousands of <u>fructose</u> molecules.

11. A few species of <u>fungi</u> are among those organisms that can break the covalent linkages in cellulose molecules.

12. The principal carbohydrate found in milk is the disaccharide <u>galactose</u>.

13. Fats, waxes, and steroids make up the group of organic compounds called <u>lipids</u>.

14. A fatty acid molecule contains an organic acid group and many <u>carboxyl</u> groups.

15. In the process of <u>dehydration synthesis</u> two molecules are bound to each other with the removal of water components.

16. Saturated fats have more <u>hydrogen</u> atoms than unsaturated fats.

17. Several examples of <u>fatty acids</u> are alanine, tryptophan, and histidine.

18. A small protein is generally called a <u>glycoside</u>.

19. Proteins are used in the construction of biological catalysts known as <u>enzymes</u>.

20. Among the human hormones composed of protein are human growth hormone and <u>adrenaline</u>.

21. All nucleic acids contain a carbohydrate molecule, <u>ammonium</u> group, and several nitrogenous bases.

22. In DNA, the nitrogenous base complementary to cytosine is <u>adenine</u>.

23. The bases cytosine, thymine, and uracil are classified as <u>purine</u> molecules.

24. The process of DNA replication in human cells occurs by a <u>conservative</u> process.

25. The basic building blocks of RNA and DNA are a series of <u>nucleotides</u>.

**Answers**

### PART A—Completion

1. organic compounds
2. elements
3. sodium
4. protons
5. atomic number
6. eight electrons
7. reduction
8. electrical charge
9. molecule
10. carbon
11. molecular weight
12. ionic bond
13. methane
14. four
15. sodium chloride
16. hydrogen atoms
17. bases
18. energy
19. hydrogen
20. monosaccharides
21. disaccharides
22. polysaccharides
23. glucose
24. milk
25. roughage
26. lipids
27. glycerol
28. unsaturated
29. adipose tissue
30. water
31. hydroxyl groups
32. lipase
33. nitrogen
34. 20
35. peptide bond
36. enzymes
37. -COOH
38. dehydration synthesis
39. cartilage
40. insulin
41. liver
42. nucleotides
43. deoxyribonucleic acid
44. ribonucleic acid
45. purines
46. guanine and cytosine
47. staircase
48. 46
49. semiconservative replication
50. deoxyribose

### PART B—Multiple Choice

| | | | | |
|---|---|---|---|---|
| 1. D | 6. D | 11. A | 16. D | 21. D |
| 2. B | 7. B | 12. C | 17. B | 22. C |
| 3. A | 8. C | 13. D | 18. D | 23. A |
| 4. C | 9. D | 14. C | 19. D | 24. C |
| 5. A | 10. B | 15. B | 20. D | 25. C |

### PART C—True/False

1. a molecule
2. electrons
3. an ion
4. daltons
5. true
6. hydrogen
7. true
8. glucose
9. isomers
10. glucose
11. bacteria
12. lactose
13. true
14. hydroxyl
15. true
16. true
17. amino acids
18. peptide
19. true
20. insulin
21. phosphate
22. guanine
23. pyrimidine
24. semiconservative
25. true

# CHAPTER 3

# CELLS AND CELL PHYSIOLOGY

The human body, like all other living things, is composed of cells. This concept, known as the **cell theory**, is a basic tenet of the science of biology. Indeed, the biology of the human body revolves about the biology of the cell (Figure 3.1).

Cells are the important consideration in separating of all living things into two major groups: **prokaryotes** and **eukaryotes**. The cells of prokaryotes lack a nucleus, while those of eukaryotes have a nucleus. In addition, prokaryotic cells lack internal cellular bodies called organelles while eukaryotic cells possess them. Prokaryotic cells do not divide by the process of mitosis, but eukaryotic cells use this process. Prokaryotes include the bacteria, while eukaryotes include plants, animals, and humans.

## CELL STRUCTURE

All cells, including human cells, have two basic features: the cytoplasm and the plasma membrane. The cytoplasm is the gel-like substance serving as the foundation of the cell. It contains the largest cellular body, the nucleus. The plasma membrane (cell membrane) is the outermost membrane separating the cell from the external environment.

## The Nucleus

With the notable exception of red blood cells, all human cells have a **nucleus**. The nucleus is composed primarily of histone protein and **deoxyribonucleic acid**, or **DNA**, which is organized into linear units called **chromosomes**. Functional segments of the chromosomes are referred to as **genes**. There are approximately 100,000 genes in the nucleus of human cells. The **histone proteins** provide a supportive framework for the DNA. The histones unite with DNA to form ultramicroscopic bodies called **nucleosomes**. The nucleosomes then coil with one another to form the final chromosome molecule.

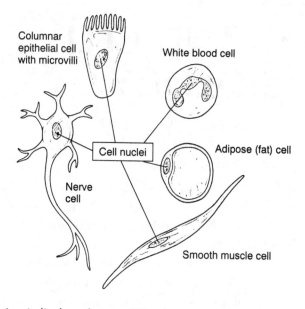

**FIGURE 3.1** *A display of some different types of cells in the human body. Note that all cells have a nucleus of varying location and size.*

The nucleus of human cells is surrounded by a membrane called the **nuclear envelope**. The nuclear envelope is a double-membrane structure consisting of two layers of phospholipid similar to the plasma membrane. Pores in the nuclear membrane allow the internal nuclear environment to communicate with the cytoplasm of the cell.

Within the nucleus are two or more dense masses referred to as **nucleoli** (singular nucleolus). The nucleolus contains **ribonucleic acid**, or **RNA**. This nucleic acid is used to construct the subunits of organelles called **ribosomes**. The subunits are later assembled into ribosomes in the cytoplasm.

## The Cytoplasm and Organelles

The **cytoplasm** is a semifluid substance representing the foundation of the cell. Within the cytoplasm are a number of microscopic bodies called **organelles** ("little organs"). Various cellular functions occur within these organelles.

An example of an organelle is the **endoplasmic reticulum**, a series of membranes extending throughout the cytoplasm of the cell (Figure 3.2). In some places the endoplasmic reticulum is studded with submicroscopic bodies called ribosomes. When ribosomes are present, the endoplasmic reticulum is referred to as **rough endoplasmic reticulum**. In other places there are no ribosomes, and this endoplasmic reticulum is said to be **smooth**. The endoplasmic reticulum is the site of protein synthesis in the cell. The ribosomes are bodies where amino acids are chemically linked together to form proteins.

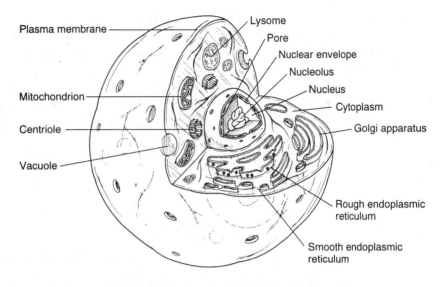

Plasma membrane
Lysome
Pore
Nuclear envelope
Nucleolus
Nucleus
Cytoplasm
Mitochondrion
Centriole
Vacuole
Golgi apparatus
Rough endoplasmic reticulum
Smooth endoplasmic reticulum

**FIGURE 3.2**   *A diagram of a "typical" human cell showing certain organelles. Some of the smaller organelles have been drawn larger than their normal relative size to show their detail.*

Another organelle is the **Golgi body**, a series of flattened sacs, usually curled at the edges. The outermost sac often bulges away to form droplike vesicles. The cell's proteins and lipids are processed in the Golgi body and packaged before moving to their final destination.

Still another organelle is the **lysosome**, a body derived from the Golgi body. The lysosome is a circular droplike sac of enzymes used in the cell's digestive processes. Enzymes break down the particles of food taken into the cell and make the products available to the cell.

The organelle where most of the cell energy is released is the **mitochondrion** (plural mitochondria). Carbohydrate molecules are broken down here, and the energy is set free. This energy is used to form molecules of adenosine triphosphate (ATP), which serves the immediate energy needs of the cell. Because they are involved in energy metabolism, the mitochondria are called the "powerhouses of the cells."

Still another organelle within the cell is the **cytoskeleton**. It is an interconnected system of fibers, threads, and interwoven molecules giving structure to the cell. The main components of the cytoskeleton are microtubules, microfilaments, and intermediate filaments (Table 3.1). All cytoskeleton components are assembled from subunits of protein.

Certain human cells contain organelles called flagella, while others have cilia. **Flagella** are long, hairlike organelles permitting movement in such cells as sperm cells. **Cilia** are shorter and more numerous than flagella. In human cells lining the nasal passages and respiratory tract, cilia wave in synchrony and move the mucus layer that has trapped foreign particles.

**TABLE 3.1** *The Structure and Function of Major Cellular Organelles*

| Organelle | Structure | Function |
|---|---|---|
| Endoplasmic reticulum | Network of interconnected membranes consisting of sacs and canals | Transports materials within the cell; provides attachment for ribosomes |
| Ribosomes | Particles composed of protein and RNA | Bodies where proteins are synthesized |
| Golgi body | Group of flattened, membranous sacs | Packages protein molecules for secretion; origin of lysosomes |
| Mitochondria | Membranous sacs with inner partitions | Site where energy released from food molecules and transformed into usable form |
| Lysosomes | Membranous sacs | Contain enzymes for intracellular digestion |
| Centrosome | Nonmembranous structure composed of two rodlike centrioles | Helps distribute chromosomes to daughter cells during cell reproduction and initiates formation of cilia |
| Cilia and flagella | Hairlike projections attached to basal bodies beneath cell membrane | Propel fluids over cellular surface and enable certain cells to move |
| Vesicles | Membranous sacs | Contain various substances after entry to the cell |
| Microfilaments and microtubules | Thin rods and tubules | Provide support to cytoplasm and help move objects within the cytoplasm; make up cytoskeleton |
| Nuclear envelope | Porous double membrane that separates nuclear contents from cytoplasm | Maintains wholeness of the nucleus and controls passage of materials between nucleus and cytoplasm |
| Nucleolus | Dense, nonmembranous body composed of protein and RNA | Contains materials to form ribosomes |
| Chromatin | Fibers composed of protein and DNA molecules | Contains genetic information for protein synthesis |
| Cell membrane | Membrane composed mainly of protein and lipid molecules | Maintains wholeness of cell and controls passage of materials into and out of cell |

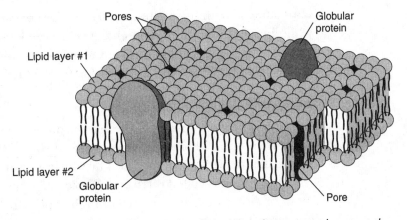

Pores

Globular protein

Lipid layer #1

Lipid layer #2

Globular protein

Pore

**FIGURE 3.3** *The fluid mosaic model of the plasma membrane at the cell border. The membrane consists of two lipid layers in which globular proteins float about.*

## The Plasma Membrane

The **plasma membrane** (also known as the **cell membrane**) lies at the border of the cells. It is composed primarily of proteins and lipids, especially **phospholipids**. The lipid occurs in two layers, a construction referred to as a bilayer. Proteins embedded in the bilayer appear to float within the lipid and the membrane is said to have a **fluid mosaic structure** (Figure 3.3). Within the membrane, proteins carry out many of the functions.

### Construction

The **phospholipids** of the plasma membrane have a phosphorous-containing polar end and a nonpolar end composed of the fatty acid chain. The polar end reacts with water and is, therefore, said to be **hydrophilic** ("loving water"), while the nonpolar end interacts with other substances that are also nonpolar, while avoiding water molecules. This second end is therefore said to be **hydrophobic** ("fearing water").

Due to these properties in phospholipids, plasma membranes share a "sandwich" structure, in which the polar ends are exposed to water at the outside and inside of the cell, and the nonpolar ends face each other within the internal portion of the membrane. Such a property in phospholipids permits the plasma membrane to form by a process of self-assembly, and to repair itself when it is torn. The lipid layer facing externally has numerous carbohydrate molecules associated with it and is called a **glycolipid**. It also contains substantial amounts of a sterol called **cholesterol**, which appears to stabilize the lipid in the membrane and makes it more fluid in character.

The proteins within the plasma membrane are both integral and peripheral proteins. **Integral proteins** generally span the entire width of the membrane, protruding on both sides and serving as channels for membrane transport, or as carriers. **Peripheral pro-**

teins attach to the membrane surface, and many act as enzymes, while helping to bring about changes in cell shape during division and during cellular contractions. Carbohydrate molecules are commonly associated with the proteins facing the environment.

## Molecular Movements

In order for the cytoplasm of the cell to communicate with the external environment, materials must move through the plasma membrane. There are several methods by which movements can occur.

One method is called **diffusion** (Table 3.2). Diffusion is the movement of molecules from an area of high concentration to one of low concentration. This movement occurs because the molecules are constantly colliding with one another, and the molecules tend to move away from where they are most concentrated. The pathway that molecules take is called the **concentration gradient**. Molecules are said to move "down the concentration gradient" because they move away from the region of high concentration. In the human lung tissues, oxygen molecules move by diffusion from the air sacs into red blood cells.

Another method for movement across the membrane is **osmosis**. Osmosis is the movement of water across a semipermeable membrane from a region of low solute concentration to a region of high solute concentration. A **semipermeable membrane** is one that lets

**TABLE 3.2** *Six Mechanisms for Molecular Movement Across the Cell Membrane*

| Mechanism | Characteristics | Example |
|---|---|---|
| Diffusion | Movement of molecules from region of high concentration to region of low concentration | Diffusion of oxygen from lung into capillaries |
| Osmosis | Diffusion of water | Reabsorption of water from kidney tubules |
| Facilitated diffusion | Diffusion assisted by carrier protein | Diffusion of glucose into red blood cells |
| Active transport | Movement of molecules from a region of low concentration to a region of high concentration assisted by carrier protein and energy from ATP | Reabsorption of salts from kidney tubules |
| Endocytosis | Membrane engulfs substance and draws it into cell in membrane-bound vesicles | Ingestion of bacterium by white blood cells |
| Exocytosis | Membrane-bound vesicle fuses with cell membrane, releasing its contents outside of cell | Release of neuro-transmitters by nerve cells |

only certain molecules (such as water molecules) pass through. A **solute** is a chemical substance dissolved in fluid (the solvent). Sodium chloride is an example of a solute.

To illustrate osmosis consider what happens when human cells are placed into a solution having a concentration of 5 percent salt. The salt concentration of the cellular cytoplasm is about 1 percent. Under these conditions, the higher concentration of solute (the salt) is outside the cell, so water moves out of the cytoplasm through the cell membrane in the direction of the higher salt concentration. The cells shrink as a result of osmosis. Since the solution has the higher solute (salt) concentration, it is called a **hypertonic** solution (Figure 3.4).

Now consider what happens when human cells are placed into a salt solution having a concentration of only 0.3 percent salt. The salt concentration of the cellular cytoplasm is still about 1 percent. Under these conditions, the higher concentration of salt (the solute) is located inside the cell, so water moves into the cytoplasm through the cell membrane in the direction of the higher salt concentration. The cells swell by osmosis. Since the outside solution has the lower salt concentration, it is called a **hypotonic** solution. If the salt concentrations were the same inside and outside the cell (about 1 percent), the solution would be an **isotonic** solution.

A third mechanism for movement across the membrane is **facilitated diffusion**. This type of diffusion is assisted by proteins present in the membrane. The proteins permit only certain molecules to pass across the membrane and encourage movement from a region of high concentration of molecules to a region of low concentration.

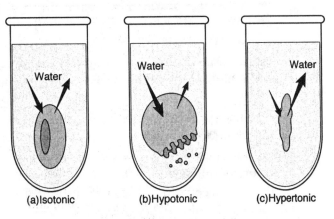

(a)Isotonic          (b)Hypotonic          (c)Hypertonic

**FIGURE 3.4**    *The process of osmosis in three different environments. (a) In an isotonic solution, the concentrations of solute are equal on both sides of the plasma membrane and no water movement occurs. (b) In a hypotonic environment, the higher solute concentration exists inside the cell, and water moves into the cell causing it to swell and explode. (c) In a hypertonic environment, the higher solute concentration exists outside the cell, and water moves out of the cell causing it to shrivel.*

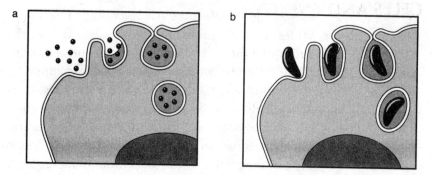

**FIGURE 3.5**   *A comparison of pinocytosis and phagocytosis. (a) In pinocytosis, the cell takes chemical substances into the cell dissolved in water (b) Phagocytosis involves the uptake of solid particles*

A fourth method for passing across the membrane is **active transport**. During active transport, proteins move chemical substances across the membrane from a region of low concentration to the region of high concentration. This movement takes place "against the concentration gradient," and it requires energy. The energy is usually derived from adenosine triphosphate (ATP). In the human body, active transport takes place in nerve cells of humans as sodium ions are transported out of the cell into the region that is already high in sodium ions.

The final mechanism for movement across the plasma membrane is **endocytosis**. During endocytosis a small patch of plasma membrane encloses particles or tiny volumes of fluid at the cell surface. The membrane enclosure then pinches off into the cytoplasm from the membrane, thereby forming a cytoplasmic **vesicle**. When endocytosis involves particulate matter, the process is called **phagocytosis**. When it involves droplets of fluid, then the process is called **pinocytosis** (Figure 3.5). Endocytosis is performed by the body's white blood cells as they remove foreign microbes from the bloodstream.

The opposite of endocytosis is **exocytosis**. During exocytosis, substances move from the interior of a cell to the external environment outside the cell. The process is used for the secretion of hormones by endocrine cells, for the release of neurotransmitters at the tip of the nerve cell, and for the secretion of mucus by cells in various organs. During exocytosis, substances contained within membranous sacs migrate to the plasma membrane. Here they fuse with the membrane, and the fused region ruptures, thereby spilling the contents of the sac into the external environment. Exocytosis is an important function in secreting cells.

# CELLS AND ENERGY

Life can exist only where molecules and cells remain organized, and organization requires energy. Physicists define **energy** as the ability to do work; in this case, the work is the continuation of cellular and human life.

Virtually every chemical reaction of the body involves a shift of energy, and usually there is a measurable loss of energy. This principle derives from a law of thermodynamics, which says that the energy in a closed system, such as a body cell, is constantly decreasing. For this reason energy must constantly be supplied to the body.

Energy must be added to most chemical reactions because compounds do not combine with one another automatically, nor do chemical compounds break apart spontaneously. To begin a chemical reaction, a type of spark referred to as the **energy of activation** is needed. For example, hydrogen and oxygen can combine to form water at room temperature, but activation energy must be provided to the chemical reaction.

Any chemical reaction in which energy is released is called an **exergonic reaction**. In an exergonic chemical reaction, the products end up with less energy than the reactants. In other chemical reactions called **endergonic reactions** energy is obtained and stored in some form.

## Enzymes

The activation energy needed to initiate an exergonic or endergonic reaction can be heat energy or chemical energy. Reactions that require activation energy can also be sparked by biological catalysts called enzymes. **Enzymes** are proteins that speed up chemical reactions while remaining unchanged. Essentially, they lower the required amount of activation energy needed for a chemical reaction.

An enzyme catalyzes only one reaction; there are thousands of different enzymes in a cell catalyzing thousands of different chemical reactions. The substance acted on by an enzyme is called the **substrate**. The products of an enzyme-catalyzed reaction are called **end products**. A key portion of the enzyme, called the **active site**, interacts with the substrate to produce the end products (Figure 3.6). With some exceptions, all enzyme names end in "-ase." For example, the enzyme that breaks down hydrogen peroxide to water and hydrogen is called catalase. Other well-known enzymes are amylase, hydrolase, peptidase, and kinase.

Enzyme reactions usually occur in milliseconds. The rate of an enzyme-catalyzed reaction depends on a number of factors, among them the concentration of the substrate, and the acidity and temperature of the environment. At higher temperatures, enzyme reactions

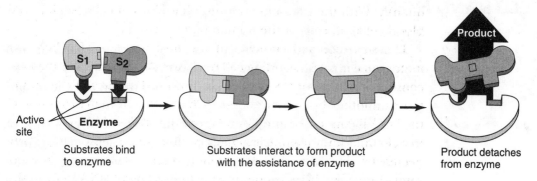

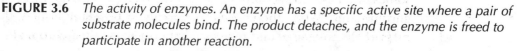

**FIGURE 3.6** *The activity of enzymes. An enzyme has a specific active site where a pair of substrate molecules bind. The product detaches, and the enzyme is freed to participate in another reaction.*

occur more rapidly, but excessive amounts of heat may cause the protein to change its structure and the enzyme to **denature** (lose its physical structure).

Enzymes work together during metabolic pathways. A **metabolic pathway** is a sequence of chemical reactions occurring in a cell. Certain metabolic pathways involve **catabolism,** the breakdown or digestion of large, complex molecules. Other metabolic pathways involve **anabolism,** the synthesis of large molecules. The pathways of metabolism are discussed in detail in Chapter 19.

## Adenosine Triphosphate (ATP)

**Adenosine triphosphate (ATP)** serves the body as the immediate energy currency for virtually all cells of the human body. The energy produced during the exergonic reactions of catabolism is stored in ATP molecules (Chapter 19).

An ATP molecule consists of three parts: a double ring of carbon and nitrogen atoms called **adenine**; a small 5-carbon carbohydrate called **ribose**; and three **phosphate units**. The phosphate units are linked together by high energy covalent bonds. When an ATP molecule is activated by an enzyme, the terminal phosphate group is released as a phosphate ion, and approximately 7.3 kilocalories of energy are released per mole of ATP (a kilocalorie is 1000 calories). This energy is made available to perform cell work.

## MITOSIS AND CELL REPRODUCTION

One of the distinguishing features of a cell is its ability to reproduce independently. In certain parts of the body, such as along the gastrointestinal tract, the cells reproduce often. In other parts of the body, such as in the nervous system, the cells reproduce less frequently. With the exception of only a few kinds of cells (such as red blood cells) all cells of the human body reproduce.

The structure and contents of the nucleus are of fundamental importance in understanding cell reproduction. The nuclear material consists of strands of DNA, which is composed of nucleotides bound to one another by covalent bonds (Chapter 2). In the chromosome, the DNA is condensed and wound around globules of histone protein to yield units known as **nucleosomes**. Millions of nucleosomes are connected by short stretches of histone protein. Stacking the nucleosomes into a coil brings about the further coiling of DNA to form the chromosome. When the chromosomes cannot be distinguished, the mass of DNA material and its associated protein is called **chromatin**.

## The Cell Cycle

The **cell cycle** is the repetition of cellular growth and reproduction (Figure 3.7). The cycle is generally divided into two major periods: interphase and mitosis. **Interphase** is the period in which the cell spends most of its time performing its unique functions. **Mitosis** is the period of the cell cycle during which the nucleus of the cell replicates and separates into two daughter cells. The actual division of the cell is **cytokinesis**.

The interphase stage of the cell cycle includes three distinctive parts: the $G_1$ phase, the S phase, and the $G_2$ phase. The **$G_1$ phase** follows mitosis and represents the time in which the cell is synthesizing its structural proteins and enzymes. The chromosomes exist as chromatin.

During the **S phase** of the cell cycle, the DNA within the nucleus replicates, although the chromosomes still cannot be seen. During this process each chromosome is faithfully copied so that by the end of the S phase, there are two chromosomes for each one present in

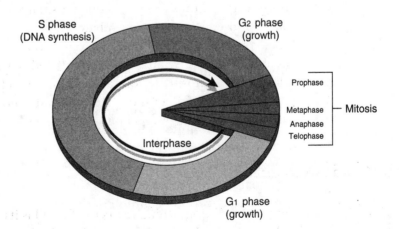

**FIGURE 3.7**    *The cell cycle. The two major periods of interphase and mitosis are shown. Three phases are contained in interphase, and four compose mitosis.*

the G$_1$ phase. In human cells, there are 46 chromosomes in the G$_1$ phase and 92 chromosomes in the S phase (see Table 3.3).

In the **G$_2$ phase**, the cell prepares for mitosis. Proteins organize themselves to form a series of fibers called the **spindle**. The spindle is constructed anew for each mitosis, then taken apart at the conclusion of the process. Spindle fibers are composed of microtubules. The nuclear material still exists as chromatin.

**TABLE 3.3** *The Cell Cycle*

| Phase | Activity |
| --- | --- |
| Interphase | G$_1$—normal cell activities<br>S—synthesis of DNA, proteins and centrioles<br>G$_2$—microtubule proteins form spindle apparatus; chromatin begins condensing |
| Mitosis | **Prophase**—duplicated chromosomes coil, nucleus and nucleolus disappear, spindle apparatus is completed, chromosomes move to center of cell<br>**Metaphase**—centromeres line up on metaphase plate<br>**Anaphase**—centromeres split, chromosomes move to opposite spindle poles<br>**Telophase**—chromosomes uncoil, nucleus and nucleoli form, spindle apparatus is dismantled, cytokinesis is completed |
| Cytokinesis | Cleavage furrow is formed by contracting microfilaments in animal cells; the cell's cytoplasm is divided by cleavage |

# Mitosis

The term **mitosis** is derived from the Latin stem "mito," meaning threads. During mitosis, (1) the nuclear material becomes visible as 92 chromosomes; (2) the chromosomes organize in the center of the cell; and (3) the chromosomes separate, and 46 chromosomes move to each new cell that is forming.

Mitosis is a continuous process comprised of a series of events. For convenience's sake and in order to denote which portion of the process is taking place, scientists divide mitosis into a series of phases called prophase, metaphase, anaphase, and telophase.

In **prophase** mitosis begins with the condensation of the chromatin material to form visible threads. Two copies of each chromosomal thread exist; the copies are called **chromatids**. The two chromatids are joined to one another at a region called the **centromere**. As prophase unfolds, the chromatids become visible in pairs, the spindle fibers form, the nucleoli disappear, and the nuclear envelope dissolves.

In human cells in prophase, microscopic bodies called **centrioles** migrate to opposite poles of the cell. When the centrioles reach the poles they are surrounded by a series of radiating microtubules called the **aster**. Spindle fibers extending out from opposite poles of the cell are also present. The chromatids attach to the spindle fibers at a structure called the **kinetochore**. The kinetochore is a region of DNA that has remained undivided. Eventually, all pairs of chromatids reach the center of the cell, an area called the **equatorial plane**. The pairs of chromatids line up across the center of the cell, and prophase comes to an end.

**Metaphase** is the stage in which all the pairs of chromatids are fully lined up in the equatorial plane, also known as the metaphase plate (Figure 3.8). In a human cell, 92 chromosomes in 46 pairs exist at the metaphase plate. Each pair is connected at the kinetochore where the spindle fiber is attached. At this point, the DNA at the kinetochore duplicates and the two chromatids separate from one another. Now they are known as chromosomes.

At the beginning of **anaphase** the chromosomes move apart from one another, each attached to a spindle fiber. The chromosomes are drawn to opposite poles of the cell by the spindle fibers and they take on a "V" shape because they are attached to the spindle fibers at their midregions. A total of 46 chromosomes move to each pole of the cell.

In **telophase** the chromosomes arrive at the opposite poles of the cell. Now the chromosomes fade from sight and form masses of chromatin. The spindle is dismantled, and its amino acids are recycled; the nucleoli reappear; and the nuclear envelope reforms.

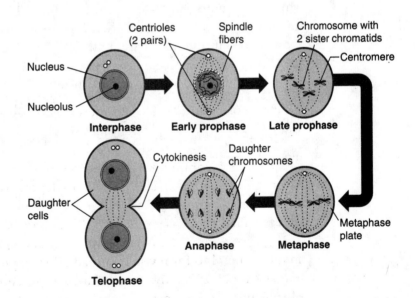

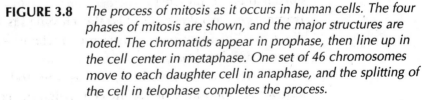

**FIGURE 3.8**    *The process of mitosis as it occurs in human cells. The four phases of mitosis are shown, and the major structures are noted. The chromatids appear in prophase, then line up in the cell center in metaphase. One set of 46 chromosomes move to each daughter cell in anaphase, and the splitting of the cell in telophase completes the process.*

# Cytokinesis

**Cytokinesis** is the process in which the cytoplasm divides and two separate cells form. In human cells, cytokinesis begins with the formation of a furrow in the equatorial plane. The cell membrane pinches into the cytoplasm, and two cells emerge. This process is often referred to as cell **cleavage**. Microfilaments contract during cleavage and assist the division of the cell into two daughter cells.

Mitosis and cytokinesis permit the entire body to grow by forming new cells. The processes also replace older cells and those that have been injured.

# PROTEIN SYNTHESIS

Proteins are used as enzymes and as structural materials in the cells. For this purpose, many proteins are retained in the cell for intracellular use. Proteins are used in microtubules and microfilaments as well as other aspects of the cytoskeleton, and they are used in the plasma membrane and other intracellular membranes such as the endoplasmic reticulum (ER). In addition, there are many specialized human proteins that are exported and function in cellular activities. For example, protein makes up the hormone insulin, the ligaments and tendons of joints, and the hair, skin, and nails of the body.

The important element of a protein molecule is the sequence in which the amino acids are linked together. This sequence is determined by a genetic code in DNA. The **genetic code** consists of the sequence of nitrogenous bases in the DNA (Chapter 2). In order for protein synthesis to occur, there are several essential materials that must be present: a supply of the 20 different amino acids, which comprise most proteins; a series of enzymes; DNA; and ribonucleic acid (RNA).

RNA carries instructions from the nuclear DNA into the cytoplasm where protein is synthesized. RNA is similar to DNA, with two exceptions; RNA contains the carbohydrate ribose rather than deoxyribose; and RNA nucleotides contain the pyrimidine uracil, rather than thymine.

## Types of RNA

In the synthesis of protein, three different types of RNA function. The first type, called **ribosomal RNA (rRNA)**, is used to manufacture ribosomes. Ribosomes are particles of RNA and protein in the cytoplasm where amino acids are linked to one another. In human cells ribosomes generally exist along the membranes of the endoplasmic reticulum.

A second important type of RNA is **transfer RNA (tRNA)**. Molecules of tRNA exist free in the cytoplasm of cells and carry amino acids to the ribosomes during protein synthesis.

The third form of RNA is **messenger RNA (mRNA)**. Messenger RNA receives the genetic code in DNA and carries the code into the cytoplasm. The genetic information is thus transferred from the DNA molecule to the mRNA molecule, which is used to synthesize a protein at a distant location.

## Transcription

One of the first processes in protein synthesis is **transcription**. In this process, a strand of mRNA is synthesized according to the nitrogenous base code of DNA. The enzyme RNA polymerase binds to one of the DNA molecules in the double helix (the other strand remains dormant) and moves along the DNA strand reading the nucleotides one by one. The enzyme selects complementary bases from available nucleotides and positions them in an mRNA molecule according to the principle of complementary base pairing (Figure 3.9). If a cytosine molecule exists on the DNA, then a guanine molecule is positioned on the RNA, and vice versa. If there is a thymine molecule on DNA, then an adenine molecule is placed in RNA, and if a thymine exists on DNA, then a uracil molecule is inserted to RNA. The chain of mRNA lengthens until a "stop" message is received.

The nucleotides of the DNA strands are read in groups of three bases called **codons**. A codon may be CGA ("cytosine-guanine-adenine") or TTA or GCT or any other combination of the four bases. The codons are transcribed to a complementary series of codons in the mRNA molecule. The synthesis of mRNA ends, and the mRNA molecule then passes through a pore in the nucleus and proceeds into the cytoplasm toward the ribosomes. Meanwhile, the DNA molecule rewinds to form a double helix.

## Translation

**Translation** is the process in which the genetic code is "translated" to an amino acid sequence in protein. The process begins with the arrival of the mRNA molecule at the ribosome. During this time, tRNA molecules have been uniting with their amino acids. The tRNA molecules now bring their amino acid molecules to the ribosomes to meet the mRNA molecule (Figure 3.10).

After it arrives at the ribosome, the mRNA molecule exposes its bases in sets of three (a codon). A tRNA molecule has an **anticodon** that complements each of the codons. When the codon of the mRNA molecule meets with its anticodon on the tRNA molecule, the latter positions its amino acid at a particular spot. Now the next

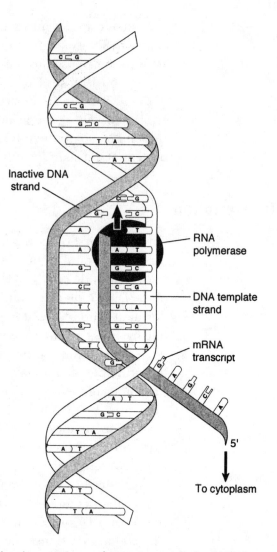

Inactive DNA strand

RNA polymerase

DNA template strand

mRNA transcript

5'

To cytoplasm

**FIGURE 3.9**    *Transcription in protein synthesis. A strand of mRNA forms according to the complementary base sequence in one strand of the DNA molecule. This synthesis transcribes the DNA message to an RNA molecule. The mRNA molecule then carries the genetic message to the cytoplasm for protein synthesis.*

codon of the mRNA is exposed, and the complementary anticodon of a tRNA molecule complements it. The amino acid carried by that tRNA molecule is positioned next to the first amino acid, and the amino acids are chemically linked to one another. The tRNA molecules then release their amino acids at the ribosome and return to the cytoplasm to search out new molecules of their amino acids.

Back at the ribosome, the next amino acid is positioned into the growing chain as the ribosome moves further down the mRNA molecule. Chain formation continues one amino acid at a time, until the protein is complete (Figure 3.11).

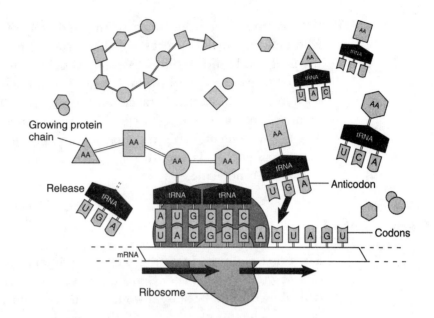

**FIGURE 3.10**   *Translation in protein synthesis. A strand of mRNA arrives at the ribosome where it is met by molecules of tRNA carrying amino acids. The tRNA molecules have three-base sequences (anticodons) that complement three-base sequences (codons) on the mRNA. This matching places the amino acids in a certain position: then they are attached to the growing protein chain. In this manner, the genetic code is translated to an amino acid sequence in a protein.*

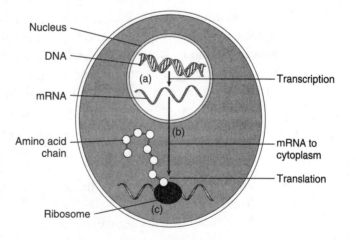

**FIGURE 3.11**   *An overview of protein synthesis showing (a) transcription in the nucleus, (b) movement of mRNA to the cytoplasm, and (c) translation at the cellular ribosome.*

Once the protein has been synthesized, it is removed from the ribosome for further processing. The protein may be modified in the Golgi body and stored in secretory vesicles before release by the cell; or it may be stored in the lysosome as a digestive enzyme, or used in the cell as a structural cellular component. The mRNA molecule is broken up and its nucleotides are returned to the nucleus. The tRNA molecules wait in the cytoplasm to unite with fresh molecules of amino acids, and the ribosome anticipates the arrival of a new mRNA molecule.

# Gene Control

The process of protein synthesis occurs at intervals followed by periods of genetic "silence." Gene expression is regulated and controlled by the cell because it would be uneconomical for the cell to be producing all its proteins all the time. For example, a digestive protein is produced when a particular food is consumed. In addition, certain cells produce only certain proteins. For example, a pancreas cell produces the hormone insulin, but a brain cell does not.

The control of gene expression occurs at several levels in the cell. For example, genes are held in control during the process of mitosis. The chromatin is compacted and tightly coiled, and this coiling regulates access to the genes.

Gene control can also occur at transcription and afterwards. In transcription, certain segments of DNA increase the activity of nearby genes and accelerate gene activity. Moreover, after transcription has taken place, the mRNA molecule is altered to regulate gene activity. It has been found, for example, that an mRNA molecule contains many useless bits of RNA called **introns**. Introns do not appear to have any genetic information for the synthesis of proteins and are found in all human cells, but not in bacterial and other simple cells. They appear to be a sort of "genetic gibberish." These bits of RNA are removed in the production of the final mRNA molecule. The remaining pieces of mRNA called **exons** are then spliced together to form the final mRNA molecule. Exons are the actual genes used to encode the proteins of the cell. They form about 5 percent of all the genetic material of a human cell, and represent the expressed portion of the human genome. In removing introns and retaining the exons, the cell alters the message received from the DNA and controls gene expression.

# REVIEW QUESTIONS

**PART A—Completion:** Add the word or words that correctly complete each of the following statements.

1. Organisms whose cells have a nucleus and organelles are known as _____ .

2. A nucleus is present in all human cells except _____ .

3. Proteins that provide a supportive framework for the DNA in chromosomes are composed of _____ .

4. The dense mass within the nucleus that contains ribonucleic acid is referred to as the _____ .

5. When ribosomes are present on the endoplasmic reticulum, the endoplasmic reticulum is said to be _____ .

6. Before being sent to their final destinations, proteins and lipids are processed in a cellular organelle called the _____ .

7. Organelles where cell energy is released from carbohydrate molecules and where ATP molecules are produced are the _____ .

8. The main components of the cytoskeleton are intermediate filaments, microfilaments, and _____ .

9. The plasma membrane at the border of the cells is composed of lipids known as _____ .

10. The movement of molecules from a region of high concentration to a region of low concentration is referred to as _____ .

11. Water molecules move across the plasma membrane from a region of low solute concentration to a region of high solute concentration in the process of _____ .

12. Facilitated diffusion takes place in the plasma membranes with the assistance of _____ .

13. When chemical substances move from a region of low concentration to a region that is already high in concentration, the movement is known as _____ .

14. Endocytosis can imply phagocytosis when particulate matter is taken into the cell, but when droplets of fluid are taken in, the process is known as _____ .

15. Those chemical reactions taking place in the body and accompanied by a release of energy are known as _____ .

16. Those chemical reactions in which energy is obtained and trapped from the environment are called _____ .

17. Proteins that speed up chemical reactions while themselves remaining unchanged are _____ .

18. With only a few exceptions, the names of all enzymes end in _____ .

19. The immediate energy currency used by virtually all cells of the human body is _____ .

20. When the temperature increases, the rate of an enzyme reaction _____ .

21. The three parts of an ATP molecule are adenine, phosphate units, and the carbohydrate known as _____ .

22. In the cell nucleus, DNA molecules are wound around globules of histone to yield _____ .

23. During the time that the chromosomes cannot be distinguished, the mass of DNA in the cell's nucleus is _____ .

24. The phase of the cell cycle during which the cell divides to form two daughter cells is called _____ .

25. The phase of interphase when the cell synthesizes its structural proteins and enzymes is _____ .

26. The phase of interphase in which the DNA replicates in preparation for mitosis is _____ .

27. During the $G_2$ phase of interphase, the cell prepares for mitosis and proteins organize themselves into a series of fibers called the _____ .

28. The copies of chromosomal threads existing in prophase are the _____ .

29. During prophase, the microscopic bodies that migrate to opposite sides of the cell are the _____ .

30. During prophase, chromatids attach to the spindle fibers at the region of DNA known as the _____ .

31. The stage of mitosis in which pairs of chromatids line up in the equatorial plane is _____ .

32. When chromosomes arrive at the opposite poles of the cell, the cell is said to be in phase of mitosis called _____ .

33. After mitosis has taken place, the cytoplasm of the cell divides to form two separate cells in the process of _____ .

34. The sequence of amino acids in a protein is determined by a message in the DNA molecule known as the _____ .

35. DNA has the base thymine, but RNA has a different base known as _____ .

36. The ultramicroscopic bodies of RNA and protein where amino acids are linked together in the cytoplasm are _____ .

37. The nucleic acid responsible for carrying instructions from nuclear DNA into the cytoplasm is _____ .

38. Transfer RNA molecules have the function of carrying to the ribosomes a series of _____ .

39. Transcription is the process in which a strand of messenger RNA is synthesized according to the base code in _____ .

40. In the synthesis of messenger RNA, the DNA strand is read in groups of three bases called _____ .

41. Once it has been synthesized, the messenger RNA molecule leaves the cell nucleus through a _____ .

42. The DNA molecule exists in the nucleus of the cell in the form of a _____ .

43. Each transfer RNA molecule has a sequence of bases that complements the sequence in the codon and is known as an _____ .

44. The chemical linkage of amino acids to form a protein according to the message in the RNA molecule takes place at the _____ .

45. The union of many hundreds or thousands of amino acids results in an organic molecule called a _____ .

46. Once the messenger RNA molecule has been used, it is broken up and its components are returned to the _____ .

47. Once a protein has been formed in the cytoplasm, it may be stored in the _____ .

48. The synthesis of messenger RNA molecules in the nucleus is brought about by the activity of _____ .

49. When one strand of the DNA double helix is functioning in protein synthesis, the other strand remains _____ .

50. In the formation of messenger RNA molecules, the cell removes useless bits of RNA called _____ .

**PART B—Multiple Choice:** Circle the letter of the item that correctly completes each of the following statements.

1. The two organic substances present in the nucleus of the cell are
   (A) fat and carbohydrate
   (B) carbohydrate and vitamins
   (C) protein and nucleic acid
   (D) minerals and vitamins

2. Genes may be defined as
   (A) carbohydrate molecules that synthesize protein
   (B) functional segments of chromosomes
   (C) protein materials that yield enzymes
   (D) chromosomes that have no exons

3. The dense mass of ribonucleic acid found within the nucleus is the
   (A) endoplasmic reticulum
   (B) nucleolus
   (C) Golgi body
   (D) ribosome

4. The endoplasmic reticulum may be described as a
   (A) sausage-shaped organelle in the cytoplasm
   (B) functional body within the nucleus
   (C) series of membranes in the cell's cytoplasm
   (D) center for energy production

5. The lysosomes of a cell are notable because they contain many
   (A) digestive enzymes
   (B) energy-laden organic compounds
   (C) vitamins and minerals
   (D) salts such as sodium chloride

6. The mitochondria of the cell are best known as the organelles where
   (A) proteins are synthesized
   (B) chromosomes gather
   (C) genes provide the genetic code for protein synthesis
   (D) energy is released from carbohydrate molecules.

7. The flagella of human cells are most closely associated with
   (A) energy production
   (B) protein synthesis
   (C) cell reproduction
   (D) cell movement

8. All the following refer to a plasma membrane of a cell except
   (A) it is composed primarily of proteins and lipids
   (B) it conforms to the fluid mosaic structure
   (C) the lipids exist in a single layer
   (D) it lies at the border of the cell

9. During the process of diffusion
   (A) molecules move from a region of high concentration to one of low concentration
   (B) molecules move through a membrane
   (C) large amounts of energy are expended
   (D) proteins assist the movement of molecules

10. If a number of human cells were placed into a beaker containing a very salty solution,
    (A) water would rush into the cells
    (B) the salt would invade the cellular cytoplasm
    (C) the cells would tend to shrink
    (D) the cells would tend to swell

11. The plasma membrane is a semipermeable membrane because it
    (A) prohibits the passage of water
    (B) lets only certain molecules through
    (C) contains chromosomes and genes
    (D) is the center for semiconservative protein synthesis

12. If a number of human cells were placed in a beaker of salt-free solution, the solution would be
    (A) isotonic
    (B) hypertonic
    (C) sinotonic
    (D) hypotonic

13. In active transport, chemical substances move from a region of low concentration to one of high concentration, and therefore they require
    (A) the functioning of genes
    (B) an expenditure of energy
    (C) the involvement of ribosomes
    (D) the involvement of the Golgi body

14. Phagocytosis is a form of endocytosis in which
    (A) the cell duplicates
    (B) gene control is exercised
    (C) the cell takes particulate matter into itself
    (D) the endoplasmic reticulum divides

15. Virtually all chemical reactions of the cell are catalyzed by
    (A) carbohydrates
    (B) minerals
    (C) sodium ions
    (D) enzymes

16. The energy is released from an ATP molecule when
    (A) the terminal phosphate group is released
    (B) the molecule binds to sodium ions
    (C) the molecule is transported to the nucleus
    (D) the molecule is combined with fat

17. All the following apply to enzymes except
    (A) enzymes are carbohydrate molecules
    (B) enzymes act on substrates
    (C) high temperatures cause enzymes to denature
    (D) enzymes work with other enzymes in metabolic pathways

18. The interphase is the phase of a cell cycle in which
    (A) the cell divides
    (B) the chromosomes line up in the equatorial plane
    (C) the cell performs its unique functions
    (D) the cellular chromosomes move to the poles

19. Chromatids, chromosomes, and chromatin are all similar to one another because
    (A) all exist as visible threads
    (B) all may be seen during interphase
    (C) all are associated with ATP molecules
    (D) all contain DNA

20. A scientist investigating the processes of telophase would be concerned with activities occurring
    (A) at the beginning of mitosis
    (B) during interphase
    (C) during the $G_2$ phase
    (D) at the end of mitosis

21. All the following are composed of RNA except
    (A) ribosomes
    (B) messenger molecules
    (C) genes
    (D) transfer molecules

22. The codon is a three-base group of nucleotides that specifies
    (A) an amino acid molecule
    (B) an enzyme molecule
    (C) a protein that facilitates diffusion
    (D) a hydrogen nucleus

23. All the following participate in the translation phase of protein synthesis except
    (A) transfer RNA molecules
    (B) DNA molecules
    (C) ribosomes
    (D) amino acids

24. Protein synthesis in the cell could not occur in the absence of
    (A) fat molecules
    (B) vitamin molecules
    (C) amino acid molecules
    (D) sterol molecules

25. A molecule of messenger RNA contains
    (A) exons but no introns
    (B) introns but no exons
    (C) ATP but no DNA
    (D) DNA but no ATP

*PART C—True/False.* **For each of the following statements, mark the letter "T" next to the statement if it is true. If the statement is false, change the <u>underlined</u> word to make the statement true.**

1. There are approximately <u>100,000</u> genes to be found in the nucleus of each human cell.

2. The nuclear envelope consists of two layers of <u>protein</u>, with pores allowing communication with the cytoplasm.

3. The nucleolus is the site of ribonucleic used to make <u>lysosomes</u>.

4. Before passages to their final destinations, the proteins and lipids of the cell are processed in the <u>mitochondrion</u>.

5. The function of the <u>lysosome</u> in the cell is to give structure to the cell.

6. Because the membrane components are constantly in a state of flux, the plasma membrane is referred to as a <u>fluid mosaic structure</u>.

7. Oxygen molecules move from the body's air sacs into its red blood cells by the process of <u>osmosis</u>.

8. If a number of body cells were placed in a solution that was free of salt, the cells would tend to <u>swell</u>.

9. If a number of body cells were placed in a solution containing the same concentration of salt as contained in the cytoplasm of the cell, the solution would be called <u>hypotonic</u>.

10. The process of facilitated diffusion is assisted by <u>carbohydrate</u> molecules present in the plasma membrane.

11. To drive the process of active transport, energy is usually obtained from a substance called <u>adenosine triphosphate</u>.

12. The processes of phagocytosis and pinocytosis are collectively known as <u>exocytosis</u>.

13. In the metabolism of cells, the activation energy is usually replaced by <u>enzymes.</u>

14. Those metabolic pathways in which large complex molecules are broken down or digested are collectively known as <u>anabolism</u>.

15. A substance can be identified as an enzyme because its name usually ends in "<u>ate</u>."

16. The $G_1$ phase of the cell cycle is the phase that immediately <u>precedes</u> mitosis.

17. During the <u>S phase</u> of the cell cycle, the DNA within the nucleus undergoes replication.

18. In the stage of mitosis known as prophase, the <u>centromeres</u> migrate to opposite sides of the cell.

19. The spindle fibers draw chromosomes to opposite poles of the cell during the stage of mitosis called <u>metaphase</u>.

20. The genetic code consists of the sequence of <u>deoxyribose molecules</u> in the DNA molecule.

21. Protein synthesis occurs in human cells along the membranes of the <u>mitochondrion</u>.

22. In the formation of messenger RNA molecules, bits of useless RNA called <u>introns</u> are eliminated.

23. A codon consists of a three-base sequence on a <u>transfer</u> RNA molecule.

24. Amino acids are brought to the ribosomes for linkage to proteins by the <u>messenger</u> RNA molecules.

25. Once the messenger RNA molecules have been used, they are broken up and their nucleotides are returned to the <u>nucleus</u>.

**Answers**

## PART A—Completion

1. eukaryotes
2. red blood cells
3. histone
4. nucleolus
5. rough
6. Golgi body
7. mitochondria
8. microtubules
9. phospholipids
10. diffusion
11. osmosis
12. proteins
13. active transport
14. pinocytosis
15. exergonic reactions
16. endergonic reactions
17. enzymes
18. ase
19. adenosine triphosphate (ATP)
20. increases
21. ribose
22. nucleosomes
23. chromatin
24. mitosis
25. $G_1$ phase
26. S phase
27. spindle
28. chromatids
29. centrioles
30. kinetochore
31. metaphase
32. telophase
33. cytokinesis
34. genetic code
35. uracil
36. ribosomes
37. RNA
38. amino acids
39. DNA
40. codons
41. pore
42. double helix
43. anticodon
44. ribosome
45. protein
46. nucleus
47. Golgi body
48. enzymes
49. dormant
50. introns

## PART B—Multiple Choice

| | | | | |
|---|---|---|---|---|
| 1. C | 6. D | 11. B | 16. A | 21. C |
| 2. B | 7. D | 12. D | 17. A | 22. A |
| 3. B | 8. C | 13. B | 18. C | 23. B |
| 4. C | 9. A | 14. C | 19. D | 24. C |
| 5. A | 10. C | 15. D | 20. D | 25. A |

## PART C—True/False

1. true
2. phospholipid
3. ribosomes
4. Golgi body
5. cytoskeleton
6. true
7. diffusion
8. true
9. isotonic
10. protein
11. true
12. endocytosis
13. true
14. catabolism
15. "ase"
16. follows
17. true
18. centrioles
19. anaphase
20. nitrogenous bases
21. endoplasmic reticulum
22. true
23. messenger
24. transfer
25. true

# CHAPTER 4

# TISSUES

Tissues are groups of cells having the same structural characteristics and performing the same functions. The human body has four basic types of tissues: epithelial tissue (epithelium), connective tissue, muscle tissue, and nervous tissue.

**Epithelial tissue**, or **epithelium**, covers the body surfaces, lines its cavities, and forms the major portions of many human glands. Certain epithelial tissues permit nutrient absorption, while others synthesize and release various secretions.

**Connective tissues** are a diverse group of tissues, each composed of cells embedded within a network of fibers called a **matrix**. The material surrounding the cells within the matrix is known as the **ground substance**. Ground substance may be hard and inflexible as in bone tissue, solid with some flexibility as in cartilage tissue, or jellylike and soft such as in adipose (fat) tissue. In some cases, such as blood, the ground substance is liquid, and the connective tissue has no fibers. Connective tissues provide support and protection to the body. They also transport materials and have a storage function.

**Muscle tissues** permit movement in the body. The structure of the tissue cells allows them to contract, and in doing so they help pump blood, move body parts, or propel food through the gastrointestinal tract.

**Nervous tissue** is composed of cells specially constructed to receive and relay signals, respond to stimuli, and coordinate conscious and unconscious activities. Nervous tissue is the principal component of the central and peripheral nervous systems (Table 4.1).

## EPITHELIAL TISSUE

Epithelial tissue exists at the body surface where it lines the skin and various organs such as the mouth, nose, and other body cavities. The tissue is found at the lining of the respiratory, reproductive, and urinary tracts. In addition, all blood vessels are lined with epithelial tissue. Cells in epithelial tissue divide by mitosis; one surface of epithelial tissue is usually free to face the air or fluid.

Epithelial tissue has no blood supply, but it is nourished by nutrients obtained from blood vessels in the underlying connective tissue. The epithelial tissue is anchored to the connective tissue by an

**TABLE 4.1** *Human Tissues and Their Functions*

| Tissue | Function |
|---|---|
| Epithelial | Provides protection and support; lines organs |
| Connective | Holds together specialized areas of human body |
| • Bone | Provides support |
| • Cartilage | Provides flexibility; absorbs shock |
| • Fibrous connective tissue | Connects muscles to bone; bone to bone |
| • Blood | Transports oxygen, nutrients, wastes, antibodies, hormones |
| Muscle | Contracts to provide body movement |
| Nerve | Conducts electrical impulses between all parts of the body and between the external environment and body |

underlying noncellular **basement membrane**. The membrane consists of glycoproteins secreted by epithelial cells. A meshwork of collagen fibers from the connective tissues is also part of the basement membrane.

# Cell Junctions

Epithelial cells are usually packed closely together and have little intercellular material. They are held together by tight junctions and adhering junctions. **Tight junctions** extend throughout the surface and around the perimeter of an epithelial cell and seal it tightly to adjacent cells (Figure 4.1). The junction is formed by fusing the cell membranes of adjacent cells with interlocking membrane lipoproteins. The intercellular space is sparse.

**Adhering junctions** between epithelial cells extend around the entire cell perimeter (**zonular adhering junctions**) or scatter about the membrane (**macular adhering junctions**). Zonular adhering junctions are also called **intermediate junctions**, because they are found between tight junctions at the cell surface and macular adhering junctions are scattered about. Macular adhering junctions are also referred to as **desmosomes,** because of their spotty occurrence. In desmosomes, a submicroscopic space separates the opposing cell membranes, and keratin fibers bind the cells together.

Still another type of junction is the gap junction. In the **gap junction**, tubular passageways and channels exist between cells, and small ions and molecules can pass from cell to cell. Smooth muscle tissue has these junctions, but epithelial tissues do not (Table 4.2).

Various functions are related to the epithelial tissues. These functions include protecting the underlying tissues from dehydration, mechanical irritation, toxic substances, and trauma. Another function

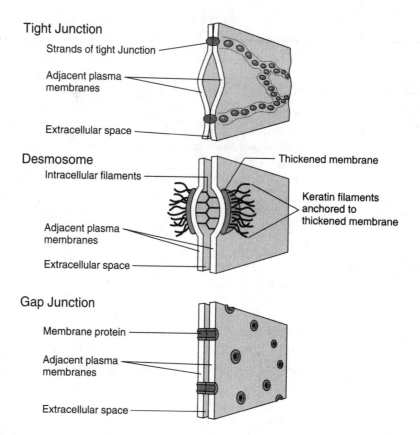

Tight Junction

Strands of tight Junction

Adjacent plasma
membranes

Extracellular space

Desmosome

Intracellular filaments

Thickened membrane

Keratin filaments
anchored to
thickened membrane

Adjacent plasma
membranes

Extracellular space

Gap Junction

Membrane protein

Adjacent plasma
membranes

Extracellular space

**FIGURE 4.1**   *Three types of junctions in human tissues. (a) Tight junctions
seal epithelial cells to one another and have fused portions of
the plasma membranes (b) Desmosomes are "spotted" seals
between cells with keratin filaments anchoring the two cells
(c) Gap junctions have tiny channels for the flow of small
ions and molecules. They are found in smooth muscle tissue.*

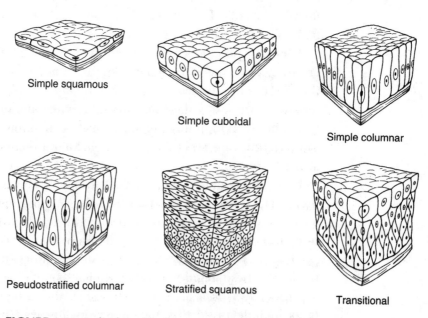

Simple squamous

Simple cuboidal

Simple columnar

Pseudostratified columnar

Stratified squamous

Transitional

**FIGURE 4.2**   *A display of various types of epithelial tissue. Different types of
epithelial tissue are found in various organs of the human body.*

**TABLE 4.2**  *A Comparison of Three Types of Intercellular Junctions*

| Type | Name | Characteristic Specializations in Cell Membranes | Width of Intercellular Space | Function |
|---|---|---|---|---|
| Organizing junctions | Tight junctions | Belts of protein that isolate parts of plasma membrane | Intercellular space disappears when the two adjacent membranes lie adjacent | Form a barrier separating surfaces of cell |
| Perimeter junctions | Adhering junctions | Buttonlike welds joining opposing cell membranes | Normal size 24 nm | Holds cells tightly together |
| Communicating junctions | Gap junctions | Channels or pores through the two cell membranes and across the inter-cellular space | Intercellular space greatly narrowed to 2 nm | Provide for electrical communication between cells and for flow of ions and small molecules |

is absorbing gases or nutrients, such as occurs in the lung or digestive system. A third function is transporting nutrients, fluids, mucus, and other particulate matter.

Still another function is secreting cell products such as enzymes, sweat, and hormones. These products are produced by epithelium known as **glandular epithelium**. The epithelium of the kidney also functions in the excretion of waste. Moreover, certain epithelial cells are specialized for sensory reception, which occurs in the ear, nose, and taste buds.

## Types of Epithelial Tissues

There are three important shapes taken by cells in epithelial tissues: flat cells are known as **squamous cells**; cube-shaped cells are called **cuboidal cells**; and tall and cylindrical cells are **columnar cells** (Figure 4.2).

Certain epithelial tissues occur in sheets consisting of a single layer. This type of epithelium is called **simple epithelium**. Other types of epithelium consist of two or more cells; this epithelium is called **stratified epithelium**. Epithelium may be considered **pseudostratified epithelium** if it appears to have more than one layer but is really a single sheet of cells having different heights. In pseudostratified epithelium all the cells touch the basement membrane. Still another form of epithelium is **transitional epithelium**, in which the cells change shape in response to mechanical stretching.

**Simple squamous epithelium** is a single layer of flat cells occuring in the lining of the blood and lymph vessels (where it is known as **endothelium**), in the lining of body cavities (where it is called **mesothelium**), in the smallest ducts of many glands, in some parts of kidney tubules, and in the terminal ducts and air sacs of the respiratory system. The main functions of this type of epithelial tissue are protection and absorption.

**Simple cuboidal epithelium** consists of cube-shaped cells having a central nucleus. This epithelial tissue is found in many glands as well as in the pigmented epithelium of the retina of the eye, the surface layer of the ovary, parts of the testis, and the anterior surface of the lens of the eye. Its functions are protection and secretion.

**Simple columnar epithelium** lines the gastrointestinal tract and is found in the fallopian tubes and uterus. It comprises portions of the respiratory tract and is found in many glands. The tissue consists of a single layer of tall cells having nuclei located close to the base of the cells. Secretion, protection, and absorption are functions associated with this epithelium. Single-celled glands, called **goblet cells**, occur in the columnar epithelium of the digestive tract. These cells secrete mucus.

**Pseudostratified columnar epithelium** lines the trachea and upper respiratory tract, and is associated with parts of the male reproductive ducts, male urethra, and a portion of the female urethra. The tissue functions in protection and secretion and consists of a single layer of cells.

**Stratified squamous epithelium** consists of many cells with columnar cells at the base, cuboidal cells at the intermediary level, and squamous cells at the free surface. The skin's epidermis contains dry, stratified squamous epithelium. Layers of squamous tissue are usually infiltrated with keratin protein. All body openings such as the mouth, anus, and openings of the urethra, ear, and vagina are lined with stratified squamous epithelium; the primary function is protection.

The ducts of sweat glands are lined with **stratified cuboidal epithelium**. This epithelium also occurs in the tubules of the testis and in the ovarian follicles. Its main function is protection. The tissue consists of layers of cuboidal cells.

**Stratified columnar epithelium** is composed of layers of tall cells. This tissue is rare in the body, and it is found in the lining of the male urethra, where it gives protection. It is also found where simple columnar epithelium meets stratified squamous epithelium, such as in the pharynx and larynx.

**Transitional epithelium** is found in the lining of the ureter and urinary bladder. Six to seven layers of cells may be seen in relaxed transitional epithelium, whereas in the distended bladder, the epithelium thins to two or three layers of cells. Protection is the principal function of this tissue (Table 4.3).

**TABLE 4.3**  *Characteristics of Epithelial Tissues*

| Epithelial Tissue | Locations | Functions |
|---|---|---|
| Simple squamous | Lining of blood vessels, lining of body cavities, part of kidney tubules | Protection, absorption |
| Simple cuboidal | Secretory portion and ducts of some glands, part of kidney tubules | Secretions, protection |
| Simple columnar | Lining of gastrointestinal tract, ducts of some glands | Absorption, protection, secretion |
| Pseudostratified columnar | Lining of trachea, upper respiratory tract, and parts of male reproductive system | Protection, secretion |
| Stratified squamous | Epidermis, lining of mouth, esophagus, and vagina | Protection, secretion, some absorption |
| Stratified cuboidal | Ducts of sweat glands | Protection |
| Stratified columnar | Part of lining of male urethra | Protection |
| Transitional | Lining of ureter and urinary bladder | Protection |

## Types of Glands

In addition to the protective and absorptive functions, the epithelium also has an important secretory function in the form of glands. Epithelium of the glands is known as **glandular epithelium**, which is specialized for secretion.

Glands are categorized as endocrine or exocrine glands. **Endocrine glands** secrete chemical regulators called hormones directly into the blood, while **exocrine glands** secrete their products into ducts. Endocrine glands are often known as **ductless glands**, while exocrine glands are known as **ducted glands**.

Exocrine glands may be classified according to the number of cells they possess. **Unicellular glands** are composed of one cell, while multicellular glands contain many cells (Table 4.4). A goblet cell typifies a unicellular gland. This gland exists within the epithelial cells of the digestive tract, where it secretes mucus. **Multicellular glands** consist of numerous cells that produce a secretion. The multicellular gland is considered **simple** if its duct does not branch; it is said to be **simple tubular** if its secretions are given off into a straight-line tube, or **coiled tubular** if the tube coils, or **branched tubular** if branches are present in the tubule, or **branched acinar** (saclike) if the secretory unit is shaped like a sac with several acini (sacs) along a duct (Figure 4.3). The multicellular gland is considered **compound** if its duct branches; it is said to be **compound tubular** if branched tubules are present; or **acinar** if saclike; or **tubuloacinar** if both tubules and sacs are present.

Exocrine glands may also be classified according to the type of secretion they produce. For example, **mucus-secreting glands** produce a slimy, viscous mucus composed of polysaccharides and proteins. Mucus is produced by goblet cells as well as epithelial cells in the respiratory, urinary, and reproductive tracts. A different type of secreting gland are the **serous-secreting glands**. Serous fluid is a watery protein material that usually contains enzymes. In some cases, the exocrine gland may produce both mucus and serous fluid. The salivary glands typify **seromucous glands** because they produce both mucus and enzymes within the serous fluid.

Exocrine glands are also classified functionally as merocrine or holocrine glands. **Merocrine glands**, such as the sweat glands and salivary glands, are glands whose cells remain intact during the secretion. **Aprocrine glands** are a subdivision of merocrine glands in which the cells release their products through the cell membranes by exocytosis. This fusion of secretory vessels with the membrane allows the cell to stay whole. Ceruminous glands, mammary glands, and certain types of sweat glands are apocrine glands (Chapter 5). **Holocrine glands** contain cells that break down to release the secretion, and thereby release the entire cell (Figure 4.4). The sebaceous glands in the skin are holocrine glands. The form of secretion is called holocrine secretion.

**TABLE 4.4** *A Comparison of Exocrine Glands*

| Type | Description | Examples |
|---|---|---|
| Unicellular | Composed of one cell; secretes the glycoprotein mucin, which combines with water to form mucus | Goblet cells are only example in humans |
| Multicellular (Simple) | Composed of many cells; duct not branched | |
| • **Tubular** | Secretory unit is a straight tube opening onto the epithelial surface; no duct is present | Intestinal glands |
| • Coiled tubular | Secretory unit is a coiled tubule; an unbranched duct conveys secretion to the surface | Sweat glands |
| • Branched tubular | Secretory unit is tubular and branched; duct may be absent | Gastric glands, uterine glands |
| • Branched acinar (alveolar) | Secretory unit is shaped like a sac, and several acini are arranged along a duct | Sebaceous (oil) glands in the skin |
| Multicellular (Compound) | Composed of many cells; duct branches | |
| • **Tubular** | Secretory unit is tubular | Liver, testes |
| • **Acinar** | Secretory unit is saclike | Salivary glands (submandibular and sublingual) |
| • **Tubuloacinar** | Secretory unit is both tubular and saclike | Pancreas, parotid salivary glands |

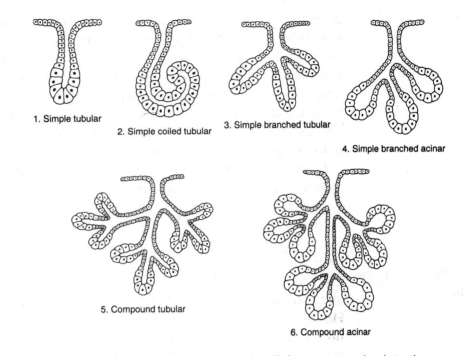

1. Simple tubular

2. Simple coiled tubular

3. Simple branched tubular

4. Simple branched acinar

5. Compound tubular

6. Compound acinar

**FIGURE 4.3**  *The various types of multicellular exocrine glands in the human body. Types 1 to 4 are "simple" because a single unbranched duct leads from the secretory cells. Types 5 and 6 are "compound" because of the branched duct. The acinar glands are saclike.*

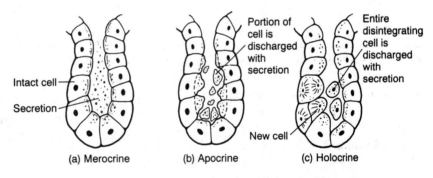

Intact cell

Secretion

Portion of cell is discharged with secretion

Entire disintegrating cell is discharged with secretion

New cell

(a) Merocrine

(b) Apocrine

(c) Holocrine

**FIGURE 4.4**  *Three types of exocrine glands classified by activity: (a) merocrine glands remain intact and permit the secretions to dissolve out of the cell; (b) apocrine glands are those in which secretions leave the cells by exocytosis; and (c) holocrine glands break down and release their secretions in disintegrating cells.*

## Types of Membranes

At several locations in the body, a sheet of epithelium and its underlying connective tissue form a structure known as a **membrane**. Membranes can be moistened by mucus or serous fluid, and therefore they are known as **mucous membranes** or **serous membranes**.

Mucous membranes line the internal surfaces of passageways leading to the outside of the body, such as in the respiratory and urinary tracts. These membranes usually consist of three layers: **epithelium** which may be simple, pseudostratified, or stratified; **lamina propia**, a connective tissue that supports the epithelium and is rich in blood; and the **muscularis mucosae**, which are various layers of circular and longitudinal smooth muscle. Mucous membranes provide protection to the organs as well as lubrication.

Serous membranes line the body cavities and cover the surfaces of organs in the ventral cavities. Serous membranes include the pleura, pericardium, and peritoneum. These membranes lubricate the organs and permit movement without friction.

# CONNECTIVE TISSUE

Connective tissue lends support to the body and binds together body parts and other tissues. The tissue consists primarily of nonliving intercellular substances produced by connective tissue cells.

There are four general types of connective tissue based on the characteristics of the matrix and ground substance. The first type is **connective tissue proper**. This tissue contains various fibers within a semifluid ground substance. The second type, **cartilage**, has a somewhat solid ground substance within which is a fibrous matrix. **Bone tissue**, the third type, also has a fibrous matrix, but the ground substance, the bone, is exceptionally solid. The fourth type, **blood**, has a fluid ground substance.

In addition to its support and connecting functions, connective tissue provides an area where blood vessels and nerves pass through. It also serves as a medium where nutrients, gases, and waste products are exchanged between cells and capillaries. The ground substance of connective tissue is a barrier against the spread of bacteria and is usually the place where defense against infection takes place.

## Connective Tissue Proper

Connective tissue proper has a ground substance that is semifluid, with several different kinds of fibers. It is subdivided into numerous different kinds of tissue, according to the types and arrangements of cells and fibers.

The first kind of connective tissue proper is **loose connective tissue**, also known as **areolar connective tissue**. This tissue consists of several types of cells embedded in a matrix of loosely arranged fibers. The cells in loose connective tissue include **fibroblasts**, which manufacture the protein fibers of the tissue and the ground substance as well as mucopolysaccharides such as hyaluronic acid. Other cells are the **macrophages**, also called **histiocytes**, which are phagocytic cells that engulf and destroy foreign agents in the tissue. In addition, there are **mast cells**, which contain cytoplasmic granules and histamine, a substance functioning during allergic reactions and inflammation. The last kind of cell are **lymphocytes**, a type of white blood cell acting in immune responses.

Loose connective tissue also contains collagen fibers, elastic fibers, and reticular fibers. **Collagen fibers** consist of the protein **collagen**, which forms fibers with high tensile strength and flexibility. The collagen may be fibrous and long-spacing collagen, which is found in the trabecular network of the eye, or segmented long-spacing collagen, which occurs in segments instead of fibers. **Elastic fibers** contain the protein **elastin**, which permits fibers to stretch and recoil quickly. **Reticular fibers** are thin delicate collagen fibers that support capillaries as well as nerve and muscle fibers. Loose connective tissue is found between the skin and muscles and beneath most epithelial linings.

The second type of connective tissue proper is **dense connective tissue** (Table 4.5). This tissue has the same components as loose

**TABLE 4.5**   *Characteristics of Connective Tissues*

| Connective Tissue | Locations | Functions |
|---|---|---|
| Loose connective tissue | Beneath the skin, between muscles, beneath most epithelial layers | Binds organs together, holds tissue fluids |
| Dense connective tissue | Tendons and ligaments | Binds organs together |
| Reticular connective tissue | Walls of liver, spleen, and lymphatic organs | Support |
| Adipose tissue | Beneath the skin, around the kidneys, behind the eyeballs, on the surface of the heart | Protection, insulation, and storage of fat |
| Elastic connective tissue | Between adjacent vertebrae, in walls of arteries and airways | Provides elastic quality |
| Pigmented connective tissue | Eyes | Store pigment |
| Hyaline cartilage | Ends of bones, nose, and rings in walls of respiratory passages (trachea, bronchi) | Support, protection, provides framework |
| Elastic cartilage | Framework of external ear and part of the larynx | Support, protection, provides framework |
| Fibrous cartilage | Between bony parts of backbone, pelvic girdle, and knee | Support, protection |
| Bone | Bones of skeleton | Support, protection, provides framework |

connective tissue, but the collagen and elastic fibers are more closely packed, and the tissue is more dense. Dense connective tissue is **regular** when the collagen fibers are arranged in parallel bundles, or **irregular** when the collagen fibers are arranged in irregular bundles. Dense regular connective tissue is found in tendons and ligaments, while dense irregular connective tissue is found over muscles, bone, cartilage, and in capsule wrappings that cover certain body organs. Sheetlike tendons called **aponeuroses** are composed of dense regular connective tissue.

The third type of connective tissue proper is **reticular connective tissue**. This tissue has delicate fibers forming a network called a reticulum; the network supports soft organs such as the spleen, lymph nodes, and liver. The fibers are synthesized by reticular cells.

The fourth type of connective tissue proper is **adipose tissue**, also known as fat tissue. In this tissue, the adipose cells expand with fat droplets and push the nucleus and cytoplasm to a thin rim around the cell's edge. The fat cell, therefore, resembles a ring. There are few fibers in the tissue. The function of the tissue is to protect, insulate against heat loss, and serve as a storage depot for fat, which is a useful energy source.

The fifth kind of connective tissue proper is **elastic connective tissue**. This tissue contains numerous branching elastic fibers arranged in parallel strands or networks. Fibroblasts are present between the fibers. The tissue is found in elastic ligaments, such as between adjacent vertebrae, in the vocal chords, and in the walls of the largest arteries.

The final kind of connective tissue proper is **pigmented connective tissue**. This tissue is found in certain structures of the eyes. It contains cells that store pigment, usually melanin pigment.

# Cartilage

Cartilage is a type of supporting connective tissue with tensile strength supported by fibers of collagen and other materials in the ground substance. These fibers are produced by cartilage cells called **chondroblasts**. Eventually the chondroblasts become trapped in their own products and revert to mature cells referred to as **chondrocytes**.

The substance of cartilage is a firm, rubberlike mixture of proteins and proteoglycans, which are protein molecules associated with carbohydrate units. Cartilage has no blood vessels and is replaced by bone in most places in the body. However, three kinds of cartilage persist in the body: hyaline cartilage, elastic cartilage, and fibrous cartilage (Figure 4.5).

**Hyaline cartilage** occurs at the ends of long bones, in the external ear, fetal skeleton, and nose, larynx, trachea, and bronchi. It is the weakest kind of cartilage as well as the most common in the body. Under the microscope, the chondrocytes can be seen in spaces called **lacunae**.

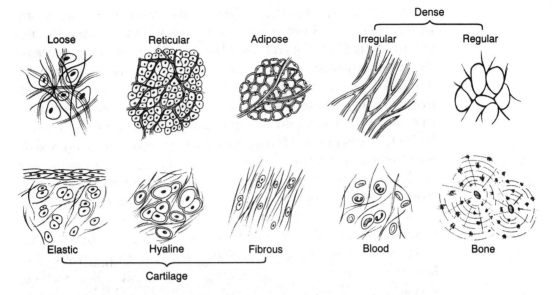

**FIGURE 4.5** *Various types of connective tissue found in the human body. The tissues differ in the types of cells they have and in the nonliving substance among the cells.*

The second kind of cartilage, **elastic cartilage**, has interlacing fibers, which are flexible. This tissue is found in the epiglottis, external ear, and Eustachian tube. The branching elastic fibers distinguish it from hyaline cartilage.

**Fibrous cartilage** is the strongest cartilage, with dense collagen fibers and a limited amount of ground substance. It is found in the pubic symphysis, skull bones, and disks between the vertebrae. Fibrous cartilage predominates in body areas that bear great amounts of weight. It has high tensile strength. Fibrous cartilage (also called fibrocartilage) also occurs in the tendinous and ligamentous insertions. It is arranged in thin, roughly parallel bundles that give it a grainy fibrous appearance. Chondrocytes are seen between the bundles of collagen.

## Bone

Bone tissue is the hardest connective tissue. It consists of cells, collagen fibers, and dense mineralized ground substance known as bone. Bone is much stronger than cartilage because bone contains inorganic salts of calcium and phosphate that make it hard and permit it to bear weight. Bone cells have rich blood supplies, while cartilage cells have no blood supply.

Various types of cells are contained in bone, including osteoblasts, osteocytes, and osteoclasts. **Osteoblasts** synthesize the components of bones, then become mature cells called **osteocytes**, which exist within lacunae in the matrix. Working with other cells, **osteoclasts** destroy bone and remodel it.

The ground substance of bone consists of an inorganic chemical substance called **hydroxyapatite**, which is a calcium-phosphorous salt that constitutes the main component of bone. **Compact (dense) bone** is solid bone present in the external portion of a long bone. **Cancelous bone (spongy bone)** has a weblike structure of thin plates of bone called **trabeculae**. This bone encloses the marrow space of long bones and is internal to the compact bone.

The basic structure of compact bone in adults is the Haversian system, or osteon. This structure and other unique structures of bone are considered in Chapter 6.

## Blood

**Blood** is a specialized type of connective tissue in which the ground substance is exclusively fluid. This fluid is not synthesized by cells contained in the fluid but rather from fluid taken into the body and then supplemented. Blood serves as a medium for the transport of nutrients, gases, and waste products to and from the cells, and it is the site where body defense usually takes place. Blood is discussed in Chapter 14.

## MUSCLE AND NERVE TISSUES

**Muscle tissue** has the ability to exert force when it contracts; this contraction generally produces movement. The cells of muscle tissue are elongated to support their contracting function. The cytoplasm is referred to as sarcoplasm, and the cellular organelles of muscle cells have specialized names as well (Chapter 8).

Muscle tissue can be attached to the skeleton, where it is called **skeletal muscle**. This type of muscle is under control by the nervous system and is **voluntary muscle**. It normally has bands surrounding it and is therefore referred to as **striated muscle** (Figure 4.6).

Another kind of muscle is not associated with the skeleton, but with the internal organs. It is found in the walls of the digestive, urinary, circulatory, and respiratory tracts. Since it has no apparent bands or striations, this muscle is known as **smooth muscle** (also called **visceral muscle**). Smooth muscle does not require nervous stimulation, and is therefore **involuntary muscle**.

The third major kind of muscle is **cardiac muscle**. This muscle is striated and not under voluntary control. It is found in the heart and in the walls of large vessels near the heart. Its branching fibers contain cells joined to one another by specialized cell junctions called **intercalated disks**.

The fourth major kind of human body tissue is **nervous tissue**. Nervous tissue serves a communication function; it receives stimuli, transports impulses, and interprets nerve impulses.

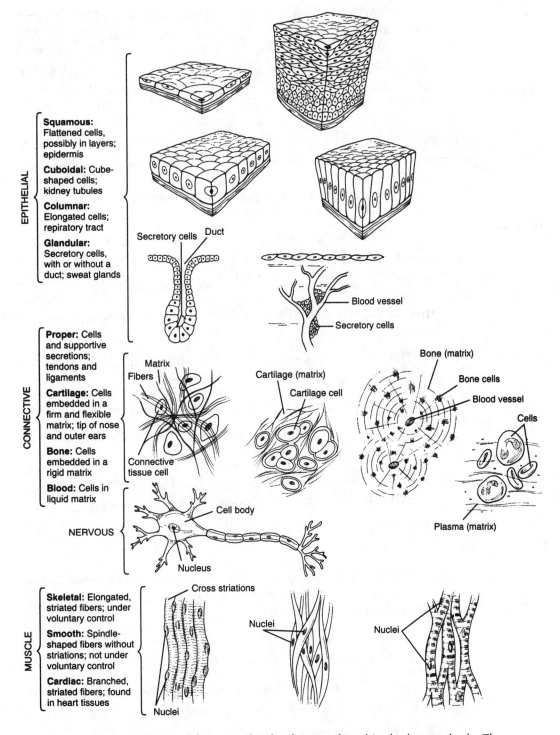

**EPITHELIAL**

**Squamous:** Flattened cells, possibly in layers; epidermis

**Cuboidal:** Cube-shaped cells; kidney tubules

**Columnar:** Elongated cells; repiratory tract

**Glandular:** Secretory cells, with or without a duct; sweat glands

Secretory cells — Duct

Blood vessel

Secretory cells

**CONNECTIVE**

**Proper:** Cells and supportive secretions; tendons and ligaments

**Cartilage:** Cells embedded in a firm and flexible matrix; tip of nose and outer ears

**Bone:** Cells embedded in a rigid matrix

**Blood:** Cells in liquid matrix

Matrix
Fibers
Connective tissue cell

Cartilage (matrix)
Cartilage cell

Bone (matrix)
Bone cells
Blood vessel
Cells
Plasma (matrix)

**NERVOUS**

Cell body
Nucleus

**MUSCLE**

**Skeletal:** Elongated, striated fibers; under voluntary control

**Smooth:** Spindle-shaped fibers without striations; not under voluntary control

**Cardiac:** Branched, striated fibers; found in heart tissues

Cross striations
Nuclei
Nuclei
Nuclei

**FIGURE 4.6**   *A summary of the many kinds of tissues found in the human body. There are four main types of tissues and several subdivisions of each.*

Nervous tissue consists of two main kinds of cells: supportive cells called **neuroglial cells** and conducting cells called **neurons**. Neurons have a distinctive shape, with numerous branches that permit stimuli to be detected and transported. **Sensory neurons** receive impulses from the external environment and transport them to the central nervous system (the brain and spinal cord), while **interneu-rons** interpret the stimuli and send out appropriate responses by means of **motor neurons**. Motor neurons carry impulses to muscles and glands for an appropriate response. The anatomy and physiology of the nervous tissue are the main topics of Chapter 10.

# REVIEW QUESTIONS

*PART A—Completion:* **Add the word or words that correctly complete each of the following statements.**

1. The tissue that covers the body surfaces, lines its cavities, and forms the major portions of many glands is called _____ .

2. The matrix of connective tissue contains both cells and _____ .

3. The major function of muscle tissue in the body is to permit _____ .

4. The cells of the nervous tissue are specially adapted to receive and relay _____ .

5. The cells in epithelial tissue divide by the process of _____ .

6. Epithelial tissue is anchored to the underlying connective tissue by a noncellular _____ .

7. Those junctions formed by fusions of cell membranes of adjacent cells with interlocking membrane proteins are _____ .

8. Adhering junctions between epithelial cells extend around the entire cell _____ .

9. A submicroscopic space separates opposing cell membranes, and keratin cells bind these cells together in macular adhering junctions, also known as _____ .

10. Epithelium-containing cells that produce products such as enzymes, sweat, and hormones are known as _____ .

11. Epithelial tissue occurring in sheets having a single layer is known as _____ .

12. In transitional epithelium, the cells change shape in response to _____ .

13. Both endothelium and mesothelium are types of _____ .

14. Epithelium consisting of a single layer of tall cells with nuclei close to the base of the cells is described as _____ .

15. The epidermis of the skin consists of epithelium known as _____ .

16. The primary function of stratified squamous epithelium in the human body is _____ .

17. The ureter and urinary bladder contain up to six or seven layers of epithelial tissue known as _____ .

18. Those glands that secrete hormones directly into the bloodstream are _____ .

19. The exocrine glands are notable because they secrete their products into _____ .

20. A goblet cell is a type of exocrine gland composed of _____ .

21. Multicellular glands consist of numerous cells and are typified by the salivary glands, mammary glands, pancreatic glands, and _____ .

22. The slimy, viscous material produced by goblet cells of the digestive tract and cells of the respiratory tract is known as _____ .

23. A serous-secreting gland produces watery material that often contains _____ .

24. Those glands that release their products through the cell membranes through the process of exocytosis are known as _____ .

25. When a cell breaks down to release its secretions, the cells are derived from glands known as _____ .

26. Those glands with branched ducts are _____ .

27. At various places in the body, membranes are formed by a sheet of epithelium and its underlying _____ .

28. In a mucous membrane, the connective tissue that supports the epithelium and contains blood vessels is _____ .

29. The pleura, pericardium, and peritoneum are examples of membranes referred to as _____ .

30. Four different types of connective tissue are established based on the characteristics of the ground substance and the _____ .

31. The protein fibers of loose connective tissue are synthesized by cells known as _____ .

32. In loose connective tissue, macrophages engulf and destroy foreign agents in the process of _____ .

33. The thin, delicate collagen fibers that support capillaries, nerve, and muscle fibers in loose connective tissue are called _____ .

34. Dense regular connective tissue contains parallel bundles of collagen fibers and is found in ligaments and _____ .

35. Dense irregular connective tissues consist of collagen fibers arranged in irregular bundles and are found as a covering over cartilage, bone, and _____ .

36. Fat tissue is a type of connective tissue proper, also known as _____ .

37. The ligaments between adjacent vertebrae, the vocal chords, and the walls of the large arteries contain a type of connective tissue proper called _____ .

38. The cells that produce cartilage are _____ .

39. Cartilage consists of a firm, rubberlike mixture of proteoglycans and _____ .

40. The weakest type of cartilage, found at the ends of the long bones and in the external ear and nose is called the _____ .

41. The epiglottis, external ear, and Eustachian tube contain a flexible type of cartilage referred to as _____ .

42. Part of the reason that bone is much stronger than cartilage is because the matrix of bone tissue contains salts of calcium and _____ .

43. Hydroxyapatite is found primarily in tissue known as _____ .

44. The components found in bone tissue are synthesized by specialized cells called _____ .

45. In adults, the basic structure of compact bone is the osteon, also known as the _____ .

46. The ground substance is exclusively fluid in the connective tissue referred to as _____ .

47. Skeletal muscle is attached to the skeleton, and since it is under control of the nervous system, it is also known as _____ .

48. The cells of skeletal muscle are surrounded by bands, and skeletal muscle is also referred to as _____ .

49. The muscle of the heart and walls of large vessels near the heart is a special form of muscle referred to as _____ .

50. Neuroglial cells and neurons are the two main kinds of cells in _____ .

*PART B—Multiple Choice:* Circle the letter of the item that correctly completes each of the following statements.

1. All the following are functions of nervous tissue except
   (A) it coordinates conscious activities
   (B) it receives and relays signals
   (C) it provides support for the body
   (D) it responds to stimuli

2. The major portions of many human glands are composed of
   (A) muscle tissues
   (B) epithelial tissues
   (C) blood tissues
   (D) nervous tissue

3. All the following systems are lined with epithelial tissue except
   (A) the reproductive system
   (B) the urinary system
   (C) the respiratory system
   (D) the endocrine system

4. Zonular junctions and macular junctions are both types of
   (A) adhering junctions
   (B) tight junctions
   (C) cartilage junctions
   (D) macrophage junctions

5. Keratin fibers bind the cells together and a submicroscopic space separates opposing cell membranes in
   (A) desmosomes
   (B) tight junctions
   (C) connective junctions
   (D) lamina propia

6. One of the important functions of epithelial tissue is to
   (A) relay nerve impulses
   (B) serve as a storage depot for fat
   (C) bear great amounts of weight in the body
   (D) absorb gases or nutrients

7. Simple squamous epithelium consists of
   (A) a single layer of tall cells
   (B) many layers of different types of cells
   (C) a single layer of flat cells
   (D) several layers of cuboidal cells

8. Epithelium in which the cells change shape in response to mechanical stretching is known as
   (A) transitional epithelium
   (B) pseudostratified epithelium
   (C) stratified epithelium
   (D) cuboidal epithelium

9. One of the principal locations of stratified squamous epithelium in the body is the
   (A) medulla of the kidney
   (B) fallopian tubes
   (C) epidermis of the skin
   (D) nerve tissue of the brain

10. Exocrine glands are those glands that
    (A) produce hormones
    (B) secrete their products directly into the bloodstream
    (C) are referred to as ductless glands
    (D) secrete their products into ducts

11. The salivary glands, mammary glands, and sweat glands are collectively referred to as
    (A) goblet glands
    (B) multicellular glands
    (C) single-celled glands
    (D) endocrine glands

12. Serous can be distinguished from mucus because serous
    (A) is more viscous
    (B) is more watery
    (C) contains no enzymes
    (D) is produced only by goblet cells

13. Those glands that release their products through the membranes of cells by the process of exocytosis are known as
    (A) exocrine glands
    (B) holocrine glands
    (C) multicellular glands
    (D) merocrine glands

14. All the following are layers of mucous membranes except
    (A) lamina propia
    (B) epithelium
    (C) cartilage
    (D) muscularis mucosae

15. Macrophages, mast cells, and lymphocytes may all be located in
    (A) blood tissue
    (B) loose connective tissue
    (C) nerve tissue
    (D) dense connective tissue

16. Reticular fibers are delicate fibers of collagen that support
    (A) nerve fibers, muscle fibers, and capillaries
    (B) bones
    (C) blood cells
    (D) the cytoskeleton of cells

17. Tendons and ligaments derive much of their strength from the fact that they are composed of
    (A) hyaline cartilage
    (B) hydroxyapatite
    (C) skeletal muscle
    (D) dense regular connective tissue

18. The cells of adipose tissue appear microscopically as
    (A) cells with many nuclei
    (B) cells with the nucleus and cytoplasm at the rim of the cell
    (C) cells with bands around them
    (D) long branching cells with several short projections and one long projection

19. The fat stored in adipose tissue is useful
    (A) for the synthesis of proteins
    (B) for the synthesis of nuclei
    (C) as an energy source
    (D) as a source of minerals

20. The most widely found type of cartilage in the body and the type that may be located in the external ear, nose, trachea, and bronchi is
    (A) elastic cartilage
    (B) hyaline cartilage
    (C) fibrous cartilage
    (D) osseus cartilage

21. Which of the following characteristics applies to cartilage?
    (A) the substance of cartilage is mainly carbohydrate
    (B) there are five different kind of cartilage in the body
    (C) cartilage has no blood vessels
    (D) cartilage is never replaced by bone in the body

22. The ground substance of bone is composed of hydroxyapatite, which is a
    (A) sodium chloride salt
    (B) potassium permanganate salt
    (C) bismuth chloride salt
    (D) calcium phosphate salt

23. The major bone-forming cells are
    (A) osteocytes
    (B) osteoclasts
    (C) osteoblasts
    (D) osteophages

24. All the following are functions of the blood except
    (A) it transports waste products from the cells
    (B) it is the site where body defense usually takes place
    (C) it is the location where carbohydrates are synthesized
    (D) it transports nutrients and gases to the cells

25. The sensory neurons of the body
    (A) receive impulses and transport them to the central nervous system
    (B) interpret the stimuli received from the environment
    (C) carry impulses from the central nervous system to the muscles
    (D) stimulate the endocrine glands to secrete hormones

*PART C—True/False:* **For each of the following statements, mark the letter "T" next to the statement if it is true. If the statement is false, change the** underlined **word to make the statement true.**

1. The type of tissue that covers the body surfaces is underlined(connective) tissue.

2. Adhering junctions in epithelial tissue extend around the perimeter of the cells and are formed by fusion of the cell membranes of adjacent cells with interlocking membrane proteins.

3. Desmosomes are a type of tight junction in which a submicroscopic space separates opposing cell membrane and keratin fibers bind the cells together.

4. Simple cuboidal epithelium is found in the pigmented epithelium of the retina of the eye, the surface layer of the ovary, and parts of the testis.

5. The trachea and upper respiratory tract as well as parts of the male reproductive ducts are lined with stratified squamous epithelium.

6. Layers of stratified cuboidal epithelium consist of layers of cuboidal cells and are found in the tubules of the testis and in follicles of the ovary.

7. The exocrine glands secrete their products into the blood for transport.

8. The slimy, viscous material known as mucus is composed primarily of protein and fat.

9. The adrenal glands produce both mucus and enzymes within serous fluid, and are therefore known as seromucous glands.

10. Merocrine glands are those glands in which the cells break down to release the product of that gland.

11. Those glands having branched ducts leading away from the gland are known as simple glands.

12. <u>Mucous</u> membranes line the body cavities and cover the surfaces of organs in the ventral cavities.

13. The pleura, pericardium, and peritoneum are examples of <u>serous</u> membranes.

14. The cells present in loose connective tissue are the <u>osteoblasts</u>.

15. Phagocytic cells that engulf foreign agents in connective tissues are the <u>macrophages</u>.

16. Collagen is a <u>polysaccharide</u> that forms fibers with high tensile strength and flexibility in loose connective tissue.

17. Dense regular connective tissue is the main component of <u>ligaments</u>, which are sheetlike tendons that bind many muscles to bones.

18. The type of connective tissue that has delicate fibers in a network and supports soft organs such as the spleen and liver is called <u>elastic</u> connective tissue.

19. Chondroblasts are the cells widely found in <u>bone tissue</u>.

20. <u>Compact</u> bone is the type of solid bone found at the external portion of long bones.

21. Blood is a specialized type of connective tissue in which the ground substance is exclusively <u>fluid</u>.

22. Smooth muscles are <u>voluntary</u> muscles.

23. Intercalated disks are specialized types of cell junctions that may be found in <u>striated</u> muscle.

24. The supportive cells found in nervous tissue are known as <u>neuroglial</u> cells.

25. <u>Motor</u> neurons are those neurons that receive impulses from the environment and transport them to the central nervous system.

**Answers**

### PART A—Completion

1. epithelium
2. ground substance
3. movement
4. signals
5. mitosis
6. basement membrane
7. tight junctions
8. perimeter
9. desmosomes
10. glandular epithelium
11. simple epithelium
12. mechanical stretching
13. simple squamous epithelium
14. simple columnar epithelium
15. stratified squamous epithelium
16. protection
17. transitional epithelium
18. endocrine glands
19. ducts
20. one cell
21. sweat glands
22. mucus
23. enzymes
24. merocrine glands
25. holocrine glands
26. compound glands
27. connective tissue
28. lamina propia
29. serous membranes
30. matrix
31. fibroblasts
32. phagocytosis
33. reticular fibers
34. tendons
35. muscles
36. adipose tissue
37. elastic connective tissue
38. chondroblasts
39. proteins
40. hyaline cartilage
41. elastic cartilage
42. phosphate
43. bone
44. osteoblasts
45. Haversian system
46. blood
47. voluntary muscle
48. striated muscle
49. cardiac muscle
50. nervous tissue

### PART B—Multiple Choice

| | | | | |
|---|---|---|---|---|
| 1. C | 6. D | 11. B | 16. A | 21. C |
| 2. B | 7. C | 12. B | 17. D | 22. D |
| 3. D | 8. A | 13. D | 18. B | 23. C |
| 4. A | 9. C | 14. C | 19. C | 24. C |
| 5. A | 10. D | 15. B | 20. B | 25. A |

### PART C—True/False

1. epithelial
2. tight
3. adhering
4. true
5. pseudostratified columnar
6. true
7. ducts
8. polysaccharide
9. salivary
10. holocrine
11. compound
12. serous
13. true
14. fibroblasts
15. true
16. protein
17. aponeuroses
18. reticular
19. cartilage
20. true
21. true
22. involuntary
23. cardiac
24. true
25. sensory

# CHAPTER 5

# THE INTEGUMENTARY SYSTEM

The integumentary system is composed of the skin and a number of skin derivatives such as hair, nails, and glands. Integument is an alternative name for skin; the skin is the major organ of the integumentary system.

As the body's largest organ, the skin accounts for about 15 percent of body weight. The skin forms a self-repairing and protective boundary between the body and the external environment.

## THE SKIN

The skin has numerous functions essential to the well-being of the body.

## Functions of the Integumentary System

The integumentary system (primarily the skin) performs several functions associated with the maintenance of homeostasis and the survival of the body. One of the most important functions is **protection**. Skin protects the body from fluid loss or gain as well as invasion by microorganisms and other mechanical and physical irritants.

The skin exerts its protective function by forming three barriers:

1. the skin surface forms a barrier to infection;
2. the outer epidermis contains a waterproof layer of the protein keratin, which is a barrier to water-soluble substances; and
3. the basement membrane at the foundation of the skin is a mechanical barrier to the area below it (Figure 5.1).

Protection is also rendered by **melanin pigment** formed in the skin. This pigment protects the skin against ultraviolet rays from the sun. The pigment is synthesized by melanocytes; the more melanin present, the darker the skin. Prevention of melanin synthesis results in albinism.

Another function of the integumentary system is **temperature regulation**. The skin permits the body to lose heat by evaporation, radiation, convection, and conduction (Chapter 21). Heat is lost in

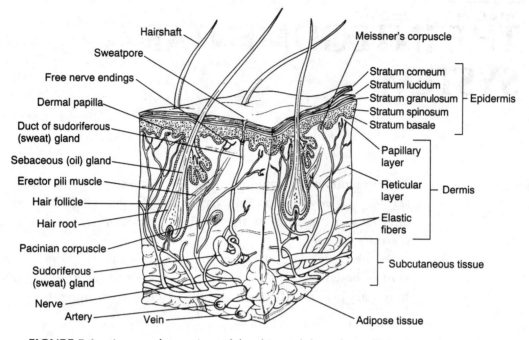

**FIGURE 5.1** *A general overview of the skin and the subcutaneous layer beneath it.*

sweat at the skin surface and from blood vessels in the skin. The skin conserves heat by reducing secretions of sweat and constricting its blood vessels.

Still another function is **excretion**. Glands located in the skin excrete water, fatty substances, and ions such as sodium ions. **Metabolism** is yet another function (Table 5.1). In the presence of ultraviolet radiation from the sun, vitamin D molecules are synthesized from precursor molecules accumulating in the skin. These precursor molecules are generally obtained from milk. The completion of vitamin D synthesis occurs in the kidney.

The integumentary system functions in the **absorption** of several substances, including the fat-soluble vitamins A, E, and K. Steroid hormones are also released by glands in the skin, then absorbed into the skin tissues.

A final function concerns **communication**. Stimuli from the environment are received by specialized skin receptors, which communicate the information to the central nervous system. Sensations such as pressure, pain, touch, and temperature are detected in this way.

**TABLE 5.1** *Functions of the Skin*

- Helps regulate the body temperature
- Protects the underlying body tissues
- Receives stimuli from the external environment
- Excretes water, salts, and certain organic compounds
- Synthesizes vitamin D
- Renders immunity via specialized cells

## Structure of the Skin

The two major layers of the skin are the **epidermis** and the **dermis**. These layers fit together in a somewhat wavy configuration. The epidermis is the outer, thinner layer of skin, while the dermis is the inner, thicker layer. The **dermal-epidermal junction** is the region where the cells of the epidermis meet the connective tissue cells of the dermis. Beneath the dermis lies a loose subcutaneous layer rich in fat and areolar tissue. This layer is called the **superficial fascia**, also known as the **hypodermis**.

### The Epidermis

The epidermis of the skin is composed of stratified squamous epithelium. There are no blood vessels in the tissue, and the cells are packed closely together. In **thin epithelium** found along most body surfaces, there are four different layers, or strata of epidermis; in **thick epithelium** such as on the palms of the hands and soles of the feet, the epidermis has five layers.

The innermost layer of epidermis lying on the basement membrane next to the dermis is the **stratum basale**, also called the **stratum germinativum**. It is a single layer of cells where cell division occurs frequently. New cells from this layer push up into the next higher layer. Many of the cells of the stratum basale are **melanocytes**, which synthesize melanin. Other cells are **keratinocytes**, the predominant epidermal cells. Still others are **Merkel cells**, which are sensitive to touch when compressed or disturbed. Merkel cells are a type of sensory receptor; other sensory receptors are located in the dermis.

The next layer of epidermis toward the body surface is the **stratum spinosum** (Figure 5.2). This layer consists of keratinocytes having a spiny appearance. The spines are points where cells adhere to one another as parts of intercellular junctions called **desmosomes** (Chapter 4). Keratinocytes synthesize the protein **keratin**, which provides a waterproof layer to the epidermis. Less frequent cell division occurs in the stratum spinosum than in the stratum basale.

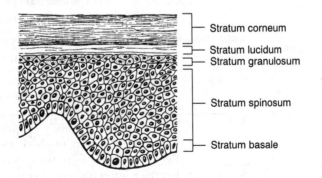

**FIGURE 5.2**   *The layers of the epidermis. The stratum lucidum is found only in the palms and soles.*

The next layer toward the surface is the **stratum granulosum**. This layer is composed of flattened cells with granules containing a substance called **keratohyalin**. Keratohyalin is used to form keratin, so the layer is rich in both keratin and keratohyalin. **Langerhans cells** related to the immune system are located in the stratum granulosum. The layer contains three to five cell layers, and as the cells die, they are replaced by keratin and keratohyalin.

The next higher layer is the **stratum lucidum**. This stratum is found only in thick epithelium. Cells in this layer are closely packed and clear, and keratinocytes are usually dead and without nuclei. The cells are filled with a transparent substance called **eleidin**, which is formed from the keratohyalin; eventually eleidin is transformed to keratin.

The uppermost layer of epidermis at the body surface is the **stratum corneum** (Table 5.2). About 25 layers of dead and dry squamous cells make up this tissue layer. Cells are constantly shed from the stratum corneum. The keratin in the cells waterproofs the cell surface, while the cells provide a barrier against infection and mechanical injury. As in the stratum spinosum, junctions called desmosomes connect the keratinocytes to one another and strengthen the layer.

The cytoplasm of most cells in the stratum corneum has been replaced by keratin. Cells in the stratum corneum are ultimately derived from cells produced in the stratum basale. As the cells move toward the body surface, they accumulate keratin and die. Thus, there is a bottom-to-top replacement of skin cells. No blood vessels and few sense receptors are found in the epidermis. A thickening of the stratum corneum is called **hyperkeratosis**.

**TABLE 5.2**  *Layers of the Epidermis*

| Stratum (Layer) | Characteristics |
| --- | --- |
| Stratum basale (stratum germinativum) | Deepest layer; single layer of cuboidal or columnar cells; site of continuous cellular reproduction; contains the only cells of the epidermis that receive nutrition; cells are constantly undergoing division and being pushed up to the body surface. |
| Stratum spinosum | Many keratinocytes with spiny appearance; some keratin. |
| Stratum granulosum | Three to five rows of flat cells; site of keratohyalin and keratin formation. |
| Stratum lucidum | Only in the thick skin of the palms and soles; consists of clear, flat, dead cells; cells contain eleidin. |
| Stratum corneum | Outermost layer of epidermis; 25 to 30 rows of flat, dead cells filled with keratin; continuously shed and replaced. |

All cells of the epidermis receive nourishment from blood vessels in the dermis. Since sensory receptors are found in the dermis, any heat, cold, pressure, or other stimulus must pass through the epidermis to reach the receptors. However, touch and pain receptors (Merkel cells) extend into the epidermis and are stimulated more easily than other receptors.

### The Dermis

The second major layer of the skin, the **dermis**, communicates closely with the epidermis by the basement membrane (Chapter 4). Specialized fibrous elements and unique polysaccharide gels also bind the two layers together. This binding provides a barrier to the passage of large molecules and certain microorganisms. Ridges from the dermis called **dermal papillae** project into the epidermis and help anchor the two layers together. Hair follicles exist within the dermal papillae.

The dermis is composed of two layers: the papillary layer, and the reticular layer. The **papillary layer** consists of loose, areolar connective tissue having macrophages, blood vessels, fibroblasts, and other cells. The layer contains many blood vessels that provide nourishment to the epidermis above. The **reticular layer** contains sebaceous (oil) glands as well as fat cells, sweat glands, and larger blood vessels. Both the reticular and papillary layers have sensory receptors for pain, pressure, temperature, and touch. The reticular layer also contains connective tissues with fibers running in multiple directions.

Much of the mechanical strength of the skin is centered in the dermis (Table 5.3). The dermis protects against injury and provides a reservoir storage area for water and electrolytes. In addition, hair follicles, sebaceous glands, muscle fibers, and many blood vessels are found in this skin layer. The blood in the dermis helps regulate body temperature.

**TABLE 5.3**   *Organs of the Integumentary System*

| Organ | Primary Functions |
| --- | --- |
| Epidermis | Protects underlying tissues |
| Dermis | Nourishes epidermis, provides strength |
| Hair Follicles | Produce hair |
| • Hairs | Provide sensation and some protection for head |
| • Sebaceous glands | Secrete lipid coating that lubricates hair shaft |
| Sweat Glands | Produce perspiration for evaporative cooling |
| Nails | Protect and stiffen distal tips of digits |
| Sensory Receptors | Provide sensations of touch, pressure, temperature, and pain |
| Superfacial Fascia | Stores fat |

Certain regions of the dermis contain both skeletal and smooth muscle fibers. For instance, there are skeletal muscles in the skin of the scalp. The muscle permits facial expressions and voluntary scalp movements. Hair follicles in the skin have smooth muscles called the **erector pili muscles**. These muscles permit the hairs to stand upright during frightful situations and intense cold.

## DERIVATIVES OF THE SKIN

Several specialized structures are derived from the skin, especially the tissue of the epidermis. These derivatives have various functions in the physiology of the body.

## Hairs

Among the skin derivatives are the **hairs** (Figure 5.3). Hairs, also known as **pili** (singular pilus), are found on the entire body surface except the eyelids, palms of the hands, soles of the feet, and lips. They are organs of sensation and protection for the skin. A **hair fiber** consists of a **shaft** extending above the skin surface and a **root** lying within a hair follicle. A **hair follicle** is a mass of epidermis, extending down into the dermis and forming a small tube.

The hair follicle swells at its base to form a **hair bulb**, which is infiltrated by connective tissues, blood vessels, and nerves to form the dermal papilla. The dermal papilla extends into the epidermis and provides nutrients to the growing hair. A sebaceous gland, an erector pili muscle, and nerve endings are associated with each hair.

Hair formation derives from activity of cells of the stratum basale found within the hair bulb. These cells produce new hair cells and push the old cells through the follicle, thereby bringing about hair growth. As the cells become keratinized, they die and form hair fibers. Blood vessels in the papilla nourish the cells in the stratum basale. Hair shafts break regularly and are continuously replaced by hair regrowth in the follicles. When hair fails to grow, the result is baldness, a condition due to genetic inheritance, scalp injury, disease, deficiencies in the diet, hormones, or drug therapy.

Hair formation begins before birth and results in extremely fine and soft hair on the fetal skin known as **lanugo**. Most lanugo is lost before birth, and stronger, pigmented hair replaces it after birth. The coarse axillary and pubic hair developing at puberty is known as **terminal hair**.

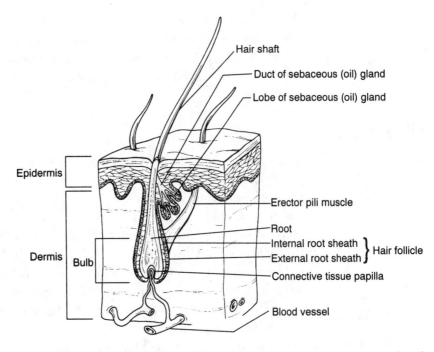

**FIGURE 5.3**   *A diagrammatic illustration of the structures associated with a hair fiber.*

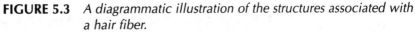
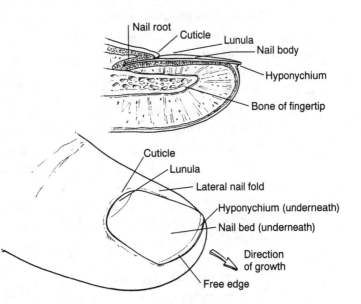

**FIGURE 5.4**   *The prominent features associated with a typical fingernail.*

## Nails

**Nails** are protective plates in the fingers and toes formed from growths of the epidermis into the dermis. The nail itself is a curved plate of keratin resting on a nail bed. The nail bed is supplied with blood vessels. At its proximal end, the nail is partially covered by the **cuticle**. A thick, white **lunula** appears near the cuticle (Figure 5.4).

Nails result from proteins produced by cells beneath the cuticle at the base of the nail. Before the cells die, they fill with keratin, and the nail forms from this keratin. Nails give structure, support, and protection to the distal tips of the digits.

## Glands

The glands found in the skin are of two types: sweat glands and sebaceous glands. **Sweat glands**, also known as **sudoriferous glands**, are the most numerous skin glands. They are subdivided as eccrine sweat glands and apocrine sweat glands.

**Eccrine sweat glands** are widely distributed over the body. They produce sweat and are the "ordinary" sweat glands that liberate sweat into ducts. The transparent watery **sweat** is important in maintaining temperature regulation and for excreting small amounts of sodium chloride and urea from the body. **Apocrine sweat glands** also liberate their secretions to a duct, but a portion of the cell is cast off as a vesicle in the process. They are larger sweat glands found primarily deep in the tissues of the groin and armpits. They secrete a white, cloudy substance that can be metabolized by bacteria to produce end products having odors. Apocrine glands are active during sexual and emotional stimuli.

Two forms of apocrine skin glands other than sweat glands are the ceruminous glands and mammary glands. **Ceruminous glands** are located in the ear canal, where they secrete **cerumin**, or ear wax. The sticky cerumin helps trap foreign substances before they penetrate deeper into the ear. **Mammary glands** are "modified" apocrine glands specialized for the production and secretion of milk, whereas ceruminous glands are "true" apocrine glands.

The second type of gland, **sebaceous glands**, secrete the substance sebum, usually into hair follicles. **Sebum** consists primarily of lipids and oils that keep the hair supple and the skin soft and pliant, while preventing excess water loss from the epidermis. Sebum also has antibacterial activity.

Sebaceous glands are simple, branched, alveolar glands (Chapter 4). They are also classified as holocrine glands, that is, they loose their secretory cells together with their secretions. Acne is a common disorder of the sebaceous glands.

## REVIEW QUESTIONS

**PART A—Completion: Add the word or words that correctly complete each of the following statements.**

1. The integumentary system is composed of the skin and a number of derivatives such as hair, nails, and _____ .

2. The largest organ of the body is the _____ .

3. Of the total body weight, the skin accounts for a percentage that approximates _____ .

4. The skin protects the body from fluid loss or gain and serves as a barrier to _____ .

5. Water-soluble substances may not enter the body because the skin contains a waterproof layer of a protein known as _____ .

6. At the foundation of the skin a barrier of protection is provided by the _____ .

7. The pigment formed in the skin is called _____ .

8. Skin pigments protect the body against radiations from the sun called _____ .

9. Skin pigments are synthesized in special cells called _____ .

10. When the body cannot produce the skin pigment melanin, the condition that results is called _____ .

11. Heat is lost at the body surface in sweat during the process of _____ .

12. The skin conserves heat by reducing its secretions of sweat and by constricting its _____ .

13. Glands in the skin release water and fatty substances in the process of _____ .

14. Ultraviolet radiation from the sun brings about the synthesis in the skin of vitamin _____ .

15. Among the vitamins absorbed in the skin are the fat-soluble vitamins E, K, and _____ .

16. The skin tissue absorbs certain steroid hormones released by the skin's _____ .

17. Environmental stimuli are received by specialized skin _____ .

18. The skin detects sensations such as pressure, touch, temperature, and _____ .

19. The epidermis and dermis fit together in the skin in a configuration that is somewhat _____ .

20. The dermis is the inner, thicker layer of the skin, while the epidermis is the outer _____ .

21. The loose subcutaneous layer of fat-rich tissue beneath the dermis is the _____ .

22. The layers of cells within the epidermis are also referred to as
    _____ .

23. The skin tissue is composed of epithelium that is stratified and
    _____ .

24. The thin epithelium of the epidermis has layers that number
    _____ .

25. The thick epithelium of the epidermis has layers that number
    _____ .

26. The innermost layer of epidermis lying on the basement membrane on the dermis is the _____ .

27. Another name for the stratum germinativum is _____ .

28. The cells that synthesize melanin are the melanocytes, which may be found in the epidermal layer called the _____ .

29. The predominant epidermal cells are the _____ .

30. The protein keratin is produced by epidermal cells known as
    _____ .

31. The epidermal cell layer lying above the stratum basale is the
    _____ .

32. Cells called Langerhans cells are related to the immune system functions and are located in the _____ .

33. Closely packed cells and dead keratinocytes are found within the layer of epidermis known as the _____ .

34. Keratohyalin eventually forms keratin, but an intermediary in the process is the transparent substance _____ .

35. At the body surface, the uppermost layer of the epidermis is the
    _____ .

36. The cells in the uppermost layer of epidermis are rich in the waterproofing substance called _____ .

37. The keratinocytes of the stratum corneum are connected to one another by junctions called _____ .

38. Most cells of the stratum corneum are ultimately derived from cells produced in the lower layer called the _____ .

39. The stratum corneum consists of about 25 layers of dead and dry cells which have been thickened by the process of _____ .

40. The cells of the epidermis receive their nourishment from blood vessels found in the _____ .

41. The epidermis receives ridges from the dermis that help anchor the two layers together and are called _____ .

42. The layer of the dermis containing fat cells, sweat glands, and larger blood vessels is the _____ .

43. The papillary layer of the dermis consists of connective tissue that is loose and _____ .

44. Skeletal muscle may be found in the dermal skin layer of the _____ .

45. The two major structures of the hair fiber are the shaft and the _____ .

46. Nerve endings are associated with each hair together with a sebaceous gland and a muscle known as _____ .

47. The nail is a protective plate consisting of the protein _____ .

48. At its proximal end, the nail is covered partially by a piece of tissue called the _____ .

49. The most numerous skin glands are the sweat glands, also known as _____ .

50. The ear canal contains a number of glands known as _____ .

**PART B—Multiple Choice:** Circle the letter of the item that correctly completes each of the following statements.

1. Which of the following pertains to the skin?
   (A) the skin accounts for about 35% of the body weight
   (B) the skin functions in support of the body organs
   (C) the skin is the largest organ
   (D) the skin has no pigments in its tissues

2. In the skin layers, the protein keratin
   (A) prevents the passage of microorganisms
   (B) serves as a barrier to ions and salts
   (C) forms a barrier to water-soluble substances
   (D) serves as the body pigment

3. The skin conserves body heat by
   (A) reducing secretions of sweat
   (B) producing heat-tolerant hormones
   (C) undergoing mitosis rapidly
   (D) producing cerumin in its ceruminous glands

4. The pigment melanin
   (A) is essential for vitamin D production
   (B) protects the skin against ultraviolet rays from the sun
   (C) is essential for synthesis of the stratum basale
   (D) is used in the synthesis of keratin

5. One of the functions of the skin is to serve in the absorption of
   (A) digestive enzymes
   (B) sodium ions
   (C) cartilage molecules
   (D) certain fat-soluble vitamins

6. Heat is lost from the body by all the following mechanisms except
   (A) evaporation
   (B) osmosis
   (C) conduction
   (D) convection

7. The superficial fascia is located
   (A) between the stratum basale and the stratum spinosum
   (B) beneath the dermis
   (C) within the stratum germinativum
   (D) near the sebaceous glands

8. Another name for the stratum germinativum is the
   (A) dermis
   (B) basement membrane
   (C) capillary layer
   (D) stratum basale

9. Intercellular junctions occurring within the stratum spinosum are referred to as
   (A) desmosomes
   (B) gap junctions
   (C) reticular junctions
   (D) papillosomes

10. Two important cells found in the stratum basale are
    (A) root cells and papillary cells
    (B) lunular cells and sweat cells
    (C) melanocytes and keratinocytes
    (D) hair cells and gland cells

11. Five layers of epithelium are found on the
    (A) legs and arms
    (B) face and neck
    (C) palms of the hands and soles of the feet
    (D) fingers and toes

12. About 25 layers of dead and dry squamous cells make up the
    (A) dermis
    (B) stratum corneum
    (C) stratum granulosum
    (D) desmosomes

13. The reticular layer of the dermis contains all the following except
    (A) fat cells
    (B) loose connective tissue
    (C) large blood vessels
    (D) sweat glands

14. In the stratum corneum, the cytoplasm of most cells has been replaced by
    (A) hair
    (B) sebum
    (C) cerumin
    (D) keratin

15. All living cells of the epidermis receive their nourishment from blood vessels located in the
    (A) basement membrane
    (B) dermis
    (C) hair cells
    (D) endocrine glands

16. The dermis and epidermis are anchored to one another by ridges known as
    (A) dermal papillae
    (B) gap junctions
    (C) hypodermis
    (D) melanocytes

17. Sensory receptors are found
    (A) in the reticular layer of the epidermis
    (B) in the papillary layer of the epidermis
    (C) in neither the reticular nor papillary layer of the dermis
    (D) in both the reticular and papillary layers of the dermis

18. Muscle fibers may be located in the
    (A) reticular layer but not the papillary layer
    (B) dermis but not the epidermis
    (C) epidermis but not the dermis
    (D) papillary layer but not the reticular layer

19. Erector pili muscles are associated with the
    (A) sebaceous glands
    (B) sweat glands
    (C) epidermis
    (D) hair follicles

20. The lanugo is the extremely fine hair associated with
    (A) the back of the hand
    (B) the fetus
    (C) the inner thigh epidermis
    (D) the back of the neck

21. The protein material of the nail results from the metabolism of cells
    (A) beneath the cuticle at the base of the nail
    (B) within the stratum germinativum
    (C) in the reticular layer of the dermis
    (D) that also produce sebum for the sebaceous glands

22. The terms eccrine and apocrine refer to two types of
    (A) ceruminous glands
    (B) basement membranes
    (C) sweat glands
    (D) nails

23. The function of cerumin is to
    (A) maintain temperature regulation at the skin surface
    (B) excrete small amounts of sodium chloride from the body
    (C) help trap foreign substances before they enter the ear
    (D) secrete milk

24. The nail consists of a
    (A) layer of stratum lucidum over a layer of stratum basale
    (B) basement membrane overlying superficial fascia
    (C) curved plate of keratin
    (D) mass of hair shafts that have fused together

25. The secretion of sebaceous glands enters the
    (A) nail
    (B) hair follicle
    (C) basement membrane
    (D) eccrine glands

*PART C—True/False:* **For each of the following statements, mark the letter "T" next to the statement if it is true. If the statement is false, change the underlined word to make the statement true.**

1. The major organ of the integumentary system is the skin.

2. The protein known as keratin prevents the passage of fat-soluble substances through the skin.

3. The melanin pigment formed in the skin protects the skin against infrared rays from the sun.

4. Heat radiates from blood vessels in the skin and is transferred to the air by the process of convection.

5. The precursor molecules necessary for the synthesis of vitamin D in the skin are generally obtained from water.

6. Sensations such as pressure, pain, touch, and temperature may be detected by receptors in the skin.

7. In thin epithelium found along most body surfaces there are <u>three</u> different layers of tissue.

8. The stratum germinativum, also known as the <u>stratum lucidum</u> lies on the basement membrane next to the dermis of the skin.

9. Keratinocytes having a spiny appearance may be found in the <u>stratum granulosum</u> of the epidermis.

10. Desmosomes are types of <u>protein globules</u> found within the cells of the stratum spinosum.

11. The substance keratohyalin is utilized by epidermal cells for the synthesis of <u>melanin</u>.

12. The thick epithelium is the only layer of the epidermis to contain the <u>stratum spinosum</u>.

13. Closely packed, clear cells and dead keratinocytes are characteristic signs of the <u>stratum lucidum</u>.

14. The layer of cells at the body surface consists of about 25 layers of dead and dry cells that are <u>cuboidal</u>.

15. The dead cells in the uppermost layer of epidermis are originally derived from cells produced in the <u>stratum lucidum</u>.

16. The epidermis has <u>numerous</u> blood vessels and few sense receptors.

17. Fibroblasts, blood vessels, and macrophages are found in the <u>reticular</u> layer of the dermis.

18. Much of the mechanical strength of the skin is centered in the <u>dermis</u>.

19. The hair follicle consists of a mass of <u>dermis</u> that has formed a small tube.

20. New hair cells are produced from cells of the <u>stratum corneum</u> located within the hair bulb.

21. The coarse axillary and pubic hair developing at puberty is called the <u>residual</u> hair.

22. At its proximal end, the nail is partially covered by the <u>cuticle</u>.

23. An alternative name for sweat glands is <u>sebaceous</u> glands.

24. <u>Apocrine</u> sweat glands secrete a white, cloudy material that can be broken down by bacteria to release odors.

25. The secretion of the sebaceous glands is emptied into the <u>ear canal</u>.

**Answers**

### PART A—Completion

1. glands
2. skin
3. 15%
4. microorganisms
5. keratin
6. basement membrane
7. melanin
8. ultraviolet rays
9. melanocytes
10. albinism
11. evaporation
12. blood vessels
13. excretion
14. D
15. A
16. glands
17. receptors
18. pain
19. wavy
20. thinner layer
21. superficial fascia
22. strata
23. squamous
24. four
25. five
26. stratum basale
27. stratum basale
28. stratum basale
29. keratinocytes
30. keratinocytes
31. stratum spinosum
32. stratum granulosum
33. stratum lucidum
34. eleidin
35. stratum corneum
36. keratin
37. desmosomes
38. stratum basale
39. hyperkeratosis
40. dermis
41. dermal papillae
42. reticular layer
43. areolar
44. scalp
45. root
46. erector pili
47. keratin
48. cuticle
49. sudoriferous glands
50. ceruminous glands

### PART B—Multiple Choice

1. C
2. C
3. A
4. B
5. D
6. B
7. B
8. D
9. A
10. C
11. C
12. B
13. B
14. D
15. B
16. A
17. D
18. B
19. D
20. B
21. A
22. C
23. C
24. C
25. B

### PART C—True/False

1. true
2. water-soluble
3. ultraviolet
4. true
5. milk
6. true
7. four
8. stratum basale
9. stratum spinosum
10. intercellular junctions
11. keratin
12. stratum lucidum
13. true
14. squamous
15. stratum basale
16. no
17. papillary
18. true
19. epidermis
20. stratum basale
21. terminal
22. true
23. sudoriferous
24. true
25. hair follicle

# CHAPTER 6

# BONES AND JOINTS

The human body contains 206 bones organized into a skeletal system (Chapter 7). Bones support the body, protect its organs, store lipids and calcium, and serve as sites of blood cell formation (Table 6.1). Joints (also called articulations) are the places where two or more bones come together, or articulate.

TABLE 6.1  *Some Functions of Bones*

| Function | Description |
|---|---|
| Movement | Maintain or change position of body parts by interacting with skeletal muscles |
| Protection | Enclose and protect the brain, lungs, and other organs |
| Support | Support and anchor muscles |
| Mineral storage | Serve as a depot for storing and withdrawing mineral ions; indirectly helps maintain body fluids and support metabolic activities. |
| Blood cell formation | Site for production of red blood cells and other blood cells |

## BONE

Bone is the hardest connective tissue in the human body. It consists of cells, collagen fibers, and dense mineralized ground substance. Various types of cells are contained in bone, and various minerals contribute to its hardness.

## Bone Classification

Bones may be classified by shape and body location. When classified by shape, bones are subdivided as flat bones, long bones, short bones, and irregular bones.

**Flat bones** consist of two thin plates of compact bone with a central region of spongy bone. This central region is called the diploe. Flat bones protect the delicate tissues of the brain and organs of the thorax. They also provide broad areas for muscle attachment. Flat bones include the skull bones, the shoulder blades (scapulae), the ribs, the breast bone (sternum), and bones of the pelvis.

**Long bones** bear the weight of the body and are longer than they are wide. Long bones have a shaft called a **diaphysis**, and two ends called **epiphyses** (singular **epiphysis**). Long bones are found in the arms and the legs.

**Short bones** resemble blocks. They include the bones of the wrists (carpals) and bones of the ankles (tarsals).

**Irregular bones** may have any number of shapes. For example, the bones of the vertebral column, the vertebrae, have a basic block shape with wings and buttresses. These projections are places where muscles, tendons, and ligaments attach, and they strengthen the connections between adjacent bones. Other irregular bones include the Wormian bones in the skull joints, the sesmoid bones in the joints, and the patellae, also called the kneecaps (Figure 6.1).

According to location, bones can be classified as components of the axial skeleton or the appendicular skeleton. The **axial skeleton** consists of all the bones that form the body's central axis. This includes the bones of the skull, the vertebral column, and the rib cage.

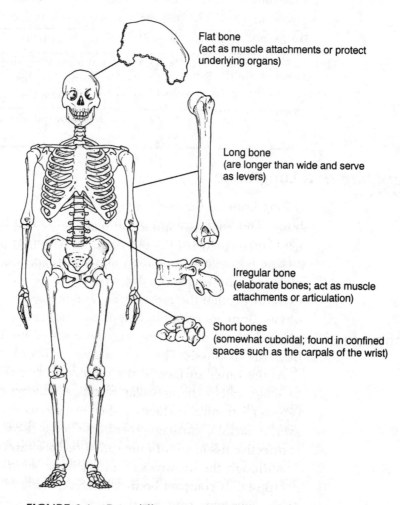

**FIGURE 6.1** *Four different types of bones and their locations.*

The **appendicular skeleton** includes bones of the upper and lower appendages, as well as bones used to attach them to the axial skeleton. The appendicular skeleton is composed of the bones of the arms and legs, hands and feet, and shoulder and hip girdles.

## Bone Tissue

To accomplish its goals, bone must be hard and rigid, yet flexible enough to bend in places. Bone gains its strength and flexibility from its chemical composition. The major component of bone is a mixture of a mineral salt called **calcium phosphate** and a minor amount of **calcium hydroxide** and **calcium carbonate** (all formed by bone-forming cells called osteoblasts). Calcium carbonate is a component of **hydroxyapatite**. Hydroxyapatite contains crystals of calcium carbonate embedded in a matrix of protein fibers. The protein is called **collagen**. Calcium carbonate adds hardness and strength to bones, while collagen fibers add to the flexibility of bones.

Bones are also the site of blood cell formation. At the center of many bones, such as the vertebrae and sternum, there is a substance called **marrow**. Marrow is the place where red and white blood cells and platelets are produced.

Bone also serves as a storage depot for calcium and phosphate. Both elements are deposited in the bone when they are plentiful in the body and released from the bone when needed in other locations in the body.

## Structure of a Long Bone

A long bone of the body is typified by the upper leg or upper arm bone. The long, straight portion of the bone is the **shaft**, or diaphysis. The two ends of the bone are the epiphyses (Figure 6.2). When a bone is growing, an active plate of cartilage called the **epiphyseal plate** is found where the diaphysis joins the wider part of the bone extremity called the **metaphysis**. The metaphysis joins with the epiphysis. Bone increases in length as cartilage is deposited at the epiphyseal plate, thereby forcing the ends of the bone further away from the diaphysis. This new cartilage is later replaced by bone.

At the outer surface of the epiphysis lies a thin layer of hyaline cartilage called the **articular cartilage**. The articular cartilage provides a frictionless surface for the bones to meet one another. Wherever it lacks a cartilage cover, the long bone is covered with a connective tissue membrane called the **periosteum**.

Although the diaphysis of a long bone is hollow, its outer wall is composed of **compact bone**, which is dense and hard. The interior of the diaphysis contains the marrow cavity where blood cells are

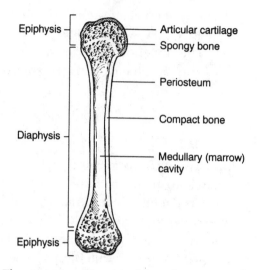

Epiphysis

Diaphysis

Epiphysis

Articular cartilage
Spongy bone

Periosteum

Compact bone

Medullary (marrow) cavity

**FIGURE 6.2**  *The structure of a typical long bone of the human body and its important parts.*

produced. In the bones of a child the marrow is red, but in adults it has been replaced by fat and is yellow.

Compact bone also covers the epiphysis. The interior is filled with **spongy bone** (cancellous bone). Spongy bone contains networks of bony plates and rods called **trabeculae** (singular trabecula). Embedded in the trabeculae are numerous spaces containing marrow. As the name implies, spongy bone is less dense than compact bone.

The marrow cavity of the bone is also known as the **medullary cavity**. This cavity consists of the cavities of spongy bone and the canals passing through the bones and is lined with a thin membrane known as the **endosteum**. Bone-forming cells called osteoblasts and bone-remodeling cells called osteoclasts are both found within the endosteum. The function of both the cells is discussed presently.

## Bone Histology

Viewed under the light microscope, compact bone displays an intricate series of concentric rings of bony tissue. The systems of rings are organized into units called **Haversian systems** (Figure 6.3). At the center of each Haversian system is a **Haversian canal**. This canal contains nerve and blood cells, which supply the bone cells within the Haversian system.

Transversing the concentric rings of a Haversian system are a system of tubules called **Volkmann's canals**. Volkmann's canals connect the bone cells with one another and link the Haversian canals. Surrounding each Haversian canal are the Haversian **lamellae**. Lamellae are the rings of the Haversian system. Spaces within the rings are the **lacunae**, which contain osteocytes.

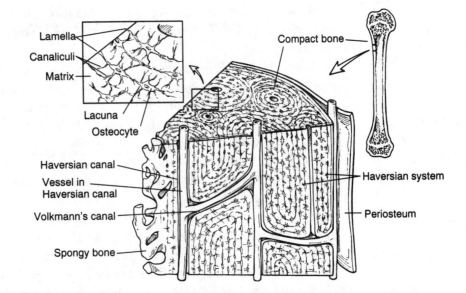

Lamella
Canaliculi
Matrix
Lacuna
Osteocyte
Compact bone
Haversian canal
Vessel in Haversian canal
Volkmann's canal
Spongy bone
Haversian system
Periosteum

**FIGURE 6.3** *The histology of bone. A section of compact bone is taken and its histology is displayed with particular reference to the Haversian system. A section of the Haversian system is magnified to show some of its fine structures.*

**Osteocytes** are inactive bone-forming cells trapped within the bone tissue they have produced. Ultramicroscopic extensions called **canaliculi** link the lacunae to one another and to the Haversian canal. The spaces between the Haversian systems are filled with material called **interstitial lamellae**. This is additional bone tissue.

The periosteum surrounds the bone and is the place where osteocytes arise. In their original form, osteocytes are highly active **osteoblasts** that secrete the protein (collagen) and hydroxyapatite of the bone. When trapped by the bone, they become osteocytes.

## Bone Formation

Approximately six weeks after fertilization, the skeleton exists as connective tissue. At this time, rods of **hyaline cartilage** develop in the shape of long bones. Flat bones exist as membranes containing fibrous connective tissue. Within these membranes and rods of hyaline cartilage the process of bone formation will occur.

Bone formation is called **ossification**. Ossification may take two forms. The first form, **intramembranous ossification**, occurs in the flat bones of the skull and takes place when osteoblasts migrate into the membranes and form clusters called **ossification centers**. Within the membranes the osteoblasts secrete the bony matrix composed of collagen, calcium phosphate, and calcium carbonate. The ossification centers are soon surrounded by bone. As the bony regions spread out, they merge to form a trabeculae of spongy bone. Red bone marrow develops in the spaces, and osteoblasts at the periosteum deposit a layer of compact bone on the surface. Bone formation is complete.

Ossification in long bones occurs by **endochondral ossification**. Blood vessels grow into the center of the rod of hyaline cartilage, and osteoblasts develop within the membrane surrounding the cartilage rod. The osteoblasts deposit compact bone as a collar around the cartilage rod. Bone formation continues at the outer surface of the rod but not in the interior areas. Thus, the interior remains as a large cavity that will contain the marrow of the bone.

While the marrow cavity is forming, the collar of compact bone is thickening and lengthening, and the cartilage rod continues to grow at the ends. A zone of cartilage remains beyond the ossification centers. Here the bone will continue to lengthen. Eventually the zone of cartilage narrows to form the epiphyseal plate, and when the epiphyseal plate turns to bone, the lengthening ceases. This process occurs usually after puberty and is under hormonal control.

Compact bone can either be woven bone or lamellar bone. **Woven bone** is bone tissue found in the embryo and young children. The bone does not show an oriented arrangement of collagen fibers characteristic of lamellar bone. **Lammellar bone** is an alternate name for compact bone. It is the dense bone that exhibits Haversian systems when viewed with the microscope.

## Bone Remodeling

The cessation of growth does not conclude the activity taking place in a bone. Indeed, bone remodeling continues throughout a person's life, controlled by the interaction of the bone forming cells (the osteoblasts) and bone-destroying cells called **osteoclasts**. Osteoclasts secrete substances that dissolve bone, and they provide the body with calcium and phosphate that may be used elsewhere in the body, such as in muscle contraction or in cell metabolism.

The activity of osteoblasts and osteoclasts normally keeps pace, and bone deposit and breakdown are in balance. In adults, however, bone breakdown may exceed deposit, and a complex condition called **osteoporosis** develops. With the loss of calcium, the bones weaken and break easily. To prevent osteoporosis, many physicians suggest numerous therapies, including increasing the intake of calcium, such as with calcium supplements.

## JOINTS AND ARTICULATIONS

Joints, or articulations, are the junctions where two or more bones come together. Different types of joints vary considerably in structure as well as function. The study of joints is called **arthrology**.

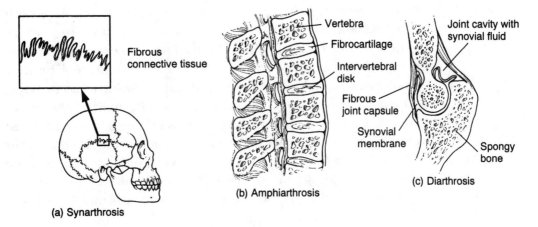

**FIGURE 6.4**   *Three basic types of joints found in the human body. (a) An immovable joint such as found between bones of the skull; (b) A slightly movable joint such as found between vertebrae in the vertebral column; (c) A freely movable joint known as a synovial joint, such as found at the hip joint. Note the structural parts of the synovial joint.*

Joints are classified according to the degree of movement they permit. Three types of joints are recognized: synarthroses, amphiarthroses, and diarthroses (Figure 6.4). **Synarthroses** are immovable joints; **amphiarthroses** are semimovable joints; and **diarthroses** are freely movable joints.

## Synarthroses

Synarthroses consist of two adjacent edges of bone separated by a very small amount of fibrous tissue. In some cases, the bones may be separated by a thin layer of cartilage. A synarthrotic joint allows no movement. Examples may be found in the skull, where the synarthroses are referred to as **sutures**. Sutures occur where the frontal bone meets the two parietal bones and where the parietal bones meet each other and the occipital bone.

Another example of a synarthrosis is a **gomphosis**. A gomphosis exists where a tooth joins its bony socket. Still another synarthrosis is a **syndesmosis**. In a syndesmosis, a fibrous membrane connects the shafts of two adjacent long bones. For example, a fibrous membrane connects the shafts of the radius and ulna in the lower arm.

## Amphiarthroses

Amphiarthroses are slightly movable joints. An amphiarthrosis consists of two adjacent bones separated by a substantial amount of cartilage. In some cases, the bones are separated by ligaments. Amphiarthrotic joints are found in two important places in the body: between the bodies of the vertebrae where the cartilage exists as a disk, and between the two pubic parts of the hip bone.

Between the bodies of vertebrae, the **intervertebral disks** are composed of fibrocartilage surrounding a gelatinous core. The disks absorb shocks and equalize the pressure between adjacent bones during body movement. Due to their slight flexibility, disks allow a limited amount of movement, such as when bending the back forward or to the side. Between the two pubic bones, the amphiarthrotic joint is called the **pubic symphysis**. A small amount of movement occurs at this joint. The **sacroiliac joint** (where the sacrum joins the ilium portion of the hip joint) is also an amphiarthrotic joint.

# Diarthroses

Diarthroses are freely movable joints. The joint consists of two bones encased within a cavity called the **synovial cavity**. Because of this cavity, the joint is often called a **synovial joint**. The articular surfaces are not directly bound to one another, and in some cases, the movement is very extensive. Examples of diarthrotic joints occur at the shoulder, elbow, wrist, hip, knee, ankle, and foot. The joints between the articular processes of the vertebrae are also diarthrotic joints.

At a synovial joint, each bone has an articular cartilage at the articulating surface. A capsule of connective tissue called the **fibrous capsule** encases the two bones and forms the cavity. The inner surface of the cavity, except at the articular surfaces, is lined by a **synovial membrane**. This membrane expresses a thick fluid called **synovial fluid**. The fluid lubricates the joint.

In certain cases, the joint cavity is divided partially or completely by **cartilaginous disks**. Two cartilaginous disks are found in each knee joint. The cartilage is crescent-shaped and is called a **semilunar meniscus**. In certain joints, the cartilage of the synovial capsule are organized into thick bundles of cords to form **ligaments**. Joints can also be stabilized by muscles connected to the bones near the joint. For example, at the shoulder, the muscles inserting to the proximal end of the humerus hold it fast in the glenoid cavity.

Several diarthrotic joints also contain closed, fluid-filled sacs called **bursae** (singular bursa). Bursae are lined with synovial membranes continuous with synovial membranes of nearby synovial cavities. Bursae are commonly found between the skin and bony prominences lying beneath the skin. For instance, there are bursae in the patella of the knee. Bursae assist the movement of tendons that pass over bones.

# Types of Diarthroses

Diarthrotic joints may be subdivided into several types depending upon their components and the type of movement they permit. One

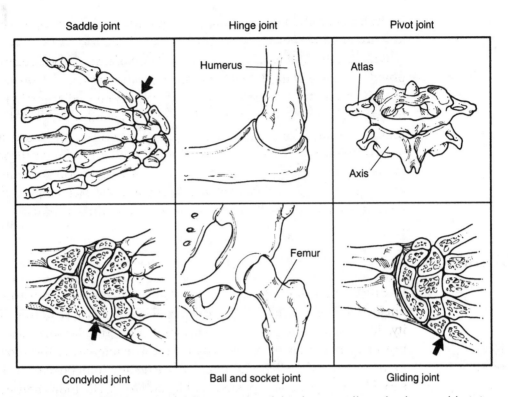

Saddle joint          Hinge joint          Pivot joint

Condyloid joint       Ball and socket joint       Gliding joint

**FIGURE 6.5**   *A display of six different types of diarthroses. All are freely movable joints.*

type of diarthrotic joint is the **hinge joint** (Figure 6.5). In a hinge joint, movement occurs in one plane only. Examples of hinge joints are at the elbow (between the humerus and ulna); and at the knee (between the femur and tibia); and at the fingers (between the phalanges).

Another type of diarthrotic joint is the **pivot joint**. A pivot joint is organized to permit rotation. In a pivot joint, the cylindrical surface of one joint rotates within a ring formed by another bone. An example of a pivot joint occurs in the neck where the first two bones of the vertebral column (the atlas and axis) come together.

A **ball-and-socket joint** is also a diarthrotic joint. This joint forms where a ball-like head fits into a cuplike cavity. A ball-and-socket joint allows the most freedom of movement. It occurs where the head of the humerus articulates with the glenoid cavity of the scapula and at the hip where the head of the femur articulates with the acetabulum (Table 6.2).

Two other types of diarthrotic joints are the condyloid joint, and the saddle joint (both are called biaxial joints). In a **condyloid joint**, the articular surfaces are oval. Rotation is not possible at this joint, but most other movements possible occur here. A condyloid joint exists where the radius meets the carpals at the wrist. In a **saddle joint**, rotation is also restricted. A saddle joint is formed where articulating bones have concave and convex surfaces. The convex surface of one bone fits the concave surface of another bone. The union of the carpal bones with one another and with the metacarpal bones of the thumb occurs at a saddle joint.

**TABLE 6.2**    *Six Types of Diarthroses*

| Diarthrosis | Description | Movement | Examples |
|---|---|---|---|
| Hinge | Spool-shaped surface fits into concave surface | In one plane about single axis, like hinged door movement | Elbow, knee, ankle, and interphalangeal joints |
| Pivot | Arch-shaped surface rotates about rounded or peglike pivot | Rotation | Between axis and atlas |
| Ball and socket | Ball-shaped head fits into concave socket | Widest range of all joints | Shoulder joint and hip joint |
| Condyloid (ellipsoidal) | Oval-shaped condyle; fits into elliptical cavity | In two planes at right angles to each other | Wrist joint (between radius and carpals) |
| Saddle | Saddle-shaped bone fits into socket that is curved in opposite direction | Same kinds of movement as condyloid joint but freer; resembles rider in saddle | Thumb, between first metacarpal and carpal (trapezium) |
| Gliding | Articulating surfaces; usually flat | Gliding, a nonaxial movement | Between carpal bones; between sacrum and ilium (sacroiliac joints) |

The final type of diarthrotic joint is a **gliding joint**. A gliding joint permits a gliding movement in a number of different directions between relatively flat, articular surfaces. The places where the articular processes of the vertebrae come together are examples of gliding joints. Gliding joints are also found between certain carpals and tarsals. Sliding and twisting movements occur at these joints.

## Joint Movements

The actions of skeletal muscles permit movements at diarthrotic joints. In most cases, one end of a muscle is attached to an immovable or fixed part on one side of a joint, and the other end of a muscle connects to the movable part at the opposite side of the joint. Contraction of the muscle exerts a pull on its movable end, and movement occurs at the joint.

There are various movements that can occur at joints. For example, assuming that one begins in the anatomical position, a **flexion** refers to the movement of the joint to reduce the angle between two bones. By contrast, **extension** is a joint movement in which the angle

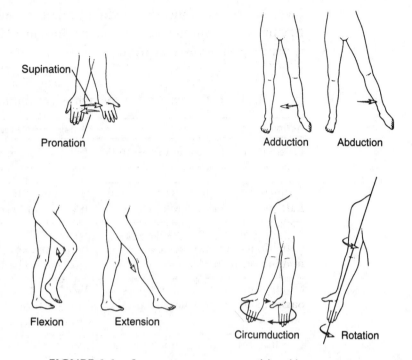

**FIGURE 6.6** *Some movements possible of human joints.*

between two bones increases. Flexion refers to the bending of a joint, while extension refers to the stretching out of a joint (Figure 6.6).

Two other joint movements are abduction and adduction. **Abduction** consists of moving a body part away from the midline of the body, while **adduction** moves a body part toward the midline. Lifting the arm horizontally is an abduction, while returning the arm to the normal position is an adduction.

Turning movements are also types of joint movements. **Rotation** is movement in which a body part moves about an axis. Twisting the head from side to side as when gesturing "no" is a rotation. Rotation can be medial if is toward the midline, or lateral if away from the midline. A special form of rotation is pronation. **Pronation** is rotation of the forearm and the hand so that the palm is turned downward. By contrast, **supination** is the forearm rotation so that the palm turns upward to the anatomical position (Table 6.3).

Raising and lowering a part of the body refers to movements called elevation and depression. **Elevation** is raising a body part, such as shrugging the shoulders, while **depression** is lowering a body part, such as drooping the shoulders. Moving a body part backward and forward illustrate retraction and protraction. **Retraction** refers to moving a body part backward, such as pulling in the chin. **Protraction** refers to moving a body part forward, such as thrusting the chin outward.

The terms eversion and inversion refer to turning movements. **Eversion** refers to turning the foot so that the sole is outward, while

**inversion** refers to turning the foot so that the sole is inward. Dorsi-flexion is bending the foot up toward the shin, while plantar flexion is bending the foot away from the shin.

**TABLE 6.3**   *Joint Movements and Antagonistic Movements*

| Action | Antagonistic Action |
|---|---|
| Flexion: the decreasing of the angle between two bones | Extension: the increasing of the angle between two bones |
| Abduction: the movement of a body part away from a midline | Adduction: the movement of a body part toward a midline |
| Medial rotation: the turning of a bone on its own axis toward the midline of the body | Lateral rotation: the turning of a bone on its own axis away from the midline of the body |
| Supination: the placing of the palm of the hand in the anatomical position | Pronation: the placing of the palm of the hand away from the anatomical position |
| Elevation: the raising of a body part | Depression: the lowering of a body part |
| Protraction: the thrusting forward of a body part | Retraction: the withdrawal of a body part |
| Dorsiflexion: the bending of the foot toward the shin | Plantar flexion: the bending of the foot away from the shin |
| Inversion: the rotation of the sole of the foot inward | Eversion: the rotation of the sole of the foot outward |

# REVIEW QUESTIONS

**PART A—Completion: Add the word or words that correctly complete each of the following statements.**

1. The skull bones and ribs and bones of the pelvis are types of bones classified as _____ .

2. Those bones that bear the weight of the body are classified as _____ .

3. The projections of irregular bones such as the vertebrae provide sites for the attachment of tendons, ligaments, and _____ .

4. The irregular bones found in the skull joints are _____ .

5. All the bones of the skull, vertebral column, and the ribcage are considered together as the _____ .

6. Two examples of short blocklike bones in the body are the bones of the wrists called carpals, and the bones of the ankles known as _____ .

7. Crystals of calcium carbonate in a matrix of protein fibers form a component of bone called _____ .

8. The major protein in the protein fibers of the bone matrix is _____ .

9. Bone serves to support the body, protect the organs, store calcium, and serve as sites of formation of _____ .

10. Bone marrow is particularly active at the center of bones such as the vertebrae and bones of the _____ .

11. The two ends of a long bone are known as _____ .

12. The thin layer of hyaline cartilage at the outer surface of the end of the bone is called the _____ .

13. Wherever it lacks a cartilage cover, the long bone is covered with a connective tissue membrane known as the _____ .

14. The interior portion of the epiphysis of the bone consists of _____ .

15. The marrow cavity in the bones of a child is usually red, but in adults, it appears _____ .

16. Spongy bone contains networks of bony plates and rods known as _____ .

17. The histological and physiological unit of the bone is the _____ .

18. The microscopic spaces that contain the osteocytes of bone are the _____ .

19. The concentric rings of a Haversian system are connected by a system of tubules called _____ .

20. The cells that secrete the protein and hydroxyapatite of bone are the _____ .

21. Bone formation takes place by a process called _____ .

22. Bone formation occurring within membranes is correctly known as _____ .

23. Bone formation taking place in the long bones occurs by the process of _____ .

24. The long bone continues to lengthen at a zone of cartilage beyond the ossification center called the _____ .

25. Bone is remodeled and dissolved by substances secreted by bone-destroying cells referred to as _____ .

26. The destruction and remodeling of bone provides the body with ions such as phosphate ions and _____ .

27. The excessive breakdown of bone may exceed its deposit in the condition _____ .

28. Semimovable joints are called _____ .

29. An immovable joint consisting of two adjacent edges of bone separated by a small amount of fibrous tissue is a _____ .

30. An example of a synarthrosis occurring in the skull is a _____ .

31. Where a tooth comes together with its bony socket, the union forms an immovable joint called a _____ .

32. An amphiarthrotic joint is found between the two pubic bones and between the bodies of the _____ .

33. A diarthrosis is a freely movable joint consisting of two bones separated by a cavity called the _____ .

34. The joints occurring at the elbow, shoulder, hip, knee, and ankle are examples of a _____ .

35. The thick fluid found within the synovial cavity of a diarthrotic joint is called _____ .

36. The cartilaginous disk found in the knee joint is referred to as the _____ .

37. Joints can be stabilized by muscles connected to bones near the joint or by thick bundles of cords known as _____ .

38. Closed, fluid-filled sacs found near diarthrotic joints are called _____ .

39. A diarthrotic joint that permits rotation is a _____ .

40. Where the head of the humerus articulates with the glenoid cavity of the scapula, there is a diarthrotic joint called a _____ .

41. Where the radius meet the carpals at the wrist, the diarthrotic joint is a _____ .

42. Where bones have concave and convex surfaces that articulate, the diarthrotic joint is known as a _____ .

43. Movement at a joint in which the angle between two bones is reduced is known as _____ .

44. When a body part is moved away from the midline of the body, the movement is referred to as _____ .

45. When a body part moves toward the midline, the movement is _____ .

46. Twisting the head from side to side, such as when gesturing "no," is the joint movement _____ .

47. Turning the hand so that the palm is upward demonstrates the movement _____ .

48. Where flexion refers to the bending of a joint, the stretching out a joint is _____ .

49. Moving a body part backward, such as pulling in the chin is a movement called _____ .

50. Moving a body part forward, such as thrusting the chin outward is _____ .

*PART B—Multiple Choice:* **Circle the letter of the item that correctly completes each of the following statements.**

1. Flat bones help protect the delicate tissues of the
   (A) abdomen and appendages
   (B) thorax and brain
   (C) spinal cord and appendages
   (D) abdomen and spinal cord

2. All the following are functions of the skeletal system except
   (A) storing lipid and calcium
   (B) serving as sites for blood cell formation
   (C) supporting the body
   (D) coordinating the body activities

3. The irregular bones of the body include the
   (A) patellae of the knee caps and Wormian bones of the skull joints
   (B) carpals and tarsals
   (C) scapulae and ribs
   (D) humerus of the upper arm and femur of the upper leg

4. All the following are bones of the axial skeleton except
   (A) bones of the ribcage
   (B) the vertebrae
   (C) bones of the skull
   (D) arm and leg bones

5. The hydroxyapatite of bone is composed of
   (A) calcium phosphate and DNA
   (B) collagen and calcium carbonate
   (C) ligaments and tendons
   (D) fibrinogen and sodium phosphate

6. Blood cell formation occurs within the marrow of bones such as
   (A) vertebrae
   (B) phalanges
   (C) temporal bones
   (D) maxillae

7. The diaphysis and epiphysis are portions of a
   - (A) rib bone
   - (B) flat bone
   - (C) pelvic bone
   - (D) long bone

8. The periosteum is a connective tissue membrane that
   - (A) is found within the marrow
   - (B) makes up part of the ligament
   - (C) covers portions of the long bone
   - (D) synthesizes the collagen of bone

9. Volkmann's canals connect the bone cells with one another in the
   - (A) epiphyseal plate
   - (B) periosteum
   - (C) Haversian system
   - (D) bone marrow

10. The principal bone-forming cells of the body are
    - (A) osteoclasts
    - (B) osteocytes
    - (C) osteoblasts
    - (D) pericytes

11. Intramembranous ossification is a type of bone formation occurring in the
    - (A) femur and humerus
    - (B) radius and ulna
    - (C) phalanges
    - (D) skull bones

12. Endochondral ossification occurs within
    - (A) bone marrow
    - (B) a rod of hyaline cartilage
    - (C) membranes
    - (D) the synovial cavity

13. Bone ceases to lengthen when the
    - (A) red marrow becomes yellow
    - (B) epiphyseal plate turns to bone
    - (C) the age of puberty is reached
    - (D) the patella forms

14. Cells that destroy bone and provide calcium to the body are known as
    - (A) calcicytes
    - (B) histiocytes
    - (C) osteoclasts
    - (D) periclasts

15. A synarthrosis is a type of joint found
    (A) in the arms and legs
    (B) in the skull
    (C) where the radius meets the femur
    (D) between  the vertebrae

16. Both a gomphosis and a syndesmosis are types of
    (A) slightly movable joints
    (B) freely movable joints
    (C) immovable joints
    (D) disjoints

17. Where the two pubic bones come together, the joint that forms
    is
    (A) amphiarthrotic
    (B) diarthrotic
    (C) gomphotic
    (D) synarthrotic

18. A diarthrotic joint occurs at all the following locations except
    (A) the knee
    (B) the elbow
    (C) the wrist
    (D) the skull

19. The function of synovial fluid is to
    (A) regulate the calcium content at the joint
    (B) lubricate a diarthrotic joint
    (C) synthesize the proteins of ligaments
    (D) manufacture red blood cells

20. An example of a hinged joint is that found at the
    (A) vertebral column
    (B) junction of atlas and axis
    (C) knee and elbow
    (D) glenoid cavity

21. Where the concave surface of one bone fits the convex surface
    of another bone, the joint is known as a
    (A) gliding joint
    (B) ball and socket joint
    (C) hinged joint
    (D) saddle joint

22. When a joint moves and reduces the angle between two bones at
    the joint, the movement is called a
    (A) protraction
    (B) flexion
    (C) pronation
    (D) supination

23. Moving a body part away from the body's midline and back to the midline represent two movements known as
    (A) pronation and supination
    (B) elevation and depression
    (C) abduction and adduction
    (D) flexion and extension

24. Pronation is a joint movement in which
    (A) the forearm is rotated so the palm is downward
    (B) the forearm is rotated so the palm faces upward
    (C) a body part such as the shoulders is raised
    (D) a body part such as the shoulders is lowered

25. Turning the foot so that the sole faces outward is a movement called
    (A) extension
    (B) protraction
    (C) eversion
    (D) abduction

*PART C—True/False:* **For each of the following statements, mark the letter "T" next to the statement if it is true. If the statement is false, change the underlined word to make the statement true.**

1. Flat bones provide protection for the brain and the organs of the abdomen.

2. The end portions of long bones are known as diaphyses.

3. Irregular bones of the body include the Wormian bones found in the skull.

4. The appendicular skeleton includes bones that form the skull, vertebral column, and rib cage.

5. The matrix of fibers found in the bone is composed of the protein albumin.

6. The place within the bone where red blood cells, white blood cells, and platelets are produced is called the bone marrow.

7. In places where a cartilage cover is lacking, the long bone is covered with a connective tissue membrane called the epiosteum.

8. The fundamental anatomical unit of the bone is called the Haversian system.

9. Bone formation is carried on by the osteoblasts, which later revert to inactive cells called osteoclasts.

10. Spaces within the rings of a Haversian system contain the osteocytes and are known as lamellae.

11. Intramembranous ossification is a type of bone formation that occurs in bones of the <u>legs</u>.

12. Endochondral ossification takes place within rods of <u>collagen</u>.

13. The lengthening of a long bone comes to an end when the epiphyseal plate turns to <u>cartilage</u>.

14. Osteoclasts dissolve bone and supply the body with phosphate and <u>calcium</u>.

15. Osteoporosis is a condition characterized by excessive <u>buildup</u> of bone.

16. Those joints that are freely movable are known as <u>synarthroses</u>.

17. Where the frontal bone of the skull meets the two parietal bones, the joint is called a <u>suture</u>.

18. A fibrous membrane connects the shafts of two adjacent long bones in a joint known as a <u>gomphosis</u>.

19. The pubic symphysis, the place where the two pubic bones come together, is an example of a <u>diarthrosis</u>.

20. The synovial fluid lubricates the joint where two <u>flat</u> bones articulate.

21. Fluid-filled sacs called <u>bursae</u> are found in the patella of the knee and between the skin and bony prominences lying beneath the skin.

22. Where the femur articulates with the acetabulum, the joint is known as a <u>ball-and-socket joint</u>.

23. A <u>condyloid joint</u> is a type of diarthrotic joint in which movement occurs in one plane only.

24. A flexion refers to the movement taking place at a joint to <u>increase</u> the angle between two bones.

25. An abduction is a joint movement in which a body part moves away from the midline, while a <u>supination</u> is a movement in which the body part moves toward the midline.

**Answers**

## PART A—Completion

1. flat bones
2. long bones
3. muscles
4. Wormian bones
5. axial skeleton
6. tarsals
7. hydroxyapatite
8. collagen
9. blood cells
10. sternum
11. epiphyses
12. articular cartilage
13. periosteum
14. spongy bone
15. yellow
16. trabeculae
17. Haversian system
18. lacunae
19. Volkmann's canals
20. osteoblasts
21. ossification
22. intramembranous ossification
23. endochondral ossification
24. epiphyseal plate
25. osteoclasts
26. calcium ions
27. osteoporosis
28. amphiarthroses
29. synarthroses
30. suture
31. gomphosis
32. vertebrae
33. synovial cavity
34. diarthrotic joint
35. synovial fluid
36. semilunar meniscus
37. ligaments
38. bursae
39. pivot joint
40. ball-and-socket joint
41. condyloid joint
42. saddle joint
43. flexion
44. abduction
45. adduction
46. rotation
47. supination
48. extension
49. retraction
50. protraction

## PART B—Multiple Choice

| | | | | |
|---|---|---|---|---|
| 1. B | 6. A | 11. D | 16. C | 21. D |
| 2. D | 7. D | 12. B | 17. A | 22. B |
| 3. A | 8. C | 13. C | 18. D | 23. C |
| 4. D | 9. C | 14. C | 19. B | 24. A |
| 5. B | 10. C | 15. B | 20. C | 25. C |

## PART C—True/False

1. thorax
2. epiphyses
3. true
4. axial
5. collagen
6. true
7. periosteum
8. true
9. osteoblasts
10. lacunae
11. skull
12. cartilage
13. bone
14. true
15. breakdown
16. diarthroses
17. true
18. syndesmosis
19. amphiarthrosis
20. long
21. true
22. true
23. hinge joint
24. reduce
25. adduction

# CHAPTER 7

# THE SKELETAL SYSTEM

The human body contains 206 bones organized into a structural framework called the **skeleton**. The bones of the skeletal system are controlled by hundreds of muscles, which in turn are controlled by impulses from the central nervous system.

Anatomically, the skeleton is divided into two major sections. The first section is the **axial skeleton**, which is composed of bones of the body's central axis (for example, the skull, the vertebral column, and the rib cage). The second section is the **appendicular skeleton**, which is composed of bones of the upper and lower appendages and bones attaching them to the axial skeleton.

## THE AXIAL SKELETON

Eighty bones make up the central axis of the human body, the axial skeleton (Figure 7.1). They include the bones of the thoracic cage, the vertebral column, and the skull. There are two major regions of the **skull**: the cranium and the facial region. Bones of the **cranium** are generally flat and tightly fused to one another. The **facial bones** support several of the sensory organs of the head including the eyes, ears, and nose.

### The Cranium

The cranium is composed of eight bones fused together at jagged, immovable joints called **sutures**. The cranial roof is formed by three bones: the **frontal bone**, which forms the forehead and anterior roof of the cranium; the **parietal bones**, which form the posterior roof of the cranium and are two arched, flattened bones joined at the midline; and the **occipital bone**, which forms the posterior cranial floor (Figure 7.2).

The frontal bone contains the **supraorbital foramen**, an opening through which nerves and blood vessels pass to the forehead. It is also the site of the **frontal sinus** where air circulates for conditioning. The occipital bone contains the **foramen magnum**, a large hole through which the spinal cord and brain are continuous. This bone also contains two rounded projections called **occipital condyles** where the base of the skull meets the top of the vertebral column.

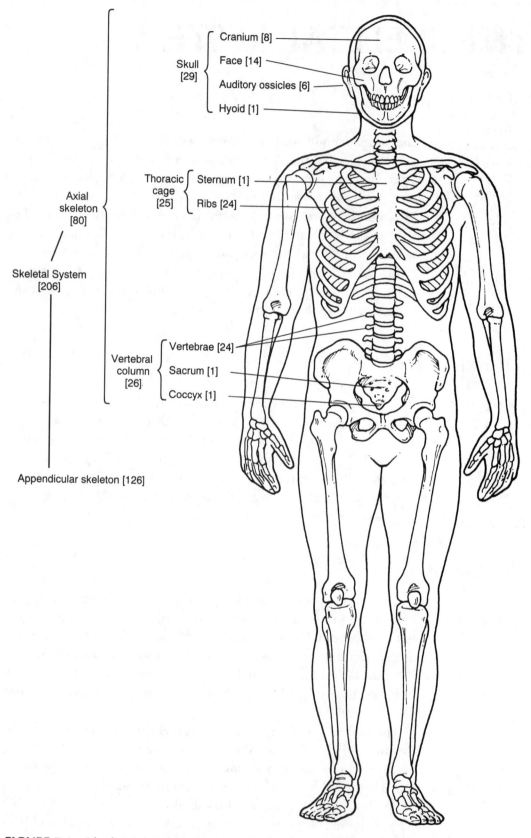

**FIGURE 7.1** *The bones of the axial skeleton in humans. Eighty bones make up this portion of the complete skeleton. Numbers in square brackets indicate number of bones for each item.*

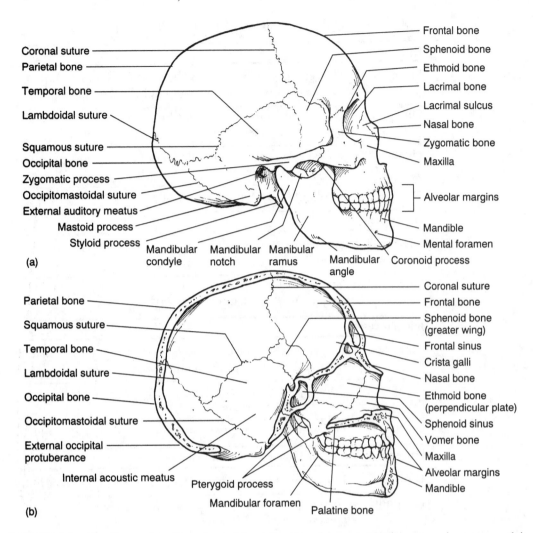

**FIGURE 7.2** *The human skull as viewed from the lateral aspect. (a) The external anatomy of the right side (b) The internal anatomy seen in a sagittal view of the left side of the skull*

The lateral walls of the cranium are formed by two **temporal bones**, which also form part of the internal floor of the cranium. An opening in the temporal bone, the **external auditory meatus,** conducts sound into the skull (Table 7.1). Sounds travel through the external auditory meatus into the inner ear located deep within the temporal bones. Below each meatus is the rounded **mastoid process**, the point of attachment for many neck muscles. A pointed **styloid process** also lies below the meatus. This is where pharyngeal and tongue muscles attach. The **mandibular fossa** is a simple depression where the temporal bone articulates with a process of the mandible. The **zygomatic process** is a projection of the temporal bone that helps form the cheekbone.

Another bone of the cranium is the **sphenoid bone**. It has a butterfly shape and forms the anterior internal floor of the cranium. The pituitary gland lies in a saddle-shaped depression in the sphenoid bone called the **sella turcica**. The superior aspect of the bone contains a slit called the **orbital fissure** for passage of blood vessels and nerves.

**TABLE 7.1** *Bone Markings and Their Descriptions*

| Openings | |
|---|---|
| Fissure | Slit between two bones through which nerves or blood vessels pass (e.g. superior orbital fissure of the sphenoid bone) |
| Foramen | Hole within a bone through which nerves or blood vessels pass (e.g. foramen magnum of the occipital bone) |
| Meatus | Tubelike passageway within a bone (e.g. auditory meatus of the temporal bone) |
| Sinus | Cavity within a bone (e.g. frontal sinus) |
| **Depressions** | |
| Fossa | Simple depression or hollowing in or on a bone (e.g. mandibular fossa of the temporal bone) |
| Sulcus | Groove that may contain a blood vessel, nerve or tendon (e.g. malleolar sulcus of the tibia) |
| **Joint Processes** | |
| Condyle | Large, convex protrusion at the end of a bone (e.g. medial or lateral condyles of the femur) |
| Head | Round protrusion separated from the rest of a bone by a neck (e.g. head of the femur) |
| Facet | Flat, smooth surface (e.g., facet of a rib) |
| **Processes for Attaching Ligaments, Tendons, and Muscles** | |
| Crest | Prominent ridge on a bone (e.g., iliac crest of the coxal bone) |
| Epicondyle | Second protrusion above a condyle (e.g., lateral epicondyles of the femur) |
| Line | Less prominent ridge on a bone (e.g., linea aspera of the femur) |
| Tubercle | Small, round protrusion (e.g., greater tubercle of the humerus) |
| Tuberosity | Large, round, and usually roughened protrusion (e.g., ischial tuberosity of the coxal bone) |
| Trochanter | Large protrusion (e.g., greater trochanter of the femur) |

A small cranial bone is the **ethmoid bone**. The ethmoid bone separates the nasal cavity from the remainder of the cranium. Olfactory nerve fibers pass from the nose to the brain through holes in the ethmoid process. These holes occur in two thin horizontal plates of bone called the **cribiform plates**. A triangular process called the **crista galli** projects upward between the cribiform plates (Figure 7.3).

Cranial bones fuse at the sutures. In the newborn, the bones have not joined, and the areas are still membranous. The membranous areas are called **fontanels** ("soft spots"). This incomplete skull development permits the skull to collapse slightly during birth. By the second year, the fontanels have ossified.

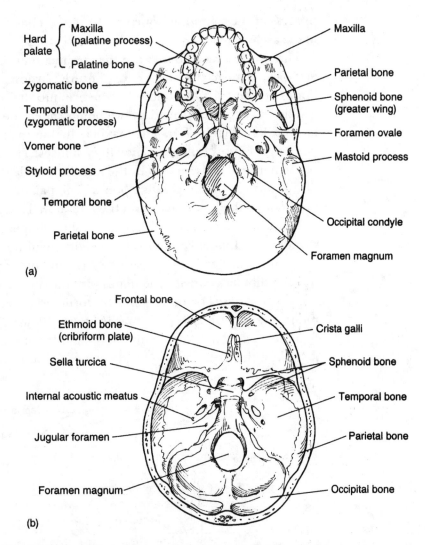

Hard palate { Maxilla (palatine process), Palatine bone }

Zygomatic bone

Temporal bone (zygomatic process)

Vomer bone

Styloid process

Temporal bone

Parietal bone

Maxilla

Parietal bone

Sphenoid bone (greater wing)

Foramen ovale

Mastoid process

Occipital condyle

Foramen magnum

(a)

Frontal bone

Ethmoid bone (cribriform plate)

Sella turcica

Internal acoustic meatus

Jugular foramen

Foramen magnum

Crista galli

Sphenoid bone

Temporal bone

Parietal bone

Occipital bone

(b)

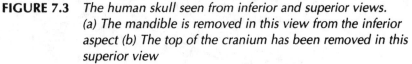

**FIGURE 7.3** *The human skull seen from inferior and superior views. (a) The mandible is removed in this view from the inferior aspect (b) The top of the cranium has been removed in this superior view*

## Facial Bones

The face of the skull is formed by 14 bones. These bones provide attachments for chewing muscles and support the other head and facial muscles.

Two bones called **nasal bones** fuse at the midline and form the bridge of the nose. The nasal cavity is divided into left and right chambers by the vertical **vomer bone**, also called the **nasal septum**. Lateral walls of the nasal cavity are formed by plates called the **inferior nasal conchae**. Mucous membranes cover the conchae and vomer bones.

Below the eye sockets, two bones support the face and form part of the cheek bones. They are called the **zygomatic bones**. Each zygomatic bone has a **temporal process** projecting to join the zygomatic process of the temporal bone and form the cheekbone at the **zygomatic arch**.

The smallest facial bones are the **lacrimal bones**. Lacrimal bones are located near the corners of the eye. Grooves in the lacrimal bones permit tears to drain from the eye into the nasal cavity.

The upper jaw is formed by two bones called the **maxillae** (singular **maxilla**) (Figure 7.4). The maxillae fuse at the midline and contain large sinuses called the **maxillary sinuses**. The anterior portion of the hard palate is formed by the maxillae, while the posterior portion is formed by two **palatine bones**. The palatine bones also form the floor of the nasal cavity and the lateral walls of the nasal cavity.

The lower jaw is formed by a horseshoe-shaped bone called the **mandible**. A hingelike joint connects the mandible to the skull. At each end, a projection extends upward. One part of the projection, the **mandibular condyle**, articulates with the mandibular fossa of the temporal bones. The other part, the **coronoid process**, is the attachment site for muscles of mastication (chewing). Because the mandible easily detaches from the remainder of the skull, it is often lost when the tissues of the head decompose after death.

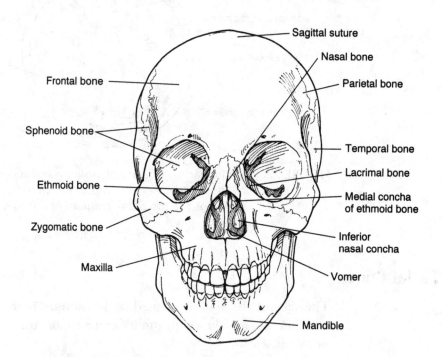

Frontal bone — Sphenoid bone — Ethmoid bone — Zygomatic bone — Maxilla

Sagittal suture — Nasal bone — Parietal bone — Temporal bone — Lacrimal bone — Medial concha of ethmoid bone — Inferior nasal concha — Vomer — Mandible

**FIGURE 7.4**  *The human skull as viewed from the anterior aspect.*

## The Vertebral Column

The **vertebral column**, also called the backbone and the spine, extends from the base of the skull downward. It is composed of a group of 26 bones called **vertebrae** (Figure 7.5). There are **7 cervical vertebrae** (in the neck), 12 are **thoracic vertebrae** (in the chest region), 5 are **lumbar vertebrae** (in the small of the back), 1 is the **sacrum** (formed by the fusion of 5 sacral vertebrae), and 1 is the **coccyx** (formed by the fusion of 4 coccygeal vertebrae).

The unfused vertebrae above the sacrum are separated by **intervertebral disks**. These disks are composed of fibrous cartilage, which absorbs shocks and permits flexibility in the vertebral column. Excessive strain can cause the disks to bulge out of shape, a condition called a **herniated disk**. A herniated disk is extremely painful because the protrusion of the disk presses on the spinal cord or a spinal nerve leaving the cord, bringing on excessive pain and numbness.

Viewed from the side, the vertebral column curves forward in the cervical and lumbar regions and backward in the thoracic and sacral regions. These curves add strength and flexibility to the column. An abnormal sideways spinal curve is called **scoliosis**. An exaggeration of the thoracic curve is known as **kyphosis** (hunchback). An exaggerated curve of the lumbar area is called **lordosis** (swayback).

Individual vertebrae vary in shape and size. Each has a **body**, the weight-bearing cylinder of bone that exists between the disks. Vertebral bodies withstand the compressions exerted by the moving body.

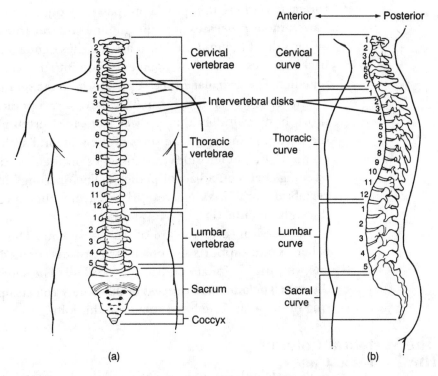

**FIGURE 7.5** *The vertebral column of the axial skeleton. (a) An anterior view of the column showing the location of the vertebrae (b) A lateral view from the right side. The curvatures of the spine are seen in this view*

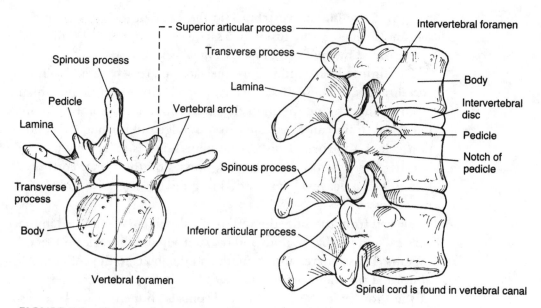

**FIGURE 7.6** *Details of a typical vertebra of the spinal column. (a) A superior view of a single vertebra showing the important processes (b) A lateral view from the right side illustrating three vertebrae in place with their associated structures.*

Each vertebral body has a **vertebral arch** extending behind the body to enclose and protect the spinal cord as it passes through the opening in the arch (Figure 7.6). This opening, called the **vertebral foramen**, is where the spinal cord passes through. The vertebrae also have **spinous processes** near the midline and **transverse processes** on either side of the midline. Both are the sites of attachment for many back muscles as well as ligaments holding the vertebral column together. The **articular processes** perform the same function.

The vertebral arch is formed in part by two **pedicles**, which are short, bony cylinders that project from the vertebral body toward the posterior and form the sides of the vertebral arch. The pedicles on the right and left sides have notches on their borders to provide openings between adjacent pedicles. The openings are called **intervertebral foramina**. Nerves arising from the spinal cord pass through the intervertebral foramina.

Two cervical vertebrae are of particular note. The first, the **atlas**, balances and supports the head. Two processes called **facets** articulate with the occipital condyles of the skull. The second vertebra is the **axis**. The **odontoid process** of this bone projects upward into the ring of the atlas. The head rotates at this joint.

## The Thoracic Cage

The **thoracic cage** is formed by the ventrally located sternum, or breastbone, plus the ribs that form a protective cage. Three recognizable parts make up the sternum: the **manubrium**, which is the

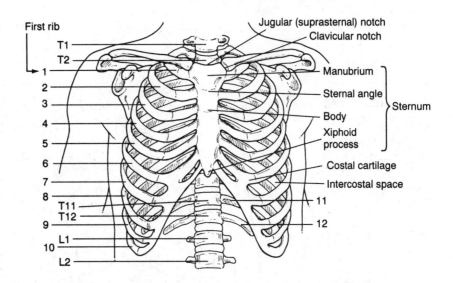

**FIGURE 7.7** *An anterior view of the thoracic cage of the axial skeleton. Ten pairs of ribs are attached to costal cartilages, and two pairs are floating ribs.*

uppermost shield-shaped portion; the **body**, which is shaped like a dagger and is attached to most ribs; and the **xiphoid process**, which forms the lowermost portion of the sternum.

Twelve pairs of **ribs** exist in the adult human (Figure 7.7). The first seven pairs are called **true ribs** because they link directly to the sternum by strips of cartilage called **hyaline costal cartilages**. These attachments provide flexibility to the ribs and absorb shocks to the chest wall. The flat sides of the ribs are called facets.

The next five pairs of ribs are called **false ribs**, so-named because they do not attach directly to the sternum. Ribs 8 to 10 have costal cartilages, but the cartilages merge with the cartilage of the seventh rib. Ribs 11 and 12 have no cartilages at all. These ribs do not attach to the sternum and are called **floating ribs**.

## THE APPENDICULAR SKELETON

The appendicular skeleton is comprised of bones of the body's appendages and bones connecting the appendages to the axial skeleton. The connecting bones are collectively known as a **girdle**. There are two girdles: the **pectoral girdle**, which connects the arms to the rib cage; and the **pelvic girdle**, which connects the legs to the sacrum (Figure 7.8).

### The Arm and Pectoral Girdle

The pectoral girdle consists of the scapula and the clavicle. The **scapula** is a large triangular bone connected to the axial skeleton by

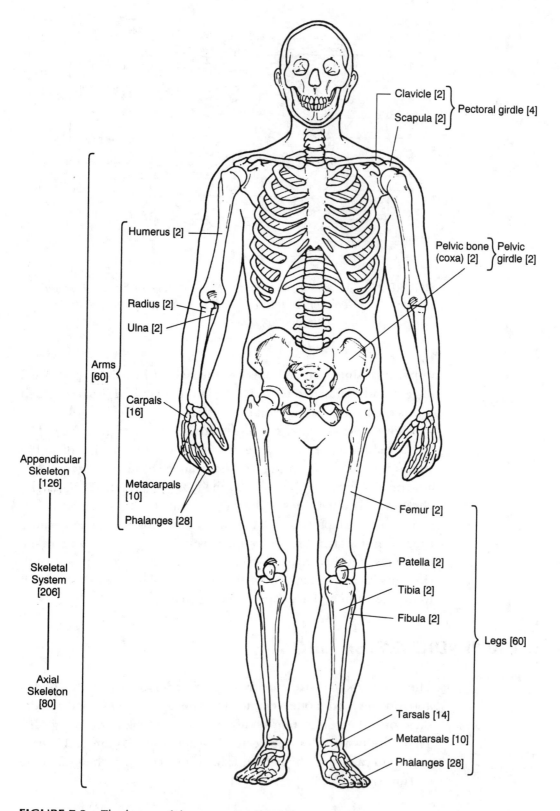

**FIGURE 7.8** *The bones of the appendicular skeleton in humans. One hundred twenty-six bones make up this portion of the complete skeleton. Numbers in square brackets indicate number of bones for each item.*

muscles and ligaments. The posterior surface of each scapula contains a bony portion called the **spine**. The spine leads to the **acromion process**, which forms the tip of the shoulder and the **coracoid process**. Both processes are muscle attachment sites. The narrow end of each scapula forms a socket to receive the upper arm bone. This socket is called the **glenoid fossa**.

The **clavicle** is a rod-shaped bone that braces the scapula against the top of the sternum. Muscles attaching to the clavicle and scapula connect the arm bones to the axial skeleton.

The upper arm bone articulating at the glenoid fossa is the **humerus**. At the upper end the humerus contains the smooth, rounded **head**. Two small, round protrusions, the **greater tubercle** and the **lesser tubercle**, lie below the head and are sites for muscles to attach. The **intertubercular groove** is the furrow lying between them. Near the center of the humerus is the roughened protrusion called the **deltoid tuberosity**, a V-shaped area where the deltoid muscle attaches. At the lower end are two **condyles** where the radius articulates, and above them are two **epicondyles** for muscle attachments. The **coronoid fossa** is a depression between the epicondyles that receives the coronoid process of the ulna. Another depression, the **olecranon fossa**, receives the olecranon process of the ulna (Figure 7.9).

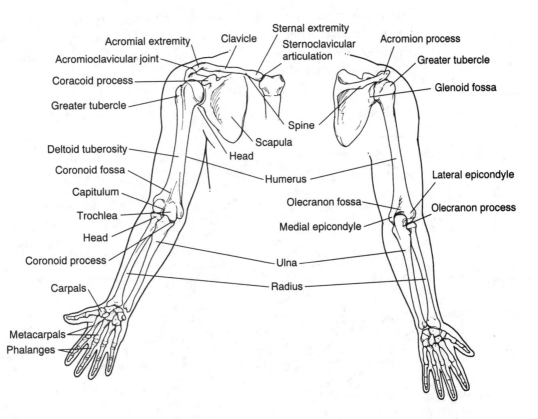

**FIGURE 7.9**    *The right arm and pectoral girdle. (a) An anterior view illustrating many important processes for articulation and muscle attachment (b) A posterior view of the right arm.*

The humerus joins with two bones of the lower arm, the **radius** and **ulna**. This union occurs at the elbow joint. The ulna, which is the medial bone of the forearm (little finger side), articulates with the humerus at the **coronoid process** and the **olecranon process**. The radius, which is the lateral bone of the forearm (thumb side), articulates with the humerus at the **head** of the radius. A process on the radius called the **radial tuberosity** is used for muscle attachments, and a **styloid process** at the lower end receives ligaments from the wrist.

The radius and ulna join at the distal end with a series of wrist bones called **carpals**. There are eight carpal bones in two rows of four each. The carpals are named the pisiform, lunate, triangular, hamate, capitate, scaphoid, trapezoid, and trapezium. Their small size and joint connections permit great flexibility of the wrist.

The carpal bones articulate with **metacarpals,** the five bones (numbered 1 to 5) in the fleshy portion of the hand. The metacarpals, in turn, articulate with finger bones called **phalanges** (singular **phalanx**). The thumb has two phalanges and each of the remaining fingers has three phalanges.

## The Pelvic Girdle and Leg

The pelvic girdle consists of two **pelvic bones**, also called the **os coxae** (singular **os coxa**). The os coxae form a basin that supports organs of the lower abdomen. They are connected to the sacrum by fibrous connective tissue.

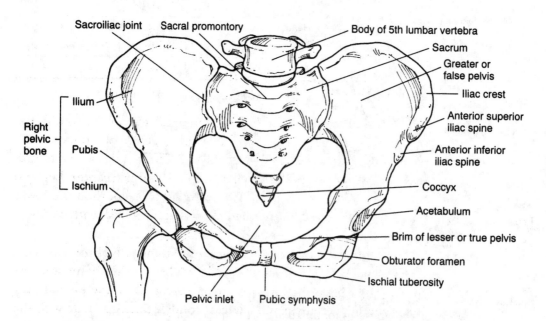
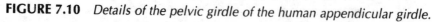

**FIGURE 7.10**  *Details of the pelvic girdle of the human appendicular girdle.*

Each pelvic bone (os coxa) is formed by the fusion of three bones: the **ilium**, the **ischium**, and the **pubis** (see Figure 7.10). In the fused pelvis, the upper region that flares is the ilium. The area on which we sit is formed by the two ischia, and the margin of this area is the **iliac crest**. In the posterior region, the ilium joins the sacrum at the **sacroiliac joint**. The ischium is the lower portion of the os coxae. It is the site of the large, round, roughened protrusion called the **ischial tuberosity**, where ligaments and leg muscles attach. A sharp projection called the **ischial spine** lies above the tuberosity. The three bones meet at a cuplike socket called **acetabulum**. At the acetabulum the rounded head of the femur—the thigh bone—articulates.

The left and right pubic bones fuse at the midline at a joint called the **pubic symphysis**. In the female, the pubic symphysis has greater flexibility to permit the pelvic bone to spread apart, thereby facilitating passage of the fetus through the birth canal (Table 7.2). Between the bodies of the pubis and ischium, a large opening called the **obturator foramen** exists for the passage of nerves and blood vessels to the leg. This is the skeleton's largest foramen.

**TABLE 7.2**   *The Female Pelvis Compared to the Male Pelvis*

| Aspect | Comparison |
| --- | --- |
| Pelvis | Female pelvic bones are lighter, thinner, and have less obvious muscular attachments; the obturator foramina and the acetabula are smaller and farther apart than in the male. |
| Pelvic cavity | Female pelvic cavity has wider diameters and is shorter, roomier and less funnel-shaped than male; the distances between the ischial spines and between the ischial tuberosities are greater than in the male. |
| Sacrum | Female sacrum is relatively wider, and the sacral curvature is bent more sharply posteriorily than in the male. |
| Coccyx | Female coccyx is more movable than male. |

The upper leg bone is called the **femur**. The femur joins with the pelvic girdle at the acetabulum and is the largest and strongest bone of the body. It has a large rounded **head**, a **neck**, and two large protrusions called the **greater trochanter** and **lesser trochanter**. Leg and buttock muscles attach here. Two rounded projections at the lower end, the **lateral condyle** and **medial condyle**, articulate with condyles of the tibia.

The femur meets the lower leg bones at the site of the kneecap, or **patella**. The lower leg bones are the tibia, or shin bone, and the fibula. The **tibia** is the larger bone and is found on the medial side. It has a **medial condyle** and a **lateral condyle** that articulate with the femur's condyles. A prominence on the ankle side called the **medial**

**malleolus** is a site for ligament attachments, and the **malleolar sulcus** is a groove where blood vessels pass. The **fibula** is the thin bone extending down the lateral portion of the leg. On the lower end is the **lateral malleolus**, where ligaments attach (Figure 7.11).

The ankle of the lower leg is formed by a series of seven **tarsals**. The names of the tarsals are the talus, calcaneus, navicular, cuboid, laterial cuneiform, medial cuneiform, and intermediate cuneiform. The tarsals join with a fleshy area in the anterior portion of the foot, formed by a series of **metatarsals**. The instep is formed by the metatarsals. The metatarsals (numbered 1 to 5) join with the toe bones called **phalanges**. As in the fingers, there are two phalanges in the large toe and three phalanges in each of the other toes.

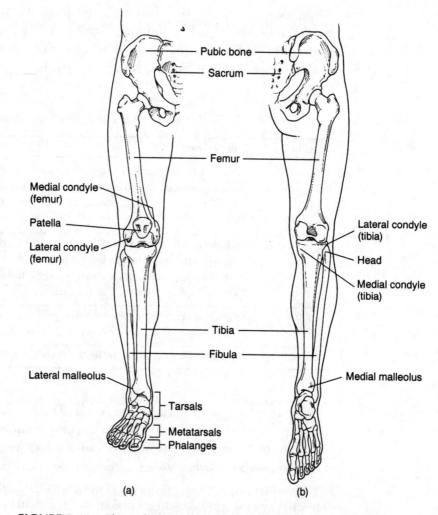

(a)

(b)

**FIGURE 7.11**   *The right leg and pelvic girdle. (a) An anterior view illustrating many important processes for articulation and muscle attachment. (b) A posterior view of the right leg.*

# REVIEW QUESTIONS

**PART A—Completion:** Add the word or words that correctly complete each of the following statements.

1. The bones of the upper and lower appendages comprise the _____ .

2. The movement of the bones is controlled by the body's _____ .

3. The number of bones in the axial skeleton is _____ .

4. The number of bones in the cranium is _____ .

5. The cranial bones are fused together at immovable joints known as _____ .

6. The forehead and anterior roof of the cranium is formed by the _____ .

7. The posterior cranial floor is formed by the _____ .

8. The occipital bone contains a large hole for passage of the spinal cord known as the _____ .

9. The bones that form the side walls of the cranium are called the _____ .

10. The mastoid process is a rounded process of the _____ .

11. The projection of the temporal bone that helps form the cheek bone is the _____ .

12. The cranial bone that has the shape of a butterfly and forms the anterior internal floor of the cranium is the _____ .

13. The two thin horizontal plates of bone in the ethmoid bone form the _____ .

14. The pituitary gland lies in a saddle-shaped depression of the sphenoid bone known as the _____ .

15. Membranous areas in the skull bones present in the newborn are known as _____ .

16. The nasal cavity is divided into left and right chambers by a vertical partition called the _____ .

17. Each zygomatic bone has a projection that helps form the cheek bone and is known as the _____ .

18. The smallest facial bones are the _____ .

19. The upper jaw is formed by two bones called _____ .

20. The lower jaw bone is shaped as a horseshoe and is called the _____ .

21. The vertebral column is composed of 26 bones known as _____ .

22. The five sacral vertebrae fuse to one another to form the _____ .

23. The vertebrae of the neck are known as _____ .

24. The unfused vertebrae of the vertebral column are separated from one another by _____ .

25. An abnormal sideways curve of the spinal column is known as _____ .

26. The weight bearing cylinder of the vertebral bone that is found between the disks is the _____ .

27. The spinal cord extends through the vertebral column by passing through openings in the vertebrae known as _____ .

28. The first vertebrae of the vertebral column is called the _____ .

29. The process of the axis that projects upward into the ring of the first vertebrae is the _____ .

30. The three recognizable parts of the sternum are the manubrium, the body, and the _____ .

31. The true ribs are attached directly to the sternum by the _____ .

32. The last two pairs of ribs do not attach to the sternum and are known as _____ .

33. The arms are connected to the ribcage by connecting bones organized as the _____ .

34. The acromion and coracoid processes are both parts of a bone called the _____ .

35. The rod shaped bone that helps connect the arm bones to the axial skeleton is the _____ .

36. The socket found in the scapula where the humerus articulates is called the _____ .

37. The area in the center of the humerus where the deltoid muscle attaches is called the _____ .

38. The coronoid fossa is a depression located between the epicondyles of the _____ .

39. The olecranon process is a marking on the _____ .

40. The hamate, capitate, and trapezoid are different types of _____ .

41. The phalanges are the bones found in the _____ .

42. The three bones of the pelvis are the ilium, ischium, and the _____ .

43. Where the ilium joins the sacrum, the joint is known as the _____ .

44. The largest and strongest bone of the human body is the _____ .

45. The large opening in the pelvic bone is referred to as the _____ .

46. The proper name for the kneecap bone is the _____ .

47. The thin leg bone extending down the lateral portion of the leg is the _____ .

48. The larger leg bone found on the medial side of the lower leg is the _____ .

49. The ankle of the lower leg is formed by a series of tarsals that number _____ .

50. The toe bones are known as _____ .

**PART B—Multiple Choice:** Circle the letter of the item that correctly completes each of the following statements.

1. The appendicular skeleton is composed of bones of the
   (A) cranium
   (B) thorax
   (C) vertebral column
   (D) upper and lower appendages

2. The cranium is composed of a series of bones
   (A) that remain unconnected
   (B) that are fused together at sutures
   (C) that contain no blood vessels
   (D) that are long and rounded at the center

3. All the following are cranial bones except the
   (A) temporal bone
   (B) frontal bone
   (C) ethmoid bone
   (D) occipital bone

4. Both the foramen magnum and the obturator foramen are
   (A) large holes in bones
   (B) nerve fibers that pass through the bones
   (C) arteries that service the bones
   (D) bones of the pelvic girdle

5. All the following are markings of the temporal bones except
   (A) the mastoid process
   (B) the external auditory meatus
   (C) the styloid process
   (D) the xyphoid process

6. The cheekbone is formed by processes of the
   - (A) occipital and ethmoid bones
   - (B) zygomatic and temporal bones
   - (C) sphenoid and nasal bones
   - (D) parietal and frontal bones

7. A saddle-shaped depression in the sphenoid bone that contains the pituitary gland is the
   - (A) manubrium
   - (B) sella turcica
   - (C) iliac crest
   - (D) crista galli

8. Both the crista galli and cribiform plates are found in the
   - (A) occipital bone
   - (B) temporal bone
   - (C) sphenoid bone
   - (D) ethmoid bone

9. Both the maxillae and the palatine bones help to form the
   - (A) zygomatic process
   - (B) sternum
   - (C) hard palate
   - (D) olecranon process

10. All the following are various types of vertebrae except
    - (A) the lumbar vertebrae
    - (B) the thoracic vertebrae
    - (C) the femoral vertebrae
    - (D) the cervical vertebrae

11. Scoliosis and kyphosis are conditions that result from
    - (A) improper formation of the sternum
    - (B) improper curvature of the spine
    - (C) failure of the skull bones to fuse
    - (D) improper formation of the leg bones

12. The coccyx and the sacrum are the names of
    - (A) lower leg bones
    - (B) facial bones
    - (C) vertebrae
    - (D) bones of the sternum

13. The atlas and the axis are the names of
    - (A) the lower arm bones
    - (B) the first two vertebrae
    - (C) the pubic bones
    - (D) the last two vertebrae

14. The manubrium is the
    (A) upper leg bone
    (B) lower wrist bone
    (C) upper bone of the sternum
    (D) lower bone of the shoulder

15. The scapula is the bone of the pectoral girdle that contains the
    (A) acromion process and coracoid process
    (B) styloid process and radial process
    (C) the scaphoid process and greater tubercle
    (D) lesser trochanter and deltoid tuberosity

16. The glenoid fossa and acetabulum are both
    (A) bones of the thoracic cage
    (B) carpals
    (C) sockets where large bones come together
    (D) processes of the radius

17. The clavicle is a rod-shaped bone of the
    (A) pelvic girdle
    (B) pectoral girdle
    (C) lumbar vertebrae
    (D) pubic bone

18. The deltoid muscle attaches to the humerus at the
    (A) deltoid tuberosity
    (B) deltoid tubercle
    (C) deltoid fossa
    (D) deltoid condyle

19. All the following are names of the carpals except
    (A) capitate
    (B) trapezoid
    (C) pisiform
    (D) cuboidal

20. All the fingers have three phalanges except the
    (A) thumb, which has two phalanges
    (B) index finger, which has one phalanx
    (C) ring finger, which has four phalanges
    (D) pinkie, which has one phalanx

21. The area of the pelvis on which we sit is formed by
    (A) the pubis and ischium
    (B) two pubic bones
    (C) the ilium and pubis
    (D) two ischia

22. All the following are markings of the femur except
    (A) the greater trochanter
    (B) medial malleolus
    (C) lesser trochanter
    (D) lateral condyle

23. All the following apply to the femur except
    (A) it joins with the pelvic girdle at the acetabulum
    (B) it has a large, rounded head
    (C) it meets the tibia and fibula at the knee joint
    (D) it is one of the smallest bones in the body

24. The thin bone extending down the lateral portion of the leg
    (A) is the fibula
    (B) contains the greater trochanter
    (C) fuses with the ischium
    (D) is the femur

25. The instep of the foot is formed by
    (A) eleven phalanges
    (B) five metatarsals
    (C) seven tarsals
    (D) the lower tibia

**PART C—True/False:** For each of the following statements, mark the letter "T" next to the statement if it is true. If the statement is false, change the underlined word to make the statement true.

1. The skeletal system is divided into two major sections called the axial skeleton and the auxiliary skeleton.

2. The cranium is composed of 16 bones fused together at immovable joints called sutures.

3. The posterior roof of the cranium is formed by two arched bones called occipital bones.

4. The foramen magnum is a large hole in the frontal bone through which the spinal cord passes and continues with the brain.

5. The external auditory meatus is an opening in the temporal bone that leads to the inner part of the eye.

6. Many neck muscles attach to the temporal bones by means of the greater trochanter.

7. The zygomatic process of the zygomatic bone is a projection that helps form the cheekbone.

8. The pituitary gland lies in a depression of the sphenoid bone called the sella turcica.

9. The cribiform plates and the <u>styloid process</u> are both processes found in the ethmoid bone.

10. Membranous areas called <u>fontanels</u> are found in the newborn skull before the bones fuse to one another.

11. The nasal cavity is divided into left and right chambers by a vertical partition called the <u>ethmoid bone</u>.

12. The lacrimal are the <u>largest</u> bones of the face.

13. The lower jaw bone has the shape of a horseshoe and is known as the <u>maxilla</u>.

14. The vertebrae of the neck are known as <u>thoracic</u> vertebrae.

15. Exaggeration of the thoracic curve of the vertebral column is known as <u>scoliosis</u>.

16. The spinous and transverse processes of the vertebrae are sites of attachment for most <u>ligaments</u> of the back.

17. The odontoid process of the <u>atlas</u> projects upward into the ring formed by the first vertebrae.

18. The lowermost portion of the sternum is formed by the <u>manubrium</u>.

19. There are <u>14</u> pairs of ribs in the adult human being.

20. The triangular bone of the pectoral girdle is called the <u>clavicle</u>.

21. The socket in the pectoral girdle that receives the upper arm bone is called the <u>acetabulum.</u>

22. The olecranon process of the <u>radius</u> articulates with the olecranon fossa of the humerus.

23. The wrist bones of the human body are referred to as <u>carpals</u>.

24. The three bones that form the pelvic bone are the ilium, <u>ischium</u>, and pubis.

25. The greater and lesser trochanters of the <u>patella</u> are the sites of attachment for many of the leg muscles.

**Answers**

## PART A—Completion

1. appendicular skeleton
2. muscles
3. eighty
4. eight
5. sutures
6. frontal bone
7. occipital bone
8. foramen magnum
9. temporal bones
10. temporal bone
11. zygomatic process
12. sphenoid bone
13. cribiform plate
14. sella turcica
15. fontanels
16. vomer bone
17. temporal process
18. lacrimal bones
19. maxillae
20. mandible
21. vertebrae
22. sacrum
23. cervical vertebrae
24. intervertebral disks
25. scoliosis
26. body
27. vertebral foramena
28. atlas
29. odontoid process
30. xiphoid process
31. hyaline costal cartilages
32. floating ribs
33. pectoral girdle
34. scapula
35. clavicle
36. glenoid fossa
37. deltoid tuberosity
38. humerus
39. ulna
40. carpals
41. fingers
42. pubis
43. sacroiliac joint
44. femur
45. obturator foramen
46. patella
47. fibula
48. tibia
49. seven
50. phalanges

## PART B—Multiple Choice

| | | | | |
|---|---|---|---|---|
| 1. D | 6. B | 11. B | 16. C | 21. D |
| 2. B | 7. B | 12. C | 17. B | 22. B |
| 3. C | 8. D | 13. B | 18. A | 23. D |
| 4. A | 9. C | 14. C | 19. D | 24. A |
| 5. D | 10. C | 15. A | 20. A | 25. B |

## PART C—True/False

1. appendicular
2. eight
3. parietal
4. occipital
5. ear
6. mastoid process
7. temporal
8. true
9. crista galli
10. true
11. vomer bone
12. smallest
13. mandible
14. cervical
15. kyphosis
16. muscles
17. axis
18. xiphoid process
19. 12
20. scapula
21. glenoid fossa
22. ulna
23. true
24. true
25. femur

# CHAPTER 8

# MUSCLE PHYSIOLOGY

Muscle tissue is one of the four basic tissues. It is distinguished from other tissues by its ability to contract and perform mechanical work. The structural unit of muscle tissue is the muscle cell, which has an elongated shape and is also called the **muscle fiber**.

Most muscles are connected to bones, and the combination of muscle and bone allows movement in the human body. Muscles are also components of other organs such as the digestive and respiratory tracts and blood vessels. Although muscles have different shapes and function, all muscle tissue can be characterized as smooth muscle, cardiac muscle, or skeletal muscle.

**Smooth ("nonbanded") muscle** consists of single contractile cells. It may be found in the walls of the digestive tract and uterus as well as in the linings of blood vessels and certain ducts. **Cardiac muscle** is also composed of single cells, but it has much more regular organization of the contracting macromolecules than found in smooth muscle cells. Cardiac muscle cells have special linkages in the walls of the heart (Table 8.1).

**TABLE 8.1**  *A Comparison of Three Types of Muscle Tissue*

| Characteristic | Skeletal Muscle | Smooth Muscle | Cardiac Muscle |
|---|---|---|---|
| Location | Attached to skeleton | Walls of stomach, intestines, etc. | Walls of heart |
| Type of control | Voluntary | Involuntary | Involuntary |
| Shape of fibers | Elongated, cylindrical, blunt ends | Elongated, spindle-shaped, pointed ends | Elongated, cylindrical fibers that branch |
| Striations | Present | Absent | Present |
| Number of nuclei per fiber | Many | One | One or two |
| Position of nuclei | Peripheral | Central | Central |
| Speed of contraction | Most rapid | Slowest | Intermediate |
| Ability to remain contracted | Least | Greatest | Intermediate |

The third type of muscle, and the most abundant type, is **striated** (**"banded"**) or **skeletal muscle**. In skeletal muscle, each muscle cell is a fused set of dozens or hundreds of cells. The muscle cells that result are usually very long, and they are called muscle fibers. A large number of these fibers make up the organ we call "muscle." Skeletal muscle is associated with the body's skeleton. It is the most researched of all muscle types, and it is the type of muscle usually meant when speaking of "muscle" in the general context.

# SKELETAL MUSCLE FUNCTION

One of the basic attributes of muscle cells and fibers is their ability to exert force, but only in one direction. Muscle cells contract by an active mechanism and relax by a passive mechanism, and contractions only occur when a stimulation has occurred.

The complex patterns of movement for locomotion require that two sets of muscles move body parts in opposite directions. Muscles thus work against each other and are said to be **antagonistic**. For example, a leg joint is bent (flexed) by flexor muscles and straightened out (extended) by extensor muscles. The muscle contractions move parts of the skeleton to which they are attached.

# Muscle Cell Structure

Skeletal (striated) muscle of the body is under **voluntary control**; it usually contracts only when stimulated by neurons that deliver impulses to it. Skeletal muscle contains bundles of muscle fibers. Each fiber has a set of 4 to 20 rodlike filaments known as **myofibrils**. The myofibrils are bathed in cytoplasm, referred to as **sarcoplasm**. Numerous **mitochondria** scattered throughout the sarcoplasm provide ATP as an energy source for the contraction of myofibrils.

Myofibrils exist in smaller units called **sarcomeres** (Figure 8.1). Each myofibril is about one or two micrometers in width, and the myofibrils may be up to 100 micrometers in length. The repetition of sarcomeres within the muscle fiber gives the muscle its characteristic striated ("banded") pattern.

A microscopic view of the sarcomere reveals that each sarcomere is composed of two types of myofilaments: **thin filaments** and **thick filaments**. The filaments run parallel to one another, and the thick filaments are composed of the protein **myosin**, while the thinner filaments are composed primarily of the protein **actin**. The point at which actin filaments from adjacent sarcomeres interweave is a dense black line called the **Z line**. The Z line bisects a relatively clear, broad stripe called the **I band**. The large dense stripe in the center of the sarcomere formed by the overlapping myosin filaments is called the **A band**. The A band is bisected by a central **H zone** containing

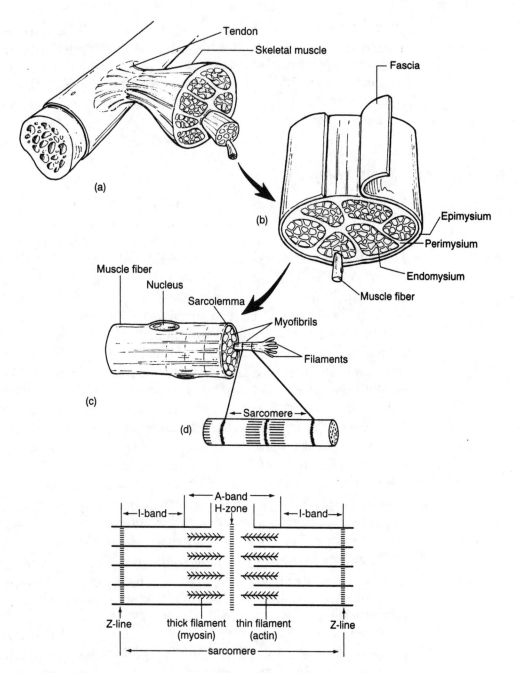

**FIGURE 8.1**  *The microscopic and submicroscopic structure of a skeletal muscle cell. (a) An entire muscle dissected through its belly (b) A cross-section through the entire muscle to show numerous muscle fibers (cells) organized to bundles (c) A single muscle fiber (cell) displaying three nuclei, the sarcolemma, and a number of sarcomeres at the end of the fiber (d) A sarcomere expanded to show the thick and thin filaments and the various bands and zones. Muscle activity takes place at the sarcomere.*

myosin filaments but no actin filaments. Repeating A and I bands account for the band pattern in the myofibrils of striated muscle.

The thin actin filaments are anchored to the Z line. During muscle contraction, opposing actin filaments slide along myosin filaments, and the two sets of actin filaments are drawn toward each other. This activity draws the attached Z lines toward one another, and the distance between the Z lines will shorten without shortening any of the filaments. As this occurs, the pale I band is reduced in size, and the broad A band approaches the Z lines. The pale H band at the center also disappears in a contracted muscle because the sets of filaments almost meet.

## The Sliding Filament Model

Research indicates that the contraction of muscle fibers occurs as thin filaments slide toward one another in the muscle fiber. This process was first proposed by Hugh E. Huxley in 1957. The myosin molecules in the thick filaments tend to bind together and remain stationary.

Molecules of myosin are composed of two polypeptide chains, each shaped like a golf club with the shafts twisted around each other and the "heads" bent to the sides at hingelike sites. Each myosin filament is surrounded by thin actin filaments so that the protruding myosin head and the actin filaments can come in contact with each other when the muscle contracts. In the **sliding filament theory**, the myosin heads act as crossbridges between actin and myosin filaments (Figure 8.2). The heads apply a power stroke similar to an oar pushing on water. The **power stroke** pushes the actin filaments inward toward the H zone. As they slide, the actin filaments pull on the Z lines and shorten the sarcomere. When the same process goes on simultaneously in thousands of sarcomeres in numerous muscle fibers, the muscle itself shortens and contracts.

Myosin crossbridges assist muscle fiber contraction because they act as enzymes. The enzymes break down **adenosine triphosphate (ATP) molecules** into adenosine diphosphate (ADP) and inorganic phosphate groups. First, ATP binds to the enzyme site on the myosin head. Then the ATP molecule is cleaved, and both ADP and phosphate remain bound to the head. The energy released by the breakdown activates the myosin head into a cocked position. In this position, the myosin head and the actin filaments bind weakly together. This binding causes the ADP and phosphate to be released, and as these substances leave, the head binds strongly to the actin molecule. Simultaneously, the head rocks forward and supplies the power stroke to the actin filament. This power stroke moves or "slides" the actin filament across the myosin filament an ultramicroscopic distance. When a new ATP molecule binds to the site, the actin is released and the cycle repeats itself.

The sliding filament cycle occurs rapidly in thousands of heads at each end of a sarcomere, so long as ATP is available to prepare the head for actin binding and provide energy for the cocking process. In so-called **red muscle**, there is a large quantity of the reddish compound **myoglobin**, where oxygen is stored. The utilization of ATP is relatively slow, and the cells can generate ATP as it is used. These

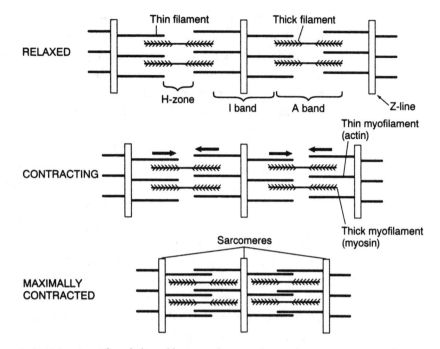

**FIGURE 8.2**   *The sliding filament theory of muscle activity as shown by two sarcomeres. (a) In the relaxed state, the thin filaments of actin are separated (b) When the muscle is contracting, the thin actin filaments are sliding toward each other as the thick myosin filaments remain stationary; the distance between the Z lines is decreasing (c) In the contracted state, the thin filaments have overlapped one another and the sarcomeres have shortened their maximum amount. The H zone has disappeared, and the I band has decreased in size dramatically. The muscle fiber is now contracted.*

properties permit red muscle to contract repeatedly and resist muscle fatigue. Red muscle is often called slow muscle or **tonic muscle**.

The other form of muscle is **white muscle**. This is sometimes called fast muscle or **twitch muscle**. White muscle has little or no myoglobin and stores little oxygen. ATP is used up quickly and cannot be replaced rapidly. Therefore white muscle becomes quickly fatigued with the build up of lactic acid.

## Triggers for Muscle Contraction

The thin actin filaments consist of two chains of actin twisted into a helix. In the grooves of this helix are molecules of a protein called **tropomyosin**. When a muscle is at rest, tropomyosin prevents myosin heads from binding to actin, probably by masking the site where binding would ordinarily occur.

Another protein called **troponin** is found at regular intervals along the actin filament. Troponin binds to both tropomyosin molecules and actin molecules. Troponin also binds to calcium ions.

Muscle cells can be triggered to contract because they initiate and propagate an action potential (a nerve impulse). A neurotransmitter such as acetylcholine is released when nerve impulses from a nerve cell reach the muscle cell at the neuromuscular junction and enter. The **neuromuscular junction** consists of a single muscle fiber and the terminal end of a single muscle cell. Although the membranes of the nerve and muscle cells are close, they do not touch. Instead, they remain separated by a fluid-filled space called the synapse. Neurotransmitters such as acetylcholine are released in this synapse. The impulse is initiated in the muscle cell and propagated over the entire cell surface. It then triggers events inside the muscle cell that culminate in the contraction.

The concentration of **calcium ions** in the cytoplasm of the muscle fiber is normally very low because calcium is continually pumped out of the cell or into the cell's mitochondria. In muscle cells there is another reservoir for calcium in a system of infoldings of the plasma membrane called the **transverse tubules**, or **T tubules**. The T tubules surround the myofibrils at the Z lines. Interwoven with the T tubules is a calcium source derived from branches of the sarcoplasmic reticulum. The sources are hollow terminal sacs.

When nerve impulses spread through the muscle cell, the T tubule membranes conduct the impulses to the terminal sacs. Immediately, calcium ions are released. The ions diffuse out of the sacs and into the fluid bathing the myosin filaments. Here they bind to sites on the troponin molecules, and the complex changes shape (Figure 8.3).

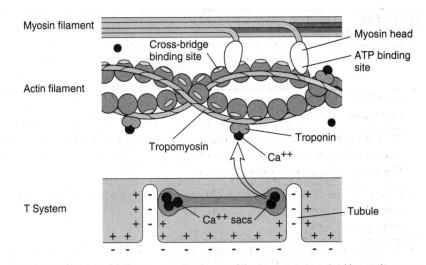

**FIGURE 8.3**   *Details of sarcomere contraction in a muscle fiber. The actin filaments contain molecules of tropomyosin in their grooves. The tropomyosin masks the actin binding sites, and troponin is attached to both actin and tropomyosin proteins. When calcium is released from reservoirs near the T tubules, it combines with troponin molecules and causes the troponin molecules to shift position, thereby revealing the actin binding sites. The actin then attaches to the globular heads of myosin molecules, using ATP molecules as a catalyst.*

Since troponin is linked to tropomyosin, the tropomyosin also shifts position and unmasks the sites on actin that bind to myosin heads. The power strokes begin. If a large quantity of calcium ions is released, then more binding to troponin takes place, and the number of sliding filaments increases. The muscle contraction becomes stronger.

### Relaxation

A muscle relaxes when nerve impulses no longer stimulate it to contract. When these impulses cease, the plasma membranes and the T tubules return to their normal states. The release of calcium ions from the sarcoplasmic reticulum comes to an end, and enzymes pump calcium ions back into the terminal sacs.

With the withdrawal of calcium, troponin reverts to its resting configuration, and tropomyosin immediately resumes covering the myosin-binding sites on the actin filaments. The actin filaments slide outward to return the sarcomere to its resting length (Table 8.2).

Although calcium ions are conserved in muscle relaxation, ATP is consumed both during contraction and relaxation. In relaxation, ATP provides energy to pump calcium ions into the terminal sacs. Muscles that lack a supply of ATP remain contracted. After death, for example, the muscle cells remain contracted in a condition called

**TABLE 8.2**   *The Events of Muscle Fiber Contraction and Relaxation*

| Muscle Fiber Contraction | Muscle Fiber Relaxation |
| --- | --- |
| 1. Stimulation occurs when acetylcholine is released from the end of a motor neuron. | 1. Cholinesterase causes acetylcholine to decompose, and muscle fiber membrane is no longer stimulated. |
| 2. Acetylcholine diffuses across gap at neuromuscular junction. | 2. Calcium ions are actively transported into the sarcoplasmic reticulum. |
| 3. Muscle fiber membrane is stimulated and a muscle impulse travels deep into the fiber through the transverse tubules. | 3. Cross-bridges between actin and myosin filaments are broken. |
| 4. Calcium ions diffuse from sarcoplasmic reticulum into the sarcoplasm and bind to troponin molecules. | 4. Actin and myosin filaments slide apart. |
| 5. Tropomyosin molecules move and expose specific sites on actin filaments. | 5. Muscle fiber lengthens as it relaxes and its resting state is reestablished. |
| 6. Cross-bridges form between actin and myosin filaments. | 6. Troponin and tropomysin molecules inhibit the interaction between actin and myosin filaments. |
| 7. Actin filaments slide inward along myosin filaments. | |
| 8. Muscle fiber shortens as contraction occurs. | |

**rigor mortis.** The actin and myosin filaments are locked together in their contracted position. After several hours, the contracted muscles relax as other degenerative processes begin to dominate.

## Graded Responses

An important functional feature of individual muscle cells is the **all-or-none response**. This means that a fiber contracts only after an impulse spreads through the cell and exceeds a certain threshold. Once the threshold has been surpassed, further increases in the intensity, rate, and duration of impulses produce a contraction that is only stronger by a very slight amount. In basic terms, the muscle fiber (cell) contracts completely or not at all.

Entire muscles do not show the all-or-none response. Rather, they display a graded response to a stimulus. A **graded response** is a variable response depending upon the number of contracting muscle cells in a muscle. If a large number of neurons conduct impulses to a muscle, many muscle fibers contract. However, if only a few fibers are stimulated, the entire muscle contracts weakly. Each nerve fiber to a muscle branches, and a single nerve fiber can supply up to 100 muscle fibers. The muscle fibers and the motor neuron that stimulates them constitute a **motor unit** (Figure 8.4).

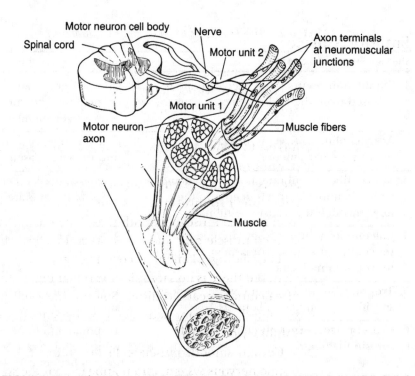

**FIGURE 8.4**    *The relationship between the nervous system and the muscular system. A motor unit consists of a motor nerve fiber and all the muscle fibers (cells) it stimulates. Two motor units are displayed. Note that a single nerve fiber services a number of muscle fibers (cells).*

The contraction of a muscle fiber is called a **twitch**. The number of twitches occurring in a muscle is called **summation**. Summation is the condition during which nerve impulses arrive at a muscle before its previous contraction has subsided. Summation results partly from the inability of the sarcoplasmic reticulum to recover all its calcium ions before new impulses arise. The strength of summated contractions is always greater than the strength of individual twitches, since twitches are responses to less-frequent stimulations. Summation can culminate in **tetanus**, a state of sustained maximum contraction. Making a fist places the muscles into a state of tetanus.

Another characteristic of muscle function is the ability to achieve muscle tone, or **tonus**. Tonus is a condition in which muscle is kept partially contracted over a long period of time. It is produced when one set of fibers, then another set, is briefly stimulated so that some part of the muscle is always contracted, although most of the muscle remains relaxed. **Normal posture** in the presence of gravity is achieved by tonus.

# OTHER ASPECTS OF MUSCLE FUNCTION

Skeletal muscle physiology has been well-studied in comparison to the physiology in smooth and cardiac muscle.

## Smooth Muscle

**Smooth muscle** tissue consists of slender, elongated cells without striations and with a single nucleus (striated cells have many nuclei). Smooth muscle cells are arranged in sheets in the wall of the colon, or as straps around certain blood vessels or ducts. The cells are linked together by collagen fibers and by **gap junctions**. The cytoplasm of smooth muscle cells has many actin filaments whose ends are inserted in the inner surface of the plasma membrane. A molecule similar to myosin is also present in smooth muscle, with heads as in striated muscle. Smooth muscle cells contract more slowly than striated muscle cells and can sustain the contraction far longer.

The sliding filament theory may also apply to smooth muscle cells, but there is no sarcoplasmic reticulum in the cells, and the role of calcium appears to differ. Smooth muscle cells usually carry out sustained, slow contractions. Typically, smooth muscle is not under voluntary control.

Certain smooth muscle cells are under the control of the autonomic nervous system, and it appears that the sympathetic nerves of this system stimulate contractions, while parasympathetic nerves inhibit contractions (Chapter 11). Smooth muscle cells also respond to hormones. Smooth muscle is often called **visceral muscle** because it is found in visceral organs. Smooth muscle has no sarcomeres, but thin and thick filaments are collected into bundles resembling myofibrils. The ratio of thick to thin filaments is 1:16, as compared to a 1:2 ratio in skeletal muscle. Smooth muscle also contains noncontractile intermediate filaments attaching to dense bodies distributed throughout the cell. Dense bodies permit the attchment of thin filaments and are counterparts to the Z lines in skeletal muscle. A strong cytoskeleton thus exists within the smooth muscle cell. The cells are surrounded by connective tissue and are bound by elastin and collagen fibers. Usually the cells are organized into sheets of muscle as in the walls of the blood vessels. Contractions of the smooth muscle in the urinary bladder, rectum, and uterus assist the organs in expelling their contents. Highly structured neuromuscular junctions do not exist in smooth muscle, but bulbous endings release their neurotransmitters into the diffuse junctions at the smooth muscle cells.

Smooth muscle is generally distinguished as single-unit muscle and multiunit muscle. **Single-unit smooth muscle** is the visceral muscle whose cells contract as a rhythmic unit and are electrically coupled to one another by gap junctions. **Multiunit smooth muscle**, by comparison, lines the airways to the lungs and large arteries as well as the erector pili muscles of the hair follicles. The fibers work independently of one another, and gap junctions are rare. The nerve endings are rich in multiunit smooth muscle, and a motor unit forms with a number of muscle fibers. The autonomic division of the nervous system supplies impulses to this muscle.

## Cardiac Muscle

Cardiac muscle is found only in the heart. Like smooth muscle cells, cardiac muscle cells have a single nucleus. Cardiac muscle, however, is striated. It is liberally supplied with mitochondria.

**Cardiac muscle cells** are often branched and form an interlocking network with neighboring cells. A T tubule system appears to be present and sarcoplasmic reticulum has been identified. The ends of cardiac muscle cells are bound firmly to one another by **intercalated disks** (Table 8.3). These are regions of cell membrane that hold cells together firmly. They are also sites where electrical current can flow easily.

Cardiac muscle tissue is not under the body's voluntary control. It receives impulses from branches of the autonomic nervous system. The intercalated disks that it contains are not found in other types of muscle tissue.

## Chemistry of Muscle Contraction

The energy utilized for contraction of a muscle fiber is derived primarily from ATP molecules. These molecules are supplied by

**TABLE 8.3**  *Cardiac Muscle Compared to Skeletal Muscle*

| Characteristic | Cardiac Muscle | Skeletal Muscle |
| --- | --- | --- |
| Nature of control | Autonomic nervous system | Somatic nervous system |
| Arrangement of fibers | Branched | Unbranched |
| Microscopic appearance | Striated | Striated |
| Nuclei per fiber | One | Many |
| Intercalated disks | Present | Absent |
| Arrangement of tubules | One per sarcomere, located at Z lines | Two per sarcomere, locates at A-I junctions |
| Duration of action potential | 150 to 100 msec | 1 to 2 msec |
| Contraction time | 150 to 300 msec | About 40 msec, but varies in different muscles |
| Absolute refractory period | 150 to 300 msec | 1 to 2 msec |

chemical reactions taking place in the numerous mitochondria lying close to the thick and thin filaments. The heads of the myosin filaments contain the enzyme ATPase, which decomposes ATP into ADP and phosphate groups and releases the energy contained in the molecule.

The supply of ATP in a muscle fiber is limited, and ATP molecules must constantly be regenerated from ADP molecules and phosphate groups. One of the sources for regenerating ATP is the substance **creatine phosphate**. Creatine phosphate contains high-energy phosphate bonds and is a storage depot for energy in the cell. When ATP is depleted, creatine phosphate breaks down and releases its energy to regenerate ATP molecules.

When muscle is very active, the supply of ATP and creatine phosphate may be exhausted (Figure 8.5). At this point, the muscle cells depend upon the breakdown of glucose as a source of energy. Glucose metabolism by cellular respiration involves glycolysis, the Krebs cycle, the electron transport system, and chemiosmosis (Chapter 19). The reactions of cellular respiration take place in the cytoplasm and mitochondria, and oxygen is required for their completion.

The oxygen needed for cellular respiration is transported by the hemoglobin in red blood cells to the muscle fibers. In muscle fibers, a pigment called myoglobin binds with oxygen molecules and stores them temporarily. As noted previously, red muscle is red because myoglobin is present. Thus, active muscle is usually red muscle. The presence of myoglobin reduces the cell's need for a continuous supply of oxygen during contraction.

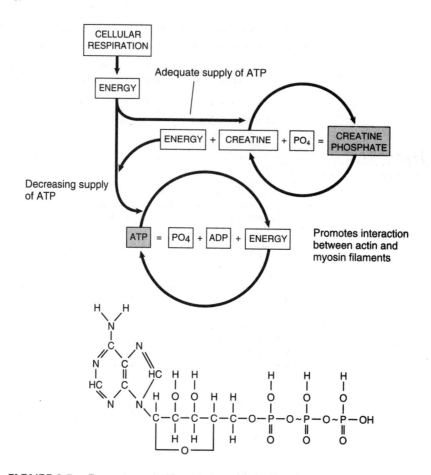

**FIGURE 8.5**    *Energy conversions in muscle cells. (a) Energy is supplied by cellular respiration to form creatine phosphate and to fuel the synthesis of ATP molecules. When ATP is broken down, its energy promotes actin-myosin activity, which leads to muscle contraction (b) The chemical structure of ATP. When the high-energy bond is broken, energy is released to stimulate muscle cell activity.*

When the muscle contracts strenuously for several minutes, oxygen cannot be supplied rapidly enough to meet the needs of the muscle cells. In this situation, the muscle cells depend on ATP derived in the anaerobic phase of cellular respiration by the process of glycolysis (Chapter 19).

During the **anaerobic reactions** of glycolysis, glucose molecules are converted through multiple steps into pyruvic acid molecules. If the cell's oxygen supply is depleted, the pyruvic acid is converted to **lactic acid**, and in the process, two molecules of ATP are produced per molecule of glucose. As lactic acid accumulates, extreme fatigue sets in, and an **oxygen debt** develops. This means that oxygen must be supplied to relieve the buildup of lactic acid. The lactic acid causes pH changes, and muscle fibers can no longer respond to stimulation.

Much of the lactic acid produced in muscle cells diffuses out of the muscle cells and is carried by the blood to the liver. The liver cells use oxygen to convert the lactic acid back into energy-yielding compounds such as glucose. Oxygen is also used for the resynthesis of ATP by cellular respiration and for the formation of creatine phosphate. The heavy breathing that one experiences after strenuous exercise is a result of the oxygen debt.

# REVIEW QUESTIONS

**PART A—Completion: Add the word or words that correctly complete each of the following statements.**

1. The type of muscle found in the walls of the digestive tract is _____ .

2. Heart muscle is also known as _____ .

3. The most abundant type of muscle in the human body is striated muscle, also known as _____ .

4. Muscles that work against each other are said to be _____ .

5. The submicroscopic filaments of a muscle cell are known as _____ .

6. The unit of muscle activity is correctly known as the _____ .

7. The thick filaments in the central portion of the sarcomere are composed of the protein _____ .

8. The thin filaments scattered throughout the sarcomere are composed of the protein _____ .

9. The thin filaments of the sarcomere in a muscle fiber are anchored to the _____ .

10. The theory that explains the contraction of muscle fibers is called the _____ .

11. During muscle contractions, as the filaments slide, they pull on the Z lines and shorten the _____ .

12. The primary source of energy for muscle contraction is the molecule known as _____ .

13. In muscle known as red muscle, much oxygen is stored in a red-pigmented compound called _____ .

14. The continual contractions of muscle fibers require a supply of the energy compound _____ .

15. The muscle known as white muscle is so-named because it has little or no _____ .

16. White muscle is not able to store much _____ .

17. White muscle is sometimes called fast muscle, also known as _____ .

18. When a muscle is at rest, myosin heads are prevented from binding to actin by a protein called _____ .

19. Muscles are triggered to contract following release of a neurotransmitter such as _____ .

20. The most important ion utilized during muscle contraction is _____ .

21. Calcium ions are stored in muscle cells within infoldings of the plasma membrane called _____ .

22. Following their release, calcium ions trigger muscle contractions by binding to the molecule _____ .

23. The binding of calcium to troponin causes a shift in the molecule _____ .

24. The shift of position of tropomyosin unmasks the sites on actin filaments that bind to the heads of _____ .

25. Muscle contractions come to an end when tropomyosin covers the myosin-binding sites on the _____ .

26. Energy is used during muscle contraction to move calcium ions into the _____ .

27. The state of continual muscle contraction following death is called _____ .

28. Muscle contracts after its threshold has been reached, and the response is called the _____ .

29. The variable response displayed by a muscle fiber is known as the _____ .

30. The number of muscle fibers contracting in a muscle depends on the reception of a certain number of _____ .

31. The contraction of a muscle fiber is known as a _____ .

32. All the muscle fibers contracting in a muscle constitute a _____ .

33. The number of twitches occurring in a muscle is known as a _____ .

34. A state of sustained maximum contraction of a muscle is a phenomenon called _____ .

35. The condition in which a muscle is kept partially contracted over a long period of time is known as _____ .

36. Where contractions of the skeletal muscles are fast, the contractions of smooth muscles are generally _____ .

37. Smooth muscle is so-named because it contains no _____ .

38. The muscle cells in smooth muscle are linked together by junctions known as _____ .

39. Where skeletal muscles are voluntary muscles, the smooth muscles are _____ .

40. Because smooth muscle is found in the internal organs, it is sometimes called _____ .

41. Cardiac muscle is striated, and it is liberally supplied by energy-yielding organelles called _____ .

42. The ends of cardiac muscle cells are firmly bound to one another by _____ .

43. Cardiac muscle is similar to smooth muscle because it is not under _____ .

44. When ATP is broken down in muscle cells, the two end products are phosphate groups and _____ .

45. One source for regenerating ATP is the high-energy molecule _____ .

46. In order for energy to be released from carbohydrate molecules in muscle cell metabolism, an essential gas is _____ .

47. When muscle is overactive, the energy for muscle contraction is derived from an anaerobic process called _____ .

48. Extreme muscle fatigue is generally due to the buildup of an acid called _____ .

49. Following strenuous activity, a person breathes deeply to satisfy an _____ .

50. Much of the lactic acid produced during strenuous muscle activity is carried from the muscle cells for metabolism to the _____ .

**PART B—Multiple Choice: Circle the letter of the item that correctly completes each of the following statements.**

1. Smooth muscle may be found in the
   (A) arms and legs
   (B) body trunk
   (C) digestive tract
   (D) head and neck

2. The only location for cardiac muscle in the human body is the
   (A) uterus
   (B) brain
   (C) lower leg
   (D) heart

3. For every muscle that acts in one direction there is another muscle that is
   (A) complementary
   (B) antagonistic
   (C) extended
   (D) flexed

4. The skeletal muscles of the body are under
   (A) voluntary control
   (B) continual stress
   (C) enzyme control
   (D) hormonal control

5. The two proteins of the myofibrils are
   (A) pepsin and peptides
   (B) lysozyme and peptidoglycan
   (C) glycogen and insulin
   (D) actin and myosin

6. The Z line, I band, A band, and H zone are all anatomical parts of the
   (A) myofibril
   (B) sarcolemma
   (C) sarcomere
   (D) thin filaments

7. The amount of space between the Z lines shortens during
   (A) ATP production
   (B) contraction
   (C) protein synthesis
   (D) sarcoplasmic release

8. The sliding filament theory was first proposed by the investigator
   (A) Hugh Corning
   (B) Charles Dowd
   (C) Hugh Huxley
   (D) Emil Von Behring

9. The energy for muscle contraction is supplied by the molecule known as
   (A) ATP
   (B) NAD
   (C) NADP
   (D) DNA

10. When muscle contraction occurs
    (A) the myosin filaments slide but the actin filaments do not
    (B) ATP is synthesized
    (C) calcium ions are not needed
    (D) the actin filaments slide while the myosin filaments do not

11. The myoglobin of red muscle is important because it
    (A) supplies ATP for muscle contraction
    (B) supplies calcium for muscle contraction
    (C) supplies oxygen for muscle contraction
    (D) removes the products of ATP breakdown during muscle contraction

12. The presence of myoglobin in red muscle permits the muscle to
    (A) resist fatigue
    (B) recovery quickly
    (C) undergo twitches
    (D) utilize NAD as an energy source

13. In a muscle at rest, the myosin heads are prevented from binding to actin
    (A) by the interference of oxygen
    (B) by the activity of iron ions
    (C) by the protein tropomyosin
    (D) by the presence of carbon dioxide

14. For the successful completion of a muscle contraction, the terminal sacs of the sarcoplasmic reticulum release
    (A) ATP, which breaks down and releases energy
    (B) calcium ions that trigger muscle contraction
    (C) lactic acid, which inhibits muscle activity
    (D) neurotransmitters, which supply impulses to muscle

15. In order for a muscle contraction to occur
    (A) lactic acid must be present
    (B) lysozyme must be present
    (C) iron ions must be available
    (D) a nerve impulse must enter the muscle fiber

16. During muscle contraction, calcium ions bind to sites
    (A) on oxygen atoms
    (B) on troponin molecules
    (C) on the endoplasmic reticulum
    (D) on the surface of the muscle cell

17. Muscle cells that lack a supply of ATP
    (A) remain contracted
    (B) utilize lactic acid instead of ATP
    (C) undergo anaerobic metabolism
    (D) are generally healthy muscle cells

18. The fact that a muscle fiber contracts completely or not at all is known as the
    (A) completion phenomenon
    (B) complementary response
    (C) theory of use and disuse
    (D) all-or-none response

19. The number of contracting muscle cells in a particular muscle represents the
    (A) muscular junction
    (B) sliding filament theory
    (C) graded response
    (D) partial phenomenon

20. The state of sustained maximum contraction by a muscle
    (A) is known as tonus
    (B) occurs most frequently in the absence of oxygen
    (C) is known as tetanus
    (D) never occurs in the muscle

21. Smooth muscle differs from striated muscle in that smooth muscle does not have
    (A) any cytoplasm
    (B) a plasma membrane
    (C) many actin filaments and no myosin
    (D) any connection to nerve fibers

22. The ends of cardiac muscle cells are bound to one another by
    (A) gap junctions
    (B) intercalated disks
    (C) nerve fibers
    (D) calcium ions

23. Cardiac muscle is
    (A) striated and involuntary
    (B) striated and voluntary
    (C) smooth and involuntary
    (D) smooth and voluntary

24. Creatine phosphate is utilized in muscle cells as a
    (A) source of protein
    (B) reservoir for amino acids
    (C) reservoir for enzymes
    (D) energy source

25. The amount of ATP produced during anaerobic reactions in a muscle cell
    (A) is about the same amount as in aerobic actions
    (B) is much less than is produced in aerobic reactions
    (C) is much higher than is produced in aerobic reactions
    (D) is produced by the reactions of the Krebs cycle and electron transport

**PART C—*True/False:* For each of the following statements, mark the letter "T" next to the statement if it is true. If the statement is false, change the <u>underlined</u> word to make the statement true.**

1. The structural unit of muscle tissue is the muscle cell, also known as the <u>muscle fiber</u>.

2. The three types of muscles are smooth muscle, skeletal muscle, and <u>epithelial</u> muscle.

3. All muscle cells are able to contract by a mechanism that is <u>active</u>.

4. The skeletal muscles of the body are under <u>voluntary</u> control.

5. The cytoplasm of the muscle cell is referred to as <u>sarcolemma</u>.

6. The thick filaments of the muscle fiber are composed of the protein <u>actin</u>.

7. The dense black line that is drawn to other dense black lines during muscle contraction is known as the <u>A band</u>.

8. During muscle contraction, there is disappearance of the <u>H band</u>.

9. During muscle contraction, crossbridges are formed by the protruding heads of <u>actin filaments</u>.

10. When a muscle is in the process of contraction, there is movement of the <u>myosin filaments</u>.

11. The major energy source for the contraction of muscle is the molecule <u>NAD</u>.

12. In order for muscle contraction to occur, there must be a ready supply of gaseous <u>carbon dioxide</u> for energy synthesis to take place.

13. In white muscle, there is little or no amount of the substance <u>myoglobin</u>.

14. Infoldings of the plasma membrane known as <u>H tubules</u> are reservoirs for the calcium needed for muscle contraction.

15. The most important ion for triggering muscle contraction is <u>chloride</u>.

16. Changes in the structure of the <u>myoglobin</u> molecule are responsible for initiating the contraction of a muscle fiber.

17. The all-or-none response indicates that a muscle fiber contracts <u>completely</u> or not at all.

18. The contraction observed in a muscle fiber is referred to as a <u>twitch</u>.

19. The state of maximum contraction observed in a muscle is referred to as <u>tonus</u>.

20. A state in which a muscle is kept partially contracted over a long period of time is called <u>tetanus</u>.

21. In smooth muscle, the cells are linked together by collagen fibers and by <u>desmosomes.</u>

22. The type of muscle found in most visceral organs is <u>skeletal</u> muscle.

23. Because of its high energy requirement, cardiac muscle is liberally supplied with <u>mitochondria</u>.

24. Intercalated disks are found only in <u>smooth</u> muscle.

25. When a person breathes deeply after strenuous muscle exercise, the oxygen brought into the body helps eliminate the <u>lactic acid</u> that has built up.

## PART A—Completion

1. smooth muscle
2. cardiac muscle
3. skeletal muscle
4. antagonistic
5. myofibrils
6. sarcomere
7. myosin
8. actin
9. Z line
10. sliding filament theory
11. sarcomere
12. ATP
13. myoglobin
14. ATP
15. myoglobin
16. oxygen
17. twitch muscle
18. tropomyosin
19. acetylcholine
20. calcium
21. transverse tubules
22. troponin
23. tropomyosin
24. myosin filaments
25. actin filaments
26. terminal sacs
27. rigor mortis
28. all-or-none response
29. graded response
30. nerve impulses
31. twitch
32. motor unit
33. summation
34. tetanus
35. tonus
36. slow
37. striations
38. gap junctions
39. involuntary
40. visceral muscle
41. mitochondria
42. intercalated disks
43. voluntary control
44. ADP
45. creatine phosphate
46. oxygen
47. glycolysis
48. lactic acid
49. oxygen debt
50. liver

## PART B—Multiple Choice

| | | | | |
|---|---|---|---|---|
| 1. C | 6. C | 11. C | 16. B | 21. C |
| 2. D | 7. B | 12. A | 17. A | 22. B |
| 3. B | 8. C | 13. C | 18. D | 23. A |
| 4. A | 9. A | 14. B | 19. C | 24. D |
| 5. D | 10. D | 15. D | 20. C | 25. B |

## PART C—True/False

1. true
2. cardiac
3. true
4. true
5. sarcoplasm
6. myosin
7. Z line
8. true
9. myosin filaments
10. actin filaments
11. ATP
12. oxygen
13. true
14. T tubules
15. calcium
16. troponin
17. true
18. true
19. tetanus
20. tonus
21. gap junctions
22. smooth
23. true
24. cardiac
25. true

# CHAPTER 9

# THE MUSCLES

Skeletal muscles are relatively long and narrow, and both ends generally attach to bones. A typical skeletal muscle has a fixed end called the **origin**, and a movable end called the **insertion**. The origin is normally close to the body's midline, and the insertion is usually distant from the body's midline. Some muscles have multiple origins and insertions. For example, the biceps brachii of the upper extremity has two origins and two insertions.

In many cases, the muscle is attached to the bone by a tendon. **Tendons** are bandlike connective tissues composed of protein. They vary in length from less than an inch to more than a foot. Muscles are also attached to bones by aponeuroses. **Aponeuroses** are sheetlike, thin tendons. Aponeuroses occur frequently in muscles that overlie the trunk (Figure 9.1).

Movements of the body occur by the actions of two or more muscles acting together. For example, the biceps brachii flexes the forearm at the elbow and brings the hand closer to the shoulder. The triceps brachii then extends the forearm and brings the hand away from the shoulder.

The muscles acting most directly and most powerfully during a given movement are called the prime movers, or **agonists**. The biceps brachii is the agonist in movements at the elbow. The muscles that act in opposition to the agonists are called **antagonists**. The triceps brachii is the antagonist to the biceps trachii.

In addition to flexors and extensors, there are abductors, adductors, levators, depressors, retractors, protractors, rotators, and sphincters (Table 9.1). Most movements in the body result from complex actions of groups of flexors, extensors, rotators, and other muscles (Figure 9.2). In this chapter, the functions of the major muscles are emphasized. Selected muscles are presented in the tables and figures, and important muscles are discussed in this chapter.

## MUSCLES OF THE BODY EXTREMITIES

The muscles of the body extremities join the arms and legs to the body trunk and regulate the movements of all parts of the extremities. They are conveniently divided in several groups as follows.

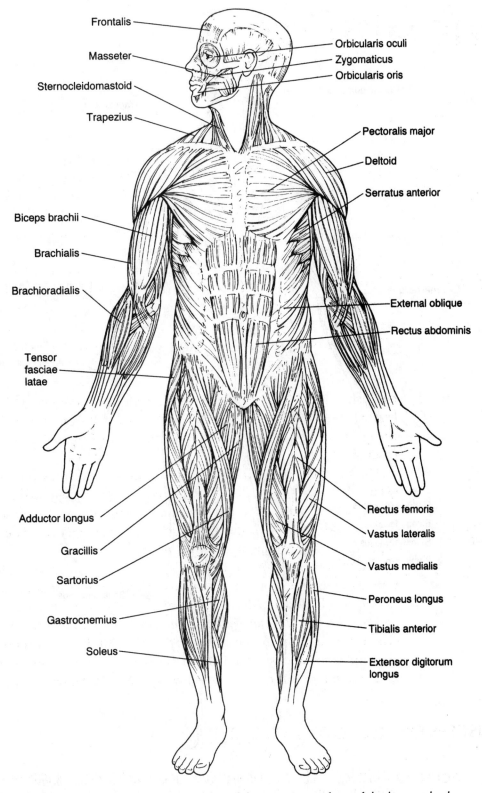

Frontalis

Masseter

Sternocleidomastoid

Trapezius

Orbicularis oculi

Zygomaticus

Orbicularis oris

Pectoralis major

Deltoid

Serratus anterior

Biceps brachii

Brachialis

Brachioradialis

External oblique

Rectus abdominis

Tensor
fasciae
latae

Adductor longus

Gracillis

Sartorius

Gastrocnemius

Soleus

Rectus femoris

Vastus lateralis

Vastus medialis

Peroneus longus

Tibialis anterior

Extensor digitorum
longus

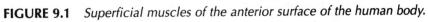

**FIGURE 9.1**   *Superficial muscles of the anterior surface of the human body.*

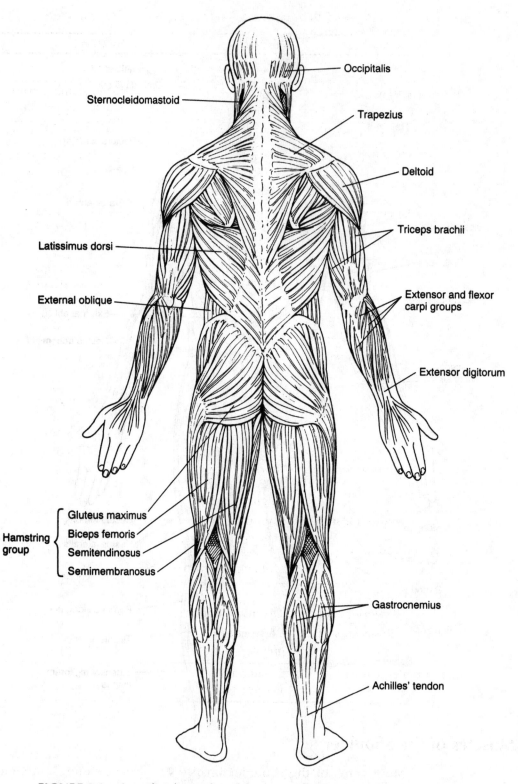

Occipitalis

Sternocleidomastoid

Trapezius

Deltoid

Triceps brachii

Latissimus dorsi

Extensor and flexor carpi groups

External oblique

Extensor digitorum

Gluteus maximus

Biceps femoris

Hamstring group

Semitendinosus

Semimembranosus

Gastrocnemius

Achilles' tendon

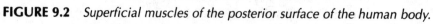

**FIGURE 9.2**  *Superficial muscles of the posterior surface of the human body.*

**TABLE 9.1**  *Selected Body Locations and Their Muscles*

| Location | Muscles | Function |
|---|---|---|
| Neck | Sternocleidomastoid | Flexes head |
| Back | Trapezius | Extends upper arm |
| | Latissimus dorsi | Extends upper arm |
| Abdominal wall | External oblique | Compresses abdomen |
| Shoulder | Deltoid | Abducts upper arm |
| | Pectoralis major | Adducts upper arm |
| | Serratus anterior | Abducts shoulder |
| Upper arm | Biceps brachii | Flexes forearm |
| | Triceps brachii | Extends forearm |
| | Brachialis | Flexes forearm |
| Forearm | Brachioradialis | Flexes forearm |
| | Pronator teres | Pronates, flexes arm |
| Buttocks | Gluteus maximus | Extends thigh |
| | Gluteus minimus | Abducts thigh |
| | Gluteus medius | Abducts thigh |
| Thigh | | |
|   Anterior surface | Quadriceps femoris group: | |
| |   Rectus femoris | Flexes thigh |
| |   Vastus lateralis | Extends leg |
| |   Vastus medialis | Extends leg |
| |   Vastus intermedius | Extends leg |
|   Medial surface | Gracilis | Adducts thigh |
| | Adductor group (brevis, longus, magnus) | |
|   Posterior surface | Hamstring group | All flex leg |
| |   Biceps femoris | |
| |   Semitendinosus | |
| |   Semimembranosus | |
| Leg | | |
|   Anterior surface | Tibialis anterior | Adducts foot |
|   Posterior surface | Gastrocnemius | Extends foot |
| | Soleus | Extends foot |
| Pelvic floor | Levator ani | Forms pelvic floor |
| | Levator coccygeus | Forms pelvic floor |

## Muscles of the Shoulder

Movements of the shoulder involve the scapula bone, sometimes called the "shoulder blade." Several muscles attach to the scapula, but only a few are considered major muscles.

The **serratus anterior** is a flat, saw-toothed muscle on the side of the chest. It originates on the outer surface of the first nine ribs, and it inserts at the costal surface of the scapula. Its function is to draw the scapula and the upper extremity outward.

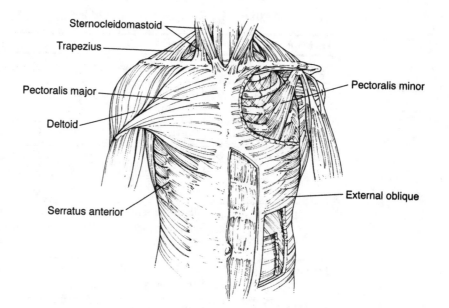

Sternocleidomastoid

Trapezius

Pectoralis major

Deltoid

Pectoralis minor

Serratus anterior

External oblique

**FIGURE 9.3** *Major muscles of the anterior surface of the chest.*

The **trapezius** is a flat, triangular muscle along the back. It has its origin on the occipital bone and spines of several vertebrae, and it inserts at both the clavicle and scapula. The trapezius muscle adducts the shoulder girdle and pulls the entire upper body extremity toward the trunk. It also rotates the scapula, so that the glenoid cavity faces upward during abduction of the arm (Figure 9.3).

The **latissimus dorsi** is a large muscle of the back, with its origin on spines of several thoracic vertebrae and its insertion by a flat tendon into the inter-tubercular groove of the humerus. The function of the latissimus dorsi is to extend the arm and to act as the antagonist to the pectoralis major.

One of the large triangular muscles on the anterior chest wall is the **pectoralis major**. This muscle originates on bones and cartilages of the thorax and clavicle and inserts into the humerus. The pectoralis major forms a significant portion of the interior wall of the axilla and flexes the arm. It works as an antagonist to the latissimus dorsi to abduct the arm.

The **deltoid** is the "shoulder pad" muscle. It originates on the acromion process of the scapula and the clavicle and it inserts into the humerus. The deltoid is a major abductor of the arm and also flexes and extends the arm. When the deltoid acts as an abductor, its antagonists are the pectoralis major and latissimus dorsi.

The **pectoralis minor** is a small, triangular muscle situated under the pectoralis major. It originates on the ribs and inserts at the coracoid process of the scapula. The pectoralis minor abducts the scapula and draws the shoulder downward; it depresses the shoulder (Figure 9.4).

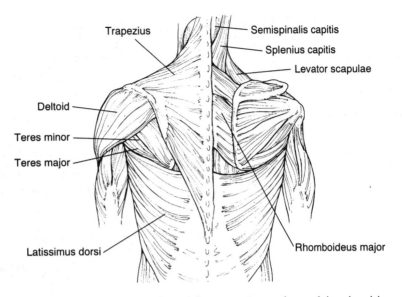

**FIGURE 9.4** *Major muscles of the posterior surface of the shoulder.*

## Muscles of the Upper Arm

Muscles of the upper arm are associated with movements of the shoulder as well as the forearm. The most superficial muscle of the upper arm is the **biceps brachii**. At its origin, this muscle has two heads, a long head and a short head, both at the scapula. The muscle inserts into the radius and into some of the connective tissue of the forearm. The biceps brachii spans two joints (the elbow and shoulder joints). At the elbow joint, it brings about flexion and supination of the forearm; at the shoulder joint, it permits flexion of the upper extremity (Table 9.2).

A large muscle on the posterior portion of the arm is the **triceps brachii**. This muscle contains three heads at its origin. Two heads are at the humerus and one is at the scapula. The muscle inserts at the olecranon process of the ulna. Its function is to extend the forearm at the elbow. The **coracobrachialis** is on the anterior surface. It extends from the scapula to the humerus and flexes and adducts the upper arm.

A fourth muscle of the upper arm is the **brachialis**. The brachialis originates on the anterior shaft of the humerus and inserts at the coronoid process of the ulna. The brachialis flexes the arm at the elbow (Figure 9.5).

## Muscles of the Forearm, Wrist, and Hand

Many muscles of the forearm are identical to those in the upper arm (for example, biceps brachii, triceps brachii, and brachialis). Another prominent muscle is the **brachioradialis**. This muscle originates at the distal lateral end of the humerus and inserts at the lateral surface of the radius. The muscle flexes the arm at the elbow.

**TABLE 9.2** *Selected Muscles of the Shoulder and Upper Arm*

| Muscle | Origin | Insertion | Innervation |
|---|---|---|---|
| Trapezius | Occipital bone | Clavicle | Spinal accessory, and certain cervical nerves |
| | Vertebrae (cervical, thoracic) | Scapula (spine and acromion) | |
| Pectoralis minor | Ribs (second to fifth) | Scapula (coracoid) | Medial and lateral anterior thoracic nerves |
| Serratus anterior | Ribs (upper eight or nine) | Scapula (anterior surface, vertebral border) | Long thoracic nerve |
| Pectoralis major | Clavicle (medial half) | Humerus (greater tubercle) | Medial and lateral anterior thoracic nerves |
| | Sternum | | |
| | Costal cartilages of true ribs | | |
| Latissimus dorsi | Vertebrae (spines of lower) | Humerus (intertubercular groove) | Thoracodorsal nerve |
| | Ilium (crest) | | |
| | Lumbodorsal fascia | | |
| Deltoid | Clavicle | Humerus (deltoid tubercle) | Axillary nerve |
| | Scapula (spine and acromion process) | | |
| Biceps brachii | Scapula (supraglenoid tuberosity) | Radius (tubercle at proximal end) | Musculocutaneous nerve |
| | Scapula (coracoid) | | |
| Brachialis | Humerus (distal half, anterior surface) | Ulna (coronoid process) | Musculocutaneous nerve |
| Triceps brachii | Scapula (infraglenoid tuberosity) | Ulna (olecranon process) | Radial nerve |
| | Humerus (posterior surface) | | |

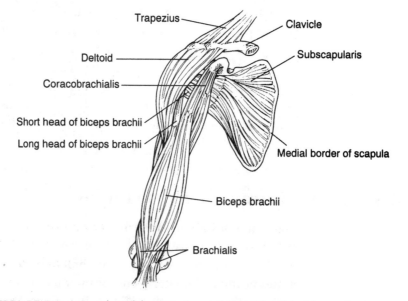

**FIGURE 9.5** *Muscles of the anterior surface of the shoulder and upper arm.*

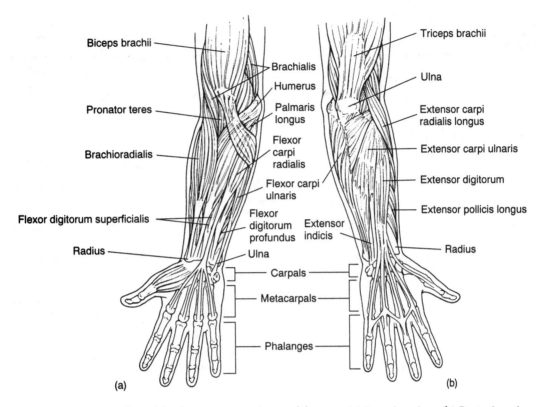

Biceps brachii

Brachialis

Humerus

Pronator teres

Palmaris longus

Flexor carpi radialis

Brachioradialis

Flexor carpi ulnaris

**Flexor digitorum superficialis**

Flexor digitorum profundus

Radius

Ulna

Triceps brachii

Ulna

Extensor carpi radialis longus

Extensor carpi ulnaris

Extensor digitorum

Extensor pollicis longus

Extensor indicis

Radius

Carpals

Metacarpals

Phalanges

(a)                                                                 (b)

**FIGURE 9.6**  *Muscles of the lower arm, wrist, and fingers. (a) Anterior view (b) Posterior view.*

The **supinator** is also a muscle of the forearm. This muscle originates at the lateral epicondyle of the humerus and inserts at the upper region of the radius. It brings about supination (rotation of the forearm so that the palm faces anteriorly). Other forearm muscles are the **pronator teres** and **pronator quadratus**. Both muscles rotate the arm medially.

Many muscles of the forearm have tendons passing to the wrist and then to the fingers. The muscles serve as powerful flexors and extensors of the fingers. Flexion of the wrist is accomplished by the **flexor carpi radialis** and **flexor carpi ulnaris**, assisted by several other muscles. Extension of the wrist is accomplished by the **extensor carpi radialis longus**, the **extensor carpi radialis brevis**, and the **extensor carpi ulnaris**, assisted by several other muscles (Figure 9.6).

These muscles also permit abduction and adduction of the wrist. The flexor muscles form a fleshy mass, part of which originates at the medial epicondyle of the humerus. The extensors originate partly from the lateral epicondyle of the humerus.

The muscles of the hand, including the fingers, can be flexed, extended, adducted, and abducted. At the base of the thumb, there are short muscles allowing the thumb to touch the tips of each of the other digits. This action, called **opposition**, is possible only in humans. The little finger also has short muscles, which deepen the palm so that the thumb and little finger can touch each other.

Movement of the digits is accomplished by several flexor and extensor muscles. A number of **dorsal interosseus muscles** permit abduction and adduction of the digits. Each finger is also supplied by long muscles from the forearm whose tendons reach the hand. The tendons are contained in tunnels called **synovial sheaths**.

## Muscles of the Thigh

Unlike the upper extremity, the lower extremity is joined to the body trunk by comparatively few muscles. An important muscle on the anterior thigh surface is the **psoas major**. The psoas major joins the lower extremity to the axial skeleton. Its origin is at the bodies of the lumbar vertebrae. It passes through the pelvis to enter the thigh and insert into the lesser trochanter of the femur. The psoas major brings about flexion of the thigh; it also flexes the trunk.

Another muscle of the anterior surface of the thigh is the **iliacus**. The iliacus originates at the iliac fossa of the ilium and inserts at the lesser trochanter of the femur together with the psoas muscle. It flexes at the hip joint, and is commonly considered together with the psoas major as one muscle called the **iliopsoas** (Figure 9.7).

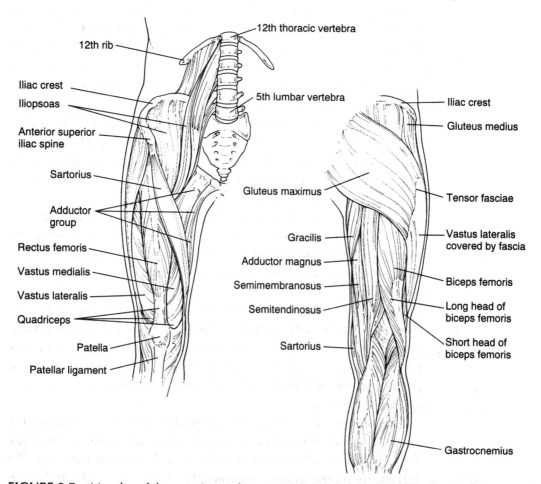

**FIGURE 9.7**   *Muscles of the anterior surface of the pelvis and thigh.*

**FIGURE 9.8**   *Muscles on the posterior surface of the right thigh.*

Also on the anterior surface of the thigh is the **sartorius**. The sartorius is a long, slender muscle that originates at the iliac spine and inserts into the medial surface of the tibia. The sartorius spans the hip and knee joints and produces movements at both. It flexes the thigh and rotates the leg medially. The muscle is so-named because it acts when one assumes the cross-legged position of a tailor (the word "sartor" means tailor).

The anterior surface of the thigh also contains the quadriceps femoris group of muscle. The quadriceps femoris group is a large, four-muscle group that constitutes the main fleshy bulk on the anterior thigh surface. The four muscles are the **rectus femoris**, the **vastus lateralis**, the **vastus medialis**, and the **vastus intermedius**. The rectus femoris has its origin at the ilium; the other three originate at the femur. Fibers from all four converge to insert at the patella and the tibial tuberosity via the patellar tendon. All are extensors of the leg.

A muscle on the posterior surface of the thigh is the **semimembranosus**. This muscle extends the thigh and flexes the leg. It originates at the ischial tuberosity and inserts at the proximal end of the tibia, thereby spanning two joints. The muscle is considered one of the three **hamstring muscles**.

Another hamstring muscle and another muscle of the posterior surface is the **semitendinosus**, a long, bandlike muscle that connects the ischium to the proximal end of the tibia (Figure 9.8). The muscle is so-named because it becomes tendenous at the middle of the thigh then continues to its insertion as a long, cordlike tendon. The semitendinosus flexes the leg and rotates it medially to extend the thigh.

The third hamstring muscle is the **biceps femoris**. This muscle has two heads, a long head and a short head. The long head is originates at the ischial tuberosity, the other head at the shaft of the femur. The biceps femoris inserts at the head of fibula and it extends the thigh and flexes and rotates the leg.

On the medial surface of the thigh, the muscles are the **adductor group** of muscles. Three muscles of this group are the **adductor magnus**, the **adductor brevis**, and the **adductor longus**. All have their origins at the body of the pubis and their insertions at the femur. They adduct the thigh.

Additional movements of the thigh are brought about by three muscles located in the buttocks. The first muscle is the **gluteus maximus**. This is the largest muscle of the buttocks. It originates at the sacrum and coccyx of the vertebral column and along the posterior ileum surface, and it inserts on the femur. Its function is to extend the leg at the hip.

The second buttocks muscle, the **gluteus medius**, originates at the ileum's lateral surface and inserts at the greater trochanter of the femur. This muscle abducts the thigh and rotates it medially. The third muscle, the **gluteus minimus**, has the same origin and insertion as the gluteus medius; it abducts the thigh.

**TABLE 9.3**   *Selected Muscles of the Pelvis and Thigh*

| Muscle | Origin | Insertion | Innervation |
|---|---|---|---|
| Iliopsoas (iliacus and psoas major) | Ilium (iliac fossa)<br>Vertebrae (bodies of twelfth thoracic to fifth lumbar) | Femur (small trochanter) | Femoral and second to fourth lumbar nerves |
| Rectus femoris | Ilium (anterior, inferior spine) | Tibia (by way of patellar tendon) | Femoral nerve |
| Gluteal group: | | | |
| Maximus | Ilium (crest and posterior surface)<br>Sacrum and coccyx (posterior surface)<br>Sacrotuberous ligament | Femur (gluteal tuberosity)<br>Iliotibial tract | Inferior gluteal nerve |
| Medius | Ilium (lateral surface) | Femur (greater trochanter) | Superior gluteal nerve |
| Minimus | Ilium (lateral surface) | Femur (greater trochanter) | Superior gluteal nerve |
| Tensor fasciae latae | Ilium (anterior part of crest) | Tibia | Superior gluteal nerve |
| Adductor group: | | | |
| Brevis | Pubic bone | Femur | Obturator nerve |
| Longus | Pubic bone | Femur | Obturator nerve |
| Magnus | Pubic bone | Femur | Obturator nerve |
| Gracilis | Pubic bone (just below symphysis) | Tibia (medial surface behind sartorius) | Obturator nerve |

# Muscles of the Lower Leg

The chief muscles of the lower leg originate in the leg bones, but also in the ankle, and many muscles have their insertion into bones of the foot. On the anterior surface of the leg lies the **tibialis anterior**. This muscle lies immediately lateral to the tibia, with its origin on the tibia and insertion in the tarsals and metatarsals. The muscle inverts the foot (turns the sole of the foot inward) and flexes the foot dorsally (turns the foot upward).

On the posterior surface of the leg lies the **gastrocnemius** (Figure 9.9). The gastrocnemius originates at the medial and lateral condyles of the femur. Like the soleus muscle, it inserts to the calcaneus (a tarsal) via the **Achilles tendon**. The gastrocnemius spans the knee and ankle joints. Its main action is to bring about plantar flexion (turning the foot downward). It also flexes the knee.

Also on the posterior surface is the **soleus** muscle. The soleus lies under the gastrocnemius, originating at the tibia and fibula and inserting into the calcaneus. The soleus permits plantar flexion of the foot (Table 9.3). The muscle pair consisting of the soleus and gastrocnemius is referred to as the **triceps surae**. The triceps surae

**TABLE 9.4**  *Selected Muscles of the Lower Leg and Foot*

| Muscle | Origin | Insertion | Innervation |
|---|---|---|---|
| Quadriceps femoris group | | | |
| Rectus femoris | Ilium | Tibia (by way of patellar tendon) | Femoral nerve |
| Vastus lateralis | Femur (linea aspera) | Tibia (by way of patellar tendon) | Femoral nerve |
| Vastus medialis | Femur | Tibia (by way of patellar tendon) | Femoral nerve |
| Vastus intermedius | Femur (anterior surface) | Tibia (by way of patellar tendon) | Femoral nerve |
| Sartorius | Os innominatum | Tibia (medial surface of shaft) | Femoral nerve |
| Hamstring group | | | |
| Biceps femoris | Ischium (tuberosity) | Fibula (head) | Hamstring nerve (branch of sciatic nerve) |
| | Femur (linea aspera) | Tibia (lateral condyle) | Hamstring nerve |
| Semitendinosus | Ischium (tuberosity) | Tibia (proximal end, medial surface) | Hamstring nerve |
| Semimembranosus | Ischium (tuberosity) | Tibia (medial condyle) | Hamstring nerve |
| Tibialis anterior | Tibia (lateral condyle of upper body) | Tarsal (first cuneiform) | Common and deep peroneal nerves |
| Gastrocnemius | Femur (condyles) | Tarsal (calcaneus by way of Achilles tendon) | Tibial nerve (branch of sciatic nerve) |
| Soleus | Tibia (underneath gastrocnemius) Fibula | Tarsal (calcaneus by way of Achilles tendon) | Tibial nerve |

shapes the posterior calf and inserts via the common Achilles tendon into the calcaneus of the heel.

One of the lateral muscles of the leg is the **peroneus longus**. This muscle originates at the upper part of the fibula and inserts to the cuneiform (a tarsal) and the first metatarsal bones. The muscle inverts the foot and supports the arch of the foot. Another lateral muscle called the **peroneus brevis** brings about eversion the foot. This muscle inserts into the fifth metatarsal.

At the ankle joint two major movements occur: flexion and extension. Dorsiflexion and plantar flexion also take place, and inversion and eversion may occur. These lateral movements take place chiefly between the tarsal bones.

The foot muscles are arranged like the hand muscles. There are many small muscles located on the plantar surface of the foot, and they insert into the bones of the toe. The muscles are grouped in rows, and each toe receives tendons traveling along the plantar aspect. On the dorsal aspect, each toe receives a tendon for extension muscles of the foot and knee.

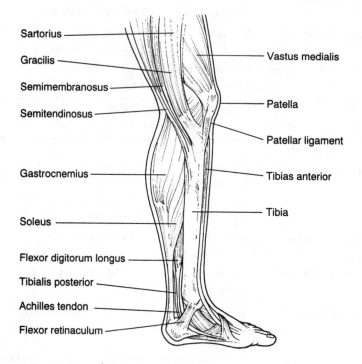

**FIGURE 9.9** *Muscles of the medial surface of the right lower leg.*

## MUSCLES OF THE HEAD AND BODY TRUNK

The muscles of the head provide facial movements and permit mastication and eye movements. The major movements of the head are flexion, extension, and rotation (as in signifying "no").

Muscles of the face differ from other muscles in the body because they move relatively light loads. For this reason, many facial muscles do not require the strong tendons found in extremity muscles. Moreover, facial muscles are different because they exist within **superficial fascia**. They have their origins in the fascia or in bones of the face, and often they insert into the skin or underlying connective tissues.

The major muscles that flex the head are the **longus coli, longus capitis**, and **rectus capitis anterior**. These muscles originate on the upper vertebrae and insert either on the occipital bone or upper cervical vertebrae. The major extensors are the **splenius capitis**, assisted by the **semispinalis capitis** and the **longissimus**. These muscles have their origins on the vertebrae and their insertions on the

occipital or temporal bone. Numerous small muscles assist these
major muscles.

## Muscles of the Face

Four powerful pairs of muscles are responsible for moving the lower
jaw (the mandible) and bringing about chewing (mastication).
These muscles have their insertions at the mandible.

The first muscle, the **temporalis**, originates at the temporal bone
and can be felt at the temple when the jaws are clenched. This muscle closes the jaw. The second muscle is the **masseter**. This muscle
originates at the zygomatic arch of the temporal and zygomatic
bones. It also closes the jaw. The third muscle, the **medial pterygoid**,
has its origin at the pterygoid plate of the sphenoid bone and adjacent parts of the maxilla and palatine bones. It inserts into the inner
surface of the ramus of the mandible. The muscle also closes the
jaw. The last muscle is the **lateral pterygoid**. This muscle moves the
jaw from side to side and protrudes the jaw and helps open the
mouth. It inserts into the neck of the mandible.

The movements of the face vary considerably, but six muscles play
a major role in facial expressions. The muscles are the **orbicularis
oculi**, **platysma**, **orbicularis oris**, **zygomaticus**, **buccinator**, and **epicranius** (Figure 9.10). Several of these muscles are paired; others are
unpaired. They are generally very thin muscles and are called cutaneous muscles because of their attachments to the skin and their
position relative to the skin. Most muscles have their origins at the
skull surface and fascia.

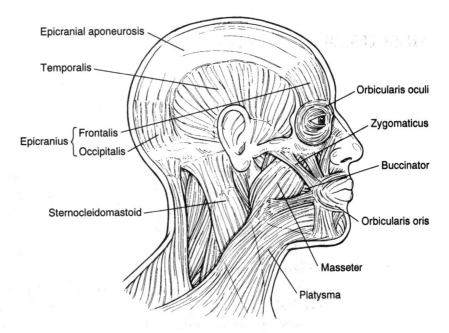

**FIGURE 9.10**   *Important muscles of the human head.*

The functions of the expression muscles vary. For example, the epicranius raises the eyebrow, while the orbicularis oculi closes the eyelid. The orbicularis oris closes the lips and encircles the mouth and is sometimes called the kissing muscle because it causes the lips to pucker. When the zygomaticus contracts, the corner of the mouth is drawn up as in smiling. When the platysma muscle contracts, the angle of the mouth is pulled downward as in pouting.

There are two sets of muscles that move the eye. The muscles that move the eyeball are called the **extrinsic muscles** of the eye. The **intrinsic muscles** change the shape of the lens (the ciliary muscle) and dilate and constrict the iris. The extrinsic muscles of the eye include the four rectus muscles that bring about elevation, rotation, adduction and abduction of the eyeball, respectively. The other important extrinsic muscles turn the eyeball downward, upward, and laterally.

## Muscles of the Neck

Many of the muscles of the neck are directly associated with movements of the head and are related to the spinal column.

A major muscle of the neck is the **sternocleidomastoid**. This muscle is located on the side of the neck. It originates on the sternum and upper surface of the clavicle, and it inserts on the mastoid process of the temporal bone. The sternocleidomastoid pulls the head to one side and toward the chest and raises the sternum. A pair of sternocleidomastoid muscles work together to flex and extend the head.

Other important muscles of the neck are the **strap muscles**. This group of long, narrow, flat muscles is located in front of the neck. The muscles cover the lateral surface of the thyroid gland and draw the larynx and hyoid bone downward.

Other muscles of the neck include the splenius capitis, semispinalis capitis, and longissimus. These muscles are also associated with the head and were noted previously.

## Muscles of the Abdomen

The anterior wall of the abdominal cavity has an extensive area containing no protective bone. Thin, expansive sheets of muscle are located in the abdominal wall to help contain the abdominal viscera and flex the trunk. A line called the **linea alba** extends in the midline of the abdominal wall from the xiphoid process to the pubic symphysis. The linea alba consists primarily of connective tissue, with no muscle, nerves, or large vessels.

A major muscle of the abdominal wall is the **rectus abdominis** (Figure 9.11). The rectus abdominis is a long, flat, straplike muscle on each side of the line alba. It connects the pubic bones to the ribs and sternum, and with other abdominal muscles, it compresses the

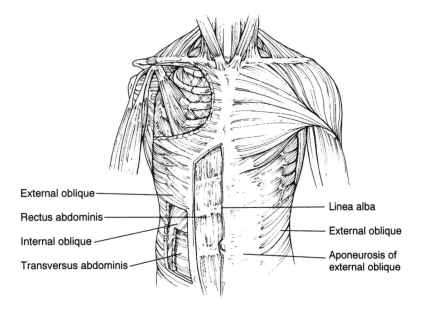

External oblique

Rectus abdominis

Internal oblique

Transversus abdominis

Linea alba

External oblique

Aponeurosis of external oblique

**FIGURE 9.11**  *Major muscles of the human abdominal wall.*

abdominal cavity and helps to flex the vertebral column. Its origin is at the crest of the pubis and its insertion is at the xyphoid process.

Another major muscle is the **external oblique**. The external oblique is a superficial and expansive sheet of muscle arising from the lower eight ribs and traveling downward and medially. It inserts by means of an aponeurosis into the linea alba and pubis. The external oblique compresses the abdominal wall.

A third abdominal muscle is the **internal oblique**. This muscle is the middle layer of the three muscle sheets lateral to the rectus abdominis. Its fibers travel almost at right angles to those of the external oblique. The internal oblique originates at the iliac crest and inserts at the cartilages of the lower ribs as well as the linea alba and pubic crest.

The deepest of the three muscles lateral to the rectus abdominus is the **transversus abdominis**. This muscle originates at the lower ribs and processes of the lumbar vertebrae and inserts at the linea alba and pubic crest. Its function is to tense the abdominal wall. In addition, it assists forced expiration by pushing the diaphragm upward, and it helps flex the body trunk.

## Muscles of Breathing

The act of breathing has two phases: inspiration, during which the chest expands and air moves into the lungs, and expiration, during which the chest relaxes and the air moves out. During inspiration, the major muscles involved include the **diaphragm** and **external intercostal muscles**. More vigorous inspiration involves muscles such as the pectoralis major and trapezius. During the expiration process, the muscles involved include the internal oblique, external oblique, transverse abdominis, and rectus abdominis. However, contraction is a largely passive process following relaxation of the muscles of inspiration.

## Muscles of the Pelvis

The muscles of the pelvis contribute to the formation of the walls and floor of the pelvis. The two major muscles that form the floor of the pelvis are the levator ani and the coccygeus.

The **levator ani** is a dual muscle having its origin in the walls of the pelvis at the pubis and ischium. Each of the two muscles inserts in the opposite muscle to form a **raphe** and create a sling (Figure 9.12). The most posterior fibers insert into the coccyx. The muscle is a prime supporter of the organs of the pelvic cavity.

The **coccygeus** is a small fan-shaped muscle that assists the levator ani. It originates at the spine of the ischium and inserts at the coccyx and helps support the pelvic organs.

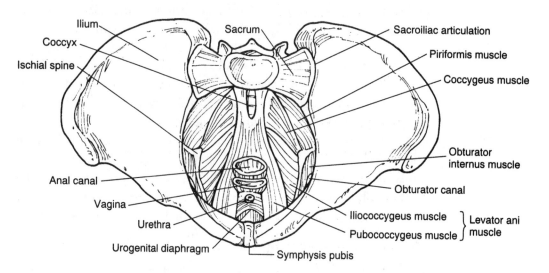

**FIGURE 9.12** *Deep muscles of the female pelvis.*

# REVIEW QUESTIONS

*PART A—Completion:* **Add the word or words that correctly complete each of the following statements.**

1. The tendons that attach many muscles to bones are composed of _____ .

2. Wide, flat, thin bands of connective tissue that also attach muscles to bones are known as _____ .

3. Muscles that are prime movers are also called _____ .

4. The muscles that act in opposition to the prime movers are known as _____ .

5. The flat, saw-toothed muscle that draws the scapula and upper extremity outward is the _____ .

6. The trapezius is a flat, triangular muscle along the back that originates on the occipital bone and spines of several vertebrae and inserts on the clavicle and the _____ .

7. The muscle that extends the arm and acts antagonistically to the pectoralis major is the _____ .

8. The shoulder pad muscle that abducts the arm and originates on the scapula and clavicle and inserts into the uterus is the _____ .

9. The most superficial muscle of the upper arm with an origin having two heads is the _____ .

10. The triceps brachii has a three-headed origin at the humerus and scapula and an insertion at the ulna in a bone marking called the _____ .

11. The upper arm muscle responsible for flexing the arm at the elbow is the _____ .

12. The supinator muscle brings about a rotation of the forearm so that the palm of the hand faces toward the _____ .

13. The muscle is rotated medially by the pronator teres and pronator _____ .

14. Flexion of the fingers is brought about by the flexor carpi radialis and the flexor carpi _____ .

15. The action allowing the thumb to touch the tips of the other digits is known as _____ .

16. Each finger is supplied by muscles from the hand and long muscles from the _____ .

17. The muscle that joins the lower extremity to the axial skeleton and brings about flexion of the thigh and flexion of the trunk is the _____ .

18. The iliacus originates at the iliac fossa of the ilium and inserts at the lesser trochanter of the _____ .

19. When one assumes the cross-legged position of a tailor, one is using the muscle known as the _____ .

20. The large, four-headed muscle on the anterior surface of the thigh is the _____ .

21. Fibers from the rectus femoris, the vastus lateralis, and two other muscles converge to insert on the _____ .

22. The hamstring muscle that extends the thigh and flexes the leg is the _____ .

23. The tendenous muscle of the hamstring group that flexes the leg and rotates it medially is the _____ .

24. The two-headed muscle originating at the ischial tuberosity and the shaft of the femur and inserting at the head of the fibula to extend the thigh and flex the leg is the hamstring muscle called the _____ .

25. The three muscles of the adductor group are the adductor magnus, the adductor brevis, and the adductor _____ .

26. The adductor group of muscles originate at the body of the pubis and insert into the femur to adduct the _____ .

27. The largest muscle of the buttocks originating partly on the ileum and inserting on the femur is the _____ .

28. The gluteus medius is a muscle that inserts at the greater trochanter of the femur and brings about the movement known as _____ .

29. The movements of dorsiflexion and inversion of the foot are both brought about by the muscle known as the _____ .

30. On the posterior surface of the leg lies a muscle that originates at the medial condyle of the femur and inserts into the calcaneus and is called the _____ .

31. Plantar flexion of the foot is provided by the muscle called the _____ .

32. The tendon joining the gastrocnemius to the calcaneus is the _____ .

33. Because the facial muscles do not move heavy loads, they do not require attachment by means of _____ .

34. Most facial muscles exist within a layer of _____ .

35. The longus coli and longus capitus both are muscles that permit the head to _____ .

36. The longissimus and semispinalis capitus both permit the movement of the head known as _____ .

37. The facial muscle that closes the jaw and is felt when the jaws are clenched is the _____ .

38. The jaw is moved from side to side and protrudes by movements of the muscle called the _____ .

39. Thin facial muscles having their attachments to the skin and their origins at the skull surface and helping to bring about facial expressions are the _____ .

40. The body muscle that raises the eyebrows is the _____ .

41. The extrinsic muscles are responsible for moving the _____ .

42. The intrinsic muscles change the shape of the eye and dilate and constrict the _____ .

43. The muscle pulling the head to one side and toward the chest while raising the sternum and flexing and extending the head is the _____ .

44. Long, narrow, flat muscles in front of the neck that cover the lateral surface of the thyroid gland and draw the larynx and hyoid downward are muscles called the _____ .

45. In the midline of the abdominal wall there is a line of connective tissue known as the _____ .

46. A straplike muscle connecting the pubic bones to the rib and sternum and compressing the contents of the abdominal cavity is the _____ .

47. The expansive sheet of muscle arising from the lower eight ribs and inserting into the pubis and linea alba to tense the abdominal wall and compress the abdomen is the _____ .

48. The major muscles involved in inspiration are the external intercostal muscles and a dome-shaped muscle known as the _____ .

49. A muscle that supports the organs of the pelvic cavity and creates a sling at the floor of the pelvic cavity is the _____ .

50. A muscle that also supports the pelvic organs originating at the spine of the ischium and inserting at the coccyx is the _____ .

**PART B—Multiple Choice:** Circle the letter of the item that correctly completes each of the following statements.

1. The movable end of the muscle is referred to as the
   (A) origin
   (B) insertion
   (C) proximum
   (D) dorsum

2. Connective tissue bands called aponeuroses connect
   (A) bones to bones
   (B) tendons to bones
   (C) bones to ligaments
   (D) muscles to bones

3. In the human body, muscles may serves as all the following except
   (A) flexors
   (B) extensors
   (C) coordinators
   (D) adductors

4. The serratus anterior is the muscle lying on the
   (A) side of the chest
   (B) above the nose
   (C) in the upper thigh
   (D) in the lower thigh

5. When the glenoid cavity faces upward during abduction of the arm, the muscle that is contracted is the
   (A) latissimus dorsi
   (B) pronator quadratus
   (C) gluteus medius
   (D) trapezius

6. The biceps brachii is so-named because it has
   (A) two heads
   (B) two motions that it performs in the body
   (C) two insertions
   (D) two muscles that it combines with

7. The brachialis, triceps brachii, and biceps brachii are all muscles of the
   (A) lower leg
   (B) upper arm
   (C) body trunk
   (D) buttocks

8. The two important muscles that rotate the arm medially are the
   (A) biceps femoris and biceps brachii
   (B) tibialis anterior and tibialis posterior
   (C) pronator teres and pronator quadratus
   (D) gluteus maximus and gluteus medius

9. The "shoulder pad" muscle into which many vaccinations are given is the
   (A) iliopsoas
   (B) deltoid
   (C) tibialis
   (D) rectus abdominis

10. The dorsal interosseus muscles permit movements of the
    (A) legs
    (B) body trunk
    (C) face
    (D) digits

11. The muscles of the thigh include the
    (A) latissimus dorsi and pectoralis major
    (B) sartorius and iliopsoas
    (C) sternocleidomastoid
    (D) diaphragm and intercostal muscles

12. The quadriceps femoris
    (A) is found in the lower arm and has three heads
    (B) lies beneath the pectoralis major and pectoralis minor
    (C) coordinates with the longus coli and longus capitus
    (D) is found near the femur and has four heads

13. The semitendinosus and semimembranosus are both considered to be
    (A) buttocks muscles
    (B) muscles that permit facial movements
    (C) hamstring muscles
    (D) muscles of the pelvis

14. The largest muscle of the buttocks is the
    (A) tibialis anterior
    (B) gluteus maximus
    (C) transversus abdominus
    (D) iliopsoas

15. The insertion point for both the soleus and the gastrocnemius is the
    (A) carpal called the navicula
    (B) last vertebrae
    (C) tarsal called the calcaneus
    (D) costal cartilage of the rib bones

16. Turning the foot downward, a movement called plantar flexion, is accomplished by the
    (A) pectoralis major and minor
    (B) peroneus longus
    (C) soleus and gastrocnemius
    (D) deltoid and trapezius

17. The muscles of the face have their origin on the bones of the face or the
    (A) superficial fascia
    (B) ligaments of the face
    (C) brain tissue
    (D) tongue

18. Among the major flexors of the head are the
    (A) temporalis and lateral pterygoid
    (B) longus capitis and longus coli
    (C) levator ani
    (D) rectus abdominis

19. When the jaws are clenched, the muscle felt at the temple is the
    (A) orbicularis oris
    (B) epicranius
    (C) extrinsic muscle
    (D) temporalis

20. The lateral pterygoid helps move the
    (A) neck up and down
    (B) body trunk when it flexes
    (C) jaw from side to side
    (D) eyebrows

21. The ciliary muscle that changes the shape of the lens, and the dilator and constrictor muscles of the iris are collectively called the
    (A) orbicularis muscles
    (B) platysma muscles
    (C) intrinsic muscles
    (D) masseter muscle

22. When the head is flexed and extended such as in saying yes and no, the muscles permitting this movement include the
    (A) masseter
    (B) soleus
    (C) sternocleidomastoid
    (D) linea alba

23. The long, flat muscle on each side of the linea alba that connects the ribs and sternum to the pubic bones is the
    (A) trapezius
    (B) quadriceps femoris
    (C) coccygeus
    (D) rectus abdominus

24. The external oblique and internal oblique lie close to the
    (A) hamstring muscles
    (B) linea alba
    (C) biceps femoris
    (D) zygomaticus

25. The external intercostal muscles are involved in the process of
    (A) digestion
    (B) excretion
    (C) movement
    (D) breathing

**PART C—True/False:** For each of the following statements, mark the letter "T" next to the statement if it is true. If the statement is false, change the <u>underlined</u> word to make the statement true.

1. The <u>origin</u> of a muscle is generally at a location distant from the body's midline.

2. Muscles are often attached to bones by protein structures referred to as <u>ligaments</u>.

3. Muscles that are prime movers of the body generally have muscles that act in opposition to them and are called <u>antagonists</u>.

4. The muscle of the forearm that brings about rotation of the forearm to face the palms anteriorally is the <u>brachioradialis</u>.

5. The pectoralis <u>major</u> is the muscle that depresses the shoulder as it abducts the scapula and draws the shoulder downward.

6. One of the muscles that flexes the arm at the elbow is the <u>deltoid</u>.

7. <u>Opposition</u> is the action in which short muscles allow the thumb to touch the tips of the other digits.

8. The muscle of the anterior surface of the thigh that acts when one assumes the cross-legged position is the <u>iliopsoas</u>.

9. The biceps femoris and semitendinosus are two of the three <u>hamstring</u> muscles.

10. Fibers from all four quadriceps femoris muscles converge to insert on the tibial tuberosity and the <u>navicula</u>.

11. The adductor group of muscles, including the adductor magnus and adductor longus, all serve to adduct the <u>shoulder</u>.

12. The major function of the gluteus maximus is to extend the leg at the <u>hip</u>.

13. The gastrocnemius is a muscle that brings about plantar flexion from its location on the <u>anterior</u> surface of the leg.

14. The insertion of the gastrocnemius to one of the tarsals takes place by means of the <u>broad ligament</u>.

15. The peroneus longus and peroneus brevis are both involved in movements of the <u>hand</u>.

16. Muscles of the face differ from other muscles of the body because they do not require attachment by <u>tendons</u>.

17. One of the major muscles involved in closing the jaw is the <u>inferior oblique</u>.

18. Muscles such as the orbicularis oris and zygomaticus are known as <u>systemic</u> muscles because of their attachments to the skin, and their closeness to the skin surface.

19. When one raises the corner of the mouth, the muscle that is functioning is the <u>trapezius</u>.

20. The <u>orbicularis oris</u> encircles the mouth and is sometimes called the kissing muscle because it causes the lips to pucker.

21. The strap muscles cover the <u>posterior</u> surface of the neck and draw the larynx downward.

22. The linea alba extends from the xiphoid process to the <u>pubic symphysis</u> and consists of connective tissue.

23. The pubic bones are connected to the ribs and sternum by the <u>diaphragm</u>.

24. The internal and external oblique are muscles that cover the <u>thoracic</u> organs.

25. Both the coccygeus and levator ani contribute to the formation of the floor and walls of the <u>pelvis</u>.

**Answers**

## PART A—Completion

1. protein
2. aponeuroses
3. agonists
4. antagonists
5. serratus anterior
6. scapula
7. latissimus dorsi
8. deltoid
9. biceps brachii
10. olecranon process
11. brachialis
12. front
13. quadratus
14. ulnaris
15. opposition
16. forearm
17. psoas major
18. femur
19. sartorius
20. quadriceps femoris
21. patella
22. semimembranosus
23. semitendinosus
24. biceps femoris
25. longus
26. thigh
27. gluteus maximus
28. abduction
29. tibialis anterior
30. gastrocnemius
31. soleus
32. Achilles tendon
33. tendons
34. superficial fascia
35. flex
36. extension
37. temporalis
38. lateral pterygoid
39. cutaneous muscles
40. epicranius
41. eyeball
42. iris
43. sternocleidomastoid
44. strap muscles
45. linea alba
46. rectus abdominis
47. external oblique
48. diaphragm
49. levator ani
50. coccygeus

## PART B—Multiple Choice

| | | | | |
|---|---|---|---|---|
| 1. B | 6. A | 11. B | 16. C | 21. C |
| 2. D | 7. B | 12. D | 17. A | 22. C |
| 3. C | 8. C | 13. C | 18. B | 23. D |
| 4. A | 9. B | 14. B | 19. D | 24. B |
| 5. D | 10. D | 15. C | 20. C | 25. D |

## PART C—True/False

1. insertion
2. tendons
3. true
4. supinator
5. minor
6. brachialis
7. true
8. sartorius
9. true
10. patella
11. thigh
12. true
13. posterior
14. Achilles tendon
15. foot
16. true
17. masseter
18. cutaneous
19. zygomaticus
20. true
21. anterior
22. true
23. rectus abdominis
24. abdominal
25. true

# CHAPTER 10

# THE NERVOUS SYSTEM: BASIC STRUCTURE AND FUNCTION

The nervous system is responsible for directing the complex processes taking place in the body's internal environment. The system also links the body to the external environment. In that sense, the nervous system permits a human being to see, hear, taste, or feel and to respond to stimuli. Without the nervous system extreme chaos would develop in the organ systems, for they would act independently of one another without reference to the body's needs. For example, muscles would not contract in any organized fashion; the body's temperature would not be regulated; the blood would not be distributed according to tissue needs; and thinking and emotions would not occur.

## BASIC ORGANIZATION

The nervous system is divided into two principal divisions: the **central nervous system (CNS)** and the **peripheral nervous system (PNS)**. The CNS consists of the brain and spinal cord and serves as a control center for the entire body (Figure 10.1). Sections of the PNS integrate incoming information and determine appropriate responses.

The PNS is composed of receptors in the sense organs and nerves that communicate between the CNS and the sense organs. The PNS contains 12 pairs of cranial nerves and 31 pairs of spinal nerves. It informs the CNS of changing conditions inside the body and at the body's surface and then transmits CNS responses to the appropriate muscles and glands that bring about adjustments.

The PNS may be further divided into the **somatic division** and the **autonomic division**. The somatic division of the PNS is concerned with changes in the external environment, while the autonomic division is concerned with changes in the internal body environment.

Both the CNS and PNS are composed of two types of nerves known as sensory and motor nerves. **Sensory**, or **afferent, nerves** transmit messages from the body's receptors to the CNS. **Motor**, or

**efferent, nerves,** transmit messages from the CNS to the muscles or other structures that respond to stimuli.

The autonomic division of the PNS has two different types of motor nerves. One type, called the **sympathetic nerves,** carry impulses to the body organs and mobilize the response to stress, such as the "flight or fight" sequence (to be discussed). The second type, called the **parasympathetic nerves,** return the organs to a quiet, calm state.

## Glial Cells

Two unique types of cells are found in the nervous system: neurons, or nerve cells, which receive and transmit biochemical information; and **glial cells,** which provide structural support for the neurons.

The glial cells are referred to collectively as the **neuroglia.** There are about 10 times as many glial cells as neurons in the nervous system. Cells of neuroglia are types of connective tissue cells.

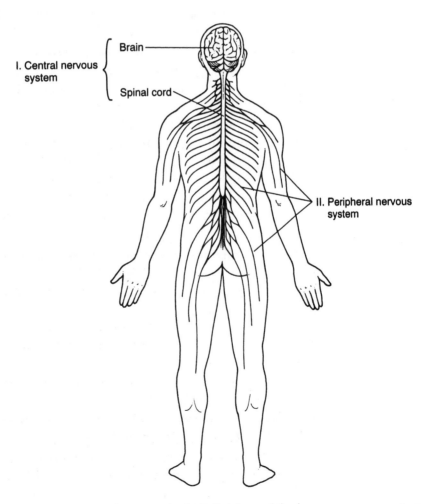

**FIGURE 10.1**    *The two principal divisions of the human nervous system.*

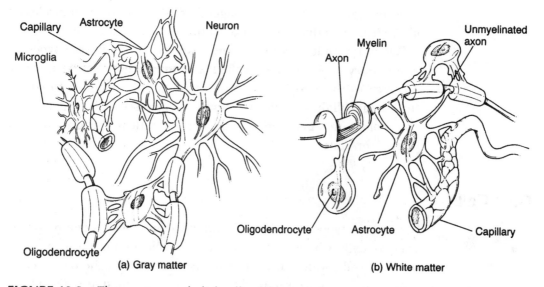

**FIGURE 10.2** *The neurons and glial cells of the nervous system. (a) The relationships of the two cell types in the gray matter of the brain and spinal cord (b) The relationships in the white matter.*

Several types of glial cells exist. One type are the **oligodendrocytes**. These cells wrap their plasma membranes about neurons and form sheaths. The sheaths are composed of fatty material called **myelin**. Another type of glial cells are the **astrocytes**, which have cytoplasm that extends itself into numerous elongated processes and gives the cells a star shape (Figure 10.2). Astrocytes help form the blood-brain barrier, which prevents or slows the flow of unwanted substances into the brain tissue. Astrocytes also help seal off damaged nerve tissue.

**Microglia** are small neuroglia cells scattered throughout the tissue of the brain and spinal cord. The cells act during responses to inflammation or injury, when they become mobile and actively phagocytize invading organisms.

**Schwann cells**, the final type, wrap themselves around neurons located outside the central nervous system. They are found in myelin sheaths covering the neurons in the PNS. Schwann cells form cellular sheaths around the myelin sheaths.

## Neurons

Neurons are cells specialized to receive and transmit information in the nervous system The **neuron** is the structural and functional unit of the nervous system. It is distinguished by the cytoplasmic extensions usually associated with the cell.

A neuron may be classified by structure or function. Structurally, neurons are described as mutlipolar neurons, bipolar neurons, and unipolar neurons. **Multipolar neurons** have many short extensions called **dendrites** and a single long extension called the **axon**. Many of the CNS neurons are of this type. **Bipolar neurons** have only one dendrite and one axon. They are located in the retina of the eye, the inner ear, and the olfactory nerves. **Unipolar neurons** have only a single extension that functions as both axon and dendrite. Most sensory neurons are unipolar (Figure 10.3).

Neurons may be classified functionally as sensory (afferent) neurons, motor (efferent) neurons, and interneurons (association neurons). **Sensory neurons** transmit information from receptors to the CNS, and **motor neurons** relay messages from the CNS to the muscles and glands. **Interneurons (association neurons)** link sensory and motor neurons to one another. Interneurons lie within the CNS, where they receive information from the sensory neurons and send out messages via the motor neurons.

The **cell body** of the neuron contains a small percentage of the cell's total volume. It houses the nucleus and many other cellular organelles including the mitochondria, Golgi bodies, and lysosomes. A characteristic of the cell body is the **Nissl body**, an accumulation of an organelle known as the rough endoplasmic reticulum (Chapter 3). Proteins are synthesized at this cellular location. The highly branched extensions of the cell body, the dendrites, are specialized to receive nerve impulses and conduct them toward the cell body. Dendritic surfaces are dotted with thousands of spines where the dendrites form junctions with other neurons.

Impulses are transmitted away from the cell body by the axon. The axon arises from a thickened part of the cell body called the **axon hillock**. The cytoplasm within the axon is called **axoplasm**, and the membrane is the **axolemma**.

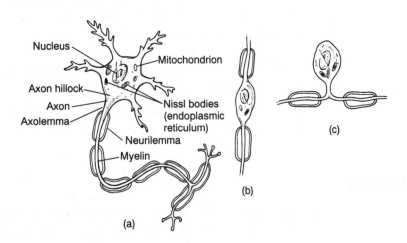

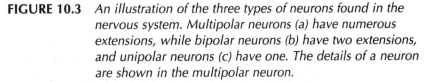

**FIGURE 10.3**    *An illustration of the three types of neurons found in the nervous system. Multipolar neurons (a) have numerous extensions, while bipolar neurons (b) have two extensions, and unipolar neurons (c) have one. The details of a neuron are shown in the multipolar neuron.*

The axon is microscopic in diameter, but it may extend several feet in length. For example, axons extending from the lower portion of the spinal cord down to the foot may be up to three feet long. Bundles of axons often travel together as a **nerve fiber**, commonly referred to as a **nerve**.

At the distal end of the axon are found thousands of microscopic branches called **axon terminals**. These branches are studded with enlargements referred to as **synaptic knobs** (sometimes called **terminal buttons**). At the synaptic knobs, nerve cells release chemical substances called **neurotransmitters**. Neurotransmitters transmit nerve impulses from a neuron to a muscle or gland or to another neuron.

## The Myelin Sheath and Neurilemma

The axons of many neurons of the PNS are covered by two sheaths: the myelin sheath and the neurilemma. The **myelin sheath** provides insulation to the axon. It is composed primarily of **myelin**, a white lipid-rich substance that is the principal component of the plasma membrane of the Schwann cell (Figure 10.4). Myelin insulates the electricity that speeds nerve impulses down the axon. Myelinated fibers conduct nerve impulses rapidly, while unmyelinated fibers conduct impulses slowly. Both myelinated and unmyelinated axons are found in the central nervous system.

The Schwann cell wraps its plasma membrane around the axon to produce the myelin sheath. Between successive Schwann cells there are gaps called the **nodes of Ranvier**. At the nodes of Ranvier, the axon is not insulated with myelin.

Myelin is responsible for the white color of the **white matter** in the brain and spinal cord. It also forms the white substance of myelinated peripheral nerves. Deterioration of patches of myelin can result in a condition called **multiple sclerosis**.

The outer sheath surrounding the axon of PNS cells is called the neurilemma. The neurilemma functions in the regeneration of injured neurons. It is formed from the bulk of the Schwann cells remaining alongside the axon outside the myelin sheaths.

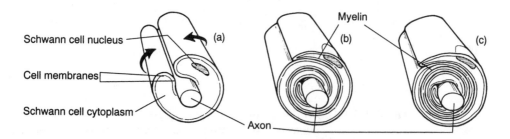

**FIGURE 10.4** *How the myelin sheath forms as the Schwann cell wraps itself around the axon in successive diagrams (a), (b), and (c). Myelin is the white, lipid-rich substance in the plasma membrane of the Schwann cell.*

## Nerves and Ganglia

A **nerve** consists of several bundles of axons, and each bundle is known as a **fascicle**. Each fascicle in a nerve is surrounded by a sheath called a **perineurium**. Fibrous connective tissue called **epineurium** surrounds the nerve and binds the fascicles to one another.

The cell bodies of neurons are generally grouped together in a mass called a **ganglion** (plural ganglia). Many ganglia exist outside the spinal cord. The axons of the cell bodies extend from these ganglia to other parts of the body.

## NERVE PHYSIOLOGY

The nervous system coordinates several activities that bring about a response to a stimulus. The first activity is **reception**, a process in which information is gathered from the external environment. The next activity is **transmission**, in which information is delivered by sensory neurons to the central nervous system. Then comes another activity called **integration**, in which an appropriate response is determined. The final activity is **response**. In response, a nerve impulse is dispatched via motor neurons to skeletal muscles or glands that will regenerate a response to the stimulus. Muscles and glands are the body's primary **effectors**.

## Nerve Activity

During nerve activity, nerve impulses travel over a sequence of neurons. The sensory neurons, interneurons, and motor neurons are generally involved. These neurons are organized into circuits called **neural circuits**. In a neural circuit, neurons are arranged so that the axon of one neuron comes close to but does not join directly with the dendrite of the next neuron in the circuit. The junction between two close neurons is called the **synapse**.

The **reflex arc** is the simplest unit of nerve activity (Figure 10.5). It is typified by the **knee-jerk reflex**, in which the patellar ligament is struck and the lower leg raises up; and by the **withdrawal reflex**, in which the finger is touched to something painful and immediately withdrawn.

A reflex arc begins when a stimulation is detected in the receptor portion at the end of a sensory neuron. A nerve impulse is generated, and the impulse travels over the sensory neuron to interneurons in the central nervous system serving as a processing center. The interneurons communicate with motor neurons, and an impulse is generated for transmission to an effector muscle or gland that will make an appropriate response. In the withdrawal reflex, for example, the finger is pulled away from the pain as the muscles contract.

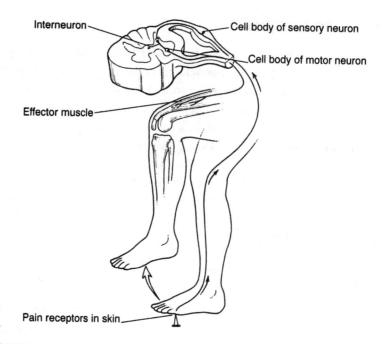

Interneuron — Cell body of sensory neuron

Cell body of motor neuron

Effector muscle

Pain receptors in skin

**FIGURE 10.5**  *A typical reflex arc as shown by the withdrawal reflex. Impulses arise at the pain receptor and travel to the spinal cord by the sensory neuron. The interneuron interprets the impulse and dispatches a response via the motor neuron to the effector muscles that withdraw the foot from the pain. Note the position of the cell body of the sensory neuron outside the spinal cord in a ganglion and the cell body of the motor neuron within the cord.*

**TABLE 10.1**  *The Components of the Reflex Arc*

| Component | Description | Function |
|---|---|---|
| Receptor | The receptor end of a dendrite or a specialized receptor cell in a sensory organ | Sensitive to an internal or external change |
| Sensory neuron | Dendrite, cell body, and axon of a sensory (afferent) neuron | Transmits nerve impulse from the receptor to the brain or spinal cord |
| Interneuron | Dendrite, cell body, and axon of a neuron within the brain or spinal cord | Serves as processing center; conducts nerve impulse from the sensory neuron to a motor neuron |
| Motor neuron | Dendrite, cell body, and axon of a motor (efferent) neuron | Transmits nerve impulse from the brain or spinal cord to an effector |
| Effector | A muscle or gland outside the nervous system | Responds to simulation by the motor neuron and produces the reflex behavioral action. |

The reflex arc is automatic and unconscious; it does not involve the brain or any mental activity. It helps maintain homeostasis in the body during such activities as sneezing, coughing, and swallowing, and it represents the most simple act that the nervous system can perform (Table 10.1).

# The Nerve Impulse

In many cases, a unit of nerve activity is initiated by stimulating a receptor at the body's surface (see Chapter 12). Among the familiar receptors are the sense organs such as the eyes, ears, nose, and taste buds. Other receptors in the skin respond to pressure, light, touch, warmth, and cold. Once the receptor has been stimulated, the neural message is transmitted to the CNS over a sensory neuron. The nature of the nerve impulse is an electrochemical event arising from changes in ion distribution in the nerve cell.

As the name implies, a **resting neuron** is not transmitting an impulse. In a resting neuron, the outer surface of the plasma membrane carries a positive charge as compared to the cytoplasm inside the membrane of the nerve cell. The resting neuron is said to be **polarized**, that is, the inside and outside regions of the membrane have different electrical charges. Electrical charges separated in this way have the potential to do work should the difference in charges disappear. The difference in potential is called the **resting potential** (Figure 10.6).

The resting potential is an imbalance in the electrical charges existing on either side of the plasma membrane. In a resting nerve cell, the resting potential is about 70 millivolts (mV). By convention this resting potential is expressed as –70mV ("negative"), because the inner surface of the plasma membrane is negatively charged relative to the positively charged region outside the membrane.

The resting potential results from an excess of positively charged ions outside the plasma membrane as compared to inside the plasma membrane. There is also a slight excess of negative ions inside the cytoplasm. Outside the plasma membrane the concentration of **sodium ions** is over 10 times greater than inside the plasma membrane. This gives the outside of the cell an overall positive charge as compared to the inside.

The ionic imbalance in a nerve cell is brought about by several factors. For example, the plasma membrane has a very efficient **sodium-potassium pump.** This pump actively transports sodium ions out of the cell and brings a small amount of potassium ions into the cell. The pump works against the concentration gradient and, therefore, much ATP is required to maintain the pump. For every three sodium ions pumped out of the cell, two potassium ions are brought into the cell. Because more positive ions are pumped out than are brought in, a positive charge develops outside the cell.

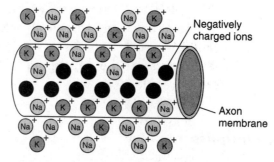

**FIGURE 10.6** *The ionic imbalance between the inside and outside environments of the resting nerve cell. The large number of negatively charged ions within the cell is attributed to protein molecules. The number of sodium ions outside the cells is greater than the number of potassium ions, thereby contributing to the overall positive charge.*

The ion imbalance is also encouraged by ions diffusing from areas of high to areas of low concentration through channels in the cell membrane. Some of these channels consist of charged protein molecules that undergo a structural change when the membrane is stimulated. Channels such as these are called **ion gates** because they act as active gates and are specific to different types of ions. Sodium gates and potassium gates are examples of ion gates.

Contributing to the ionic imbalance are large numbers of negatively charged proteins within the cytoplasm of the neuron. Organic phosphate ions are also negatively charged, and their accumulation within the cytoplasm adds further to the negative charge. Large negatively charged ions also exist within the cell and add to the ionic imbalance.

A nerve impulse is called an **action potential**. When an action potential is generated, an electrical, chemical, or mechanical stimulus alters the resting potential by increasing the permeability of the plasma membrane to sodium. The sodium gates open, and the membrane of the neuron depolarizes. During depolarization, the resting potential drops to about -55mV, the **threshold level** of a nerve impulse. At that point the sodium ion channels open, and sodium ions flow rapidly into the cell. The flow continues for a thousandth of a second, then the channels close. The resting potential continues to drop to zero at this point and overshoots to about +35 mV ("positive"), so that a momentary reversal in polarity takes place.

The action potential is sufficiently strong to depolarize the adjacent area of the membrane, and that area depolarizes the next area, and so on. A **wave of depolarization** thus spreads like a chain reaction from area to adjacent area. The wave "travels" down the neuron at a constant velocity; the wave is the action potential. It is the nerve impulse.

Once the action potential has shot down the axon, the membrane begins to **repolarize** (Figure 10.7). The sodium gates close and the potassium gates open, causing potassium to move out of the membrane. The leakage of potassium returns the exterior region to a positive state and repolarizes the membrane. The entire depolarization and repolarization can occur in less than one millisecond.

The return to the normal resting condition requires that sodium be pumped out of the neuron, and this pumping occurs over the next few seconds. In its depolarized state, the neuron is **refractory**; that is, it cannot transmit another action potential unless the stimulation is intense. However, once sodium ions have been pumped out and the ionic imbalance is once again established, the nerve cell is free to undergo a new action potential. Thus, the nerve cell conforms to the **all-or-none law**: a stimulus strong enough to depolarize the neuron to its critical threshold results in a nerve impulse; a stronger stimulus results in the same impulse; the neuron either propagates the nerve impulse or it does not; there is no variation in the strength of an impulse.

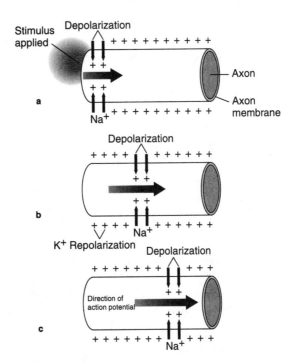

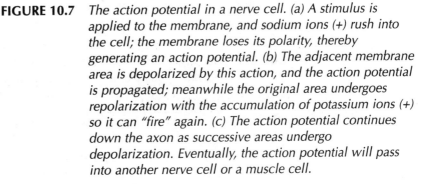

**FIGURE 10.7**    *The action potential in a nerve cell. (a) A stimulus is applied to the membrane, and sodium ions (+) rush into the cell; the membrane loses its polarity, thereby generating an action potential. (b) The adjacent membrane area is depolarized by this action, and the action potential is propagated; meanwhile the original area undergoes repolarization with the accumulation of potassium ions (+) so it can "fire" again. (c) The action potential continues down the axon as successive areas undergo depolarization. Eventually, the action potential will pass into another nerve cell or a muscle cell.*

# The Synapse

The **synapse** is the junction between two neurons or between a neuron and its effector muscle or gland. When the synapse exists between a neuron and a muscle cell, it is called a **neuromuscular junction,** or **motor end plate.** The space itself is the **synaptic cleft.** When an impulse reaches the end of an axon it is unable to jump the synaptic cleft. A chemical mechanism is needed to bridge the gap and conduct the message across to the next neuron or muscle or gland.

On arriving at the synaptic knobs at the axon's end, the impulse stimulates the release of chemical substances called **neurotransmitters.** Neurotransmitters swiftly diffuse across the synaptic cleft and change the permeability of the next neuron. If sufficient neurotransmitters are available, the dendrite of the next neuron is depolarized, and a nerve impulse is propagated.

Neurotransmitters are continually synthesized in the synaptic knobs of axons. They are stored in membrane-bound sacs called **synaptic vesicles** within the synaptic knobs. When a nerve impulse reaches the synaptic knob, calcium ions from the surrounding tissue pass into the axon terminal and encourage the synaptic vesicles to release their contents into the synaptic cleft. This process occurs by exocytosis from the presynaptic membrane (Figure 10.8).

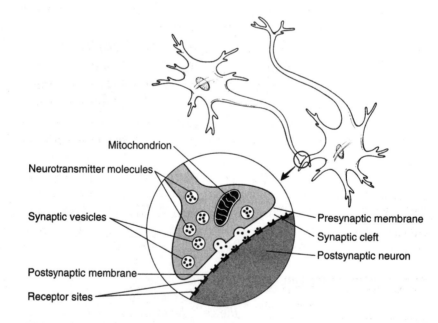

**FIGURE 10.8** *Activity at the synapse. When an action potential reaches the end of the axon, neurotransmitter molecules (such as acetylcholine) are released from synaptic vesicles by exocytosis at the presynaptic membranes. The molecules move across the synaptic cleft (space) and react with receptor sites on the surface of the postsynaptic membrane. This reaction generates an action potential in the postsynaptic neuron.*

After diffusing across the synaptic cleft, neurotransmitters combine with receptor sites on membranes of the dendrites of the next neuron (the postsynaptic membrane). These receptors are protein molecules that form ion channels. The channels then open and permit ions to pass through the membrane into the dendrite. If ion passage is sufficiently intense, depolarization takes place, and a new nerve impulse is generated.

More than 50 different kinds of neurotransmitters are known to exist. One neurotransmitter extensively studied is **acetylcholine**. Acetylcholine is released from neurons that innervate skeletal muscles, and it triggers muscle contraction. Acetylcholine is released by certain neurons in the autonomic division in the PNS and by certain neurons in the brain. After acetylcholine has combined with its receptors, the excess is removed by an enzyme called **cholinesterase**.

Another well known neurotransmitter is **norepinephrine**. Norepinephrine is released by sympathetic neurons (Chapter 11) and many neurons in the brain and spinal cord. Other neurotransmitters are **epinephrine** and **dopamine**. The three compounds belong to a class of organic substances called **catecholamines**.

Still other neurotransmitters include serotonin, glutamate, glycine, and endorphins (Table 10.2). Different neurotransmitters play different roles in the body (some are inhibitors), and many may be identical to hormones released in the endocrine glands. Different neurotransmitters are produced in different tissues of the body; for instance, glutamate is produced in the cerebral cortex, while glycine is released in the spinal cord. Certain neurotransmitters excite the second neuron and lead to depolarization and nerve impulses called **excitatory postsynaptic potentials** (**EPSPs**). Other neurotransmitters inhibit the development of nerve impulses in the second neuron by decreasing the membrane permeability to sodium ions. This inhibition leads to **inhibitory postsynaptic potentials** (**IPSPs**).

**TABLE 10.2**  *Properties of Several Neurotransmitters*

| Neurotransmitter | Location | Actions |
| --- | --- | --- |
| Acetylcholine | Neuromuscular junctions, autonomic nervous system, and brain | Excites muscles, decreases heart rate, and relays various signals in the autonomic nervous system and the brain |
| Norepinephrine | Sympathetic nervous system and brain | Regulates activity of visceral organs and some brain functions |
| Dopamine | Brain | Involved in control of certain motor functions |
| Serotonin | Brain and spinal cord | May be involved in mental functions, circadian rhythms, and sleep and wakefulness |
| Gamma-aminobutyric acid | Brain and spinal cord | Inhibits various neurons |
| Glycine | Spinal cord | Inhibits various neurons |

# REVIEW QUESTIONS

**PART A—Completion:** Add the word or words that correctly complete each of the following statements.

1. The nervous system has two principal divisions called the central nervous system and the _____ .

2. The central nervous system consists of the brain and the _____ .

3. The peripheral nervous system is composed of sensory receptors that are located in the _____ .

4. There are 31 pairs of spinal nerves and 12 pairs of _____ .

5. Nerve impulses from the central nervous system are transmitted to the glands and the _____ .

6. The two divisions of the peripheral nervous system are the somatic division and the _____ .

7. Sensory nerves are also known as _____ .

8. Motor nerves are also known as _____ .

9. The common name given to the nerve cell is the _____ .

10. Supporting cells of the nervous system are called _____ .

11. Those glial cells that wrap their plasma membranes about neurons and form sheaths are called _____ .

12. Those glial cells with cytoplasm extended into elongated processes are called _____ .

13. Glial cells that phagocytize invading microorganisms are called _____ .

14. The glial cells that wrap themselves around neurons outside the central nervous system are the _____ .

15. In the peripheral nervous system, Schwann cells form cellular sheaths around the _____ .

16. The neuron is the structural and functional unit of the _____ .

17. Neurons with many dendrites and a single long axon are known as _____ .

18. Neurons with only one dendrite and one axon are known as _____ .

19. Most sensory neurons are neurons described as _____ .

20. Neurons that transmit information from receptors to the central nervous system are called afferent neurons, or _____ .

21. Motor neurons relay impulses from the central nervous system to the glands or the _____ .

22. Motor neurons are also known as _____ .

23. The neurons that link sensory and motor neurons to one another are association neurons, also called _____ .

24. The nucleus and organelles are contained in that part of the neuron called the _____ .

25. Proteins are synthesized at an organelle of the cell body known as the _____ .

26. Impulses are conducted away from the cell body by a long extension called the _____ .

27. A nerve fiber is composed of bundles of _____ .

28. At the ends of axons are located thousands of microscopic branches called _____ .

29. Chemical substances released at the synaptic knobs are referred to as _____ .

30. The covering that provides insulation to the axons is the _____ .

31. Between successive Schwann cells are gaps called the _____ .

32. Deterioration of patches of myelin can result in a condition called _____ .

33. The outer sheath that surrounds the axon is called the _____ .

34. A nerve consists of several bundles of axons in which each bundle is known as a _____ .

35. The cell bodies of neurons are often grouped together in a mass referred to as a _____ .

36. The first activity in the body's response to a stimulus is called _____ .

37. A neuron that is not transmitting a nerve impulse is a _____ .

38. Because the regions inside and outside the membrane of a resting neuron have opposite electrical charges, the resting neuron is said to be _____ .

39. The difference in electrical potential in a resting neuron is called the _____ .

40. Outside the plasma membrane of a resting neuron there is a high concentration of _____ .

41. Outside the plasma membrane of a resting neuron, the electrical charge is _____ .

42. To maintain the sodium-potassium pump, energy must be supplied from the molecule _____ .

43. Another name for the nerve impulse is the _____ .

44. Once the nerve impulse has moved down the axon, the neuron membrane must _____ .

45. In its depolarized state, the neuron is said to be _____ .

46. The same nerve impulse will develop in a neuron regardless of the strength of the stimulus, and this is the _____ .

47. The junction where two neurons come together is the _____ .

48. The space within the synapse that must be filled by neurotransmitters is the _____ .

49. A neurotransmitter that has been extensively studied is _____ .

50. A well-known neurotransmitter released by neurons of the sympathetic system and by brain neurons is _____ .

**PART B—Multiple Choice: Circle the letter of the item that correctly completes each of the following statements.**

1. The peripheral nervous system is composed of
   (A) brain and cranial nerves
   (B) sensory receptors and nerves
   (C) brain and spinal cord
   (D) spinal cord and sensory receptors

2. The brain and spinal cord are components of the
   (A) peripheral nervous system
   (B) autonomic nervous system
   (C) sensory nervous system
   (D) central nervous system

3. The sympathetic and parasympathetic nerves are nerves of the
   (A) central nervous system
   (B) sensory nervous system
   (C) autonomic nervous system
   (D) cranial nervous system

4. All the following are types of neuroglia except
   (A) astrocytes
   (B) microglia
   (C) oligodendrocytes
   (D) lymphocytes

5. Schwann cells are located in the neurons
   (A) within the Nissl body
   (B) in the myelin sheaths
   (C) in astrocytes only
   (D) within the nucleus of the cell body

6. Within the nervous system there are
   (A) more glial cells than neurons
   (B) more neurons than glial cells
   (C) the same number of neurons and glial cells
   (D) no glial cells

7. Bipolar neurons have
   (A) a single Nissl body and a single Golgi body
   (B) two ribosomes per cell body
   (C) one axon and one dendrite
   (D) two origins in the sense receptors

8. All the following are true of interneurons (association neurons) except
   (A) they have no axons
   (B) they are found in the central nervous system
   (C) they link sensory and motor neurons
   (D) they receive information from sensory neurons

9. The function of dendrites is to
   (A) interpret nerve impulses
   (B) synthesize proteins
   (C) house the cell nucleus
   (D) conduct nerve impulses to the cell body

10. Bundles of axons generally travel together as
    (A) nerve fibers
    (B) dendrites
    (C) neurilemmas
    (D) microglia

11. Neurotransmitters are released by neurons
    (A) in the cell body
    (B) at the terminal knobs of dendrites
    (C) at the synaptic knobs
    (D) within the Golgi bodies

12. The axon is not insulated with myelin
    (A) at the cell body
    (B) at the nodes of Ranvier
    (C) in the brain
    (D) in the autonomic nervous system

13. Multiple sclerosis is accompanied by
    (A) deterioration of patches of myelin
    (B) absence of axon terminals
    (C) inability to release neurotransmitters
    (D) absence of Schwann cells

14. The perineurium and epineurium are associated with the
    (A) neurilemma
    (B) dendrites
    (C) nerves
    (D) neuroglia

15. The primary effectors of nerve activity in the body are the
    (A) neurons
    (B) bones and glands
    (C) axons and dendrites
    (D) glands and muscles

16. The synapse is an area that occurs
    (A) between the cell body and axons
    (B) only at the sensory receptors
    (C) between dendrites and cell bodies
    (D) between two neurons

17. In a resting neuron, the inner surface of the plasma membrane
    (A) carries a positive charge
    (B) is uncharged
    (C) carries a negative charge
    (D) carries both a positive and negative charge

18. A resting neuron
    (A) is polarized
    (B) has no myelin sheath
    (C) has axons but no dendrites
    (D) has no cytoplasm

19. A nerve impulse is the same thing as the
    (A) Nissl body
    (B) action potential
    (C) resting potential
    (D) synaptic potential

20. The ions that maintain the ionic imbalance in a resting neuron are
    (A) sulfur and boron
    (B) oxygen and carbon
    (C) beryllium and radon
    (D) potassium and sodium

21. When stimulated, the membrane of the neuron
    (A) contracts
    (B) undergoes depolarization
    (C) expands
    (D) begins to synthesize protein

22. The neuron repolarizes after a nerve impulse passes by the leakage of
    (A) carbon isotopes
    (B) hydrogen ions
    (C) oxygen atoms
    (D) potassium ions

23. A synapse occurring between a neuron and a muscle cell is called a
    (A) desmosome
    (B) gap junction
    (C) neuromuscular junction
    (D) synovial junction

24. All the following are possible neurotransmitters except
    (A) pitressin
    (B) norepinephrine
    (C) acetylcholine
    (D) dopamine

25. Once a neurotransmitter has been utilized in a synapse, it is
    (A) left in place
    (B) broken down
    (C) converted to an enzyme
    (D) converted to potassium ions

**PART C—True/False:** For each of the following statements, mark the letter "T" next to the statement if it is true. If the statement is false, change the underlined word to make the statement true.

1. The two main divisions of the nervous system are the peripheral nervous system and the <u>outer</u> nervous system.

2. Afferent nerves are also called <u>motor</u> nerves.

3. The two main types of cells in the nervous system are nerve cells and <u>glial</u> cells.

4. The <u>oligodendrocytes</u> are glial cells that have long processes and help form the blood-brain barrier.

5. Neuroglia cells provide <u>support</u> to the nerve cells.

6. Neurons with a single extension functioning as both an axon and dendrite are called <u>bipolar</u> neurons.

7. Motor neurons transmit impulses from the <u>peripheral</u> nervous system to the muscles and the glands.

8. The function of the association neurons is to link sensory neurons to <u>afferent</u> neurons.

9. The nucleus and mitochondria of a neuron are found in the <u>axon</u>.

10. Nerve cell extensions that are specialized to receive nerve impulses are the <u>axons</u>.

11. A nerve fiber is really a bundle of <u>dendrites</u>.

12. Chemical substances called <u>neurotransmitters</u> are released by the nerve cells at the synaptic knobs.

13. The nodes of Ranvier are places on the <u>dendrite</u> where there is no myelin.

14. The white matter of the brain is due to the white color of the <u>cytoplasm</u> that surrounds the nerve cell.

15. A <u>ganglion</u> is a mass of cell bodies of several neurons.

16. The place where an axon comes close to but does not join a dendrite is called a <u>synergism</u>.

17. A resting neuron is polarized because of the difference in electrical charges on either side of its <u>membrane.</u>

18. Outside the plasma membrane of a resting neuron the concentration of <u>hydrogen</u> ions is ten times greater than inside the membrane.

19. Energy to power the sodium-potassium pump is derived from <u>NAD</u> molecules within the cytoplasm of the neuron.

20. To stimulate a nerve impulse, a mechanical stimulus alters the resting potential by increasing the permeability of the <u>nuclear membrane</u>.

21. A wave of depolarization in a nerve cell is the same as the <u>nerve impulse</u>.

22. The same impulse will be generated in a nerve cell regardless of the size of the stimulus once the threshold has been reached. This is called the <u>threshold</u> law.

23. Without neurotransmitters such as <u>glucose</u>, a nerve impulse could not be propagated across the synapse.

24. The process of <u>endocytosis</u> accounts for the release of neurotransmitter into the synaptic cleft.

25. The three neurotransmitters of the catecholamine group are dopamine, epinephrine, and <u>norepinephrine</u>.

**Answers**

## PART A—Completion

1. peripheral nervous system
2. spinal cord
3. sense organs
4. cranial nerves
5. muscles
6. autonomic division
7. afferent neurons
8. efferent neurons
9. neuron
10. glial cells
11. oligodendrocytes
12. astrocytes
13. microglia
14. Schwann cells
15. axons
16. nervous system
17. multipolar neurons
18. bipolar neurons
19. unipolar
20. sensory neurons
21. muscles
22. efferent neurons
23. interneurons
24. cell body
25. Nissl body
26. axon
27. axons
28. axon terminals
29. neurotransmitters
30. myelin sheath
31. nodes of Ranvier
32. multiple sclerosis
33. neurilemma
34. fascicle
35. ganglion
36. reception
37. resting neuron
38. polarized
39. resting potential
40. sodium ions
41. positive
42. ATP
43. action potential
44. repolarize
45. refractory
46. all-or-none law
47. synapse
48. synaptic cleft
49. acetylcholine
50. norepinephrine

## PART B—Multiple Choice

| | | | | |
|---|---|---|---|---|
| 1. B | 6. B | 11. C | 16. D | 21. B |
| 2. D | 7. C | 12. B | 17. C | 22. D |
| 3. C | 8. A | 13. A | 18. A | 23. C |
| 4. D | 9. D | 14. C | 19. B | 24. A |
| 5. B | 10. A | 15. D | 20. D | 25. B |

## PART C—True/False

1. central
2. sensory
3. true
4. astrocytes
5. true
6. unipolar
7. central
8. motor
9. cell body
10. dendrites
11. axons
12. true
13. axon
14. myelin
15. true
16. synapse
17. true
18. sodium
19. ATP
20. plasma membrane
21. true
22. all-or-none
23. acetylcholine
24. exocytosis
25. true

# CHAPTER 11

# THE NERVOUS SYSTEM: ORGANIZATION

For purposes of study, the human nervous system is divided into two major subdivisions: the central nervous system and the peripheral nervous system.

The **central nervous system (CNS)** consists of the brain and spinal cord. These organs are composed primarily of the cell bodies and axons of interneurons. The skull and vertebral column surround and protect the brain and spinal cord, respectively. The sense organs are located near the brain, and this proximity reduces the distance that nerve impulses travel for interpretation (Table 11.1).

The **peripheral nervous system (PNS)** is composed mainly of the axons and dendrites of sensory and motor neurons. The cell bodies of these neurons are located within or near the central nervous system. The axons and dendrites usually extend away from the central nervous system as "nerves." Most nerves are mixed because both sensory and motor elements are present. The peripheral nervous system informs the central nervous system of stimuli received from the external environment and carries appropriate responses to the glands and muscles.

**TABLE 11.1**  *Organization of the Human Nervous System*

| Central Nervous System | Peripheral Nervous System |
| --- | --- |
| Brain | Cranial nerves arising from the brain<br>1. somatic fibers connecting to the skin and skeletal muscles<br>2. autonomic fibers connecting to visceral organs |
| Spinal cord | Spinal nerves arising from the spinal cord<br>1. somatic fibers connecting the skin and skeletal muscles<br>2. autonomic fibers connecting to visceral organs |

## THE CENTRAL NERVOUS SYSTEM

The central nervous system is the main interpretation center for the human body. It is made up of the brain and spinal cord, which are continuous with one another.

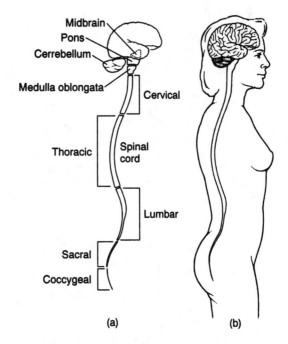

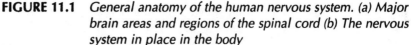

**FIGURE 11.1**  *General anatomy of the human nervous system. (a) Major brain areas and regions of the spinal cord (b) The nervous system in place in the body*

## The Spinal Cord

In the average adult, the **spinal cord** is a white cord of nerve tissue approximately 18 inches in length. It passes downward from the brain and extends through the bony tunnel formed by vertebrae (Figure 11.1). The spinal cord is continuous with the brain, and anatomically, it begins where the nerve tissue leaves the cranial cavity at the level of the foramen magnum of the occipital bone. The spinal cord tapers and terminates near the intervertebral disc separating the first and second lumbar vertebrae.

The outside portion of the spinal cord is white (**"white matter"**) due to the accumulation of myelin sheaths of nerve fibers. The inner material of the cord is gray (**"gray matter"**) because it is composed mainly of the cell bodies of neurons, and cell bodies have no myelin sheaths.

The spinal cord is surrounded and protected by a group of three membranes called **meninges**. The meninges are composed of three layers: the dura mater, the arachnoid mater, and the pia mater. The **dura mater** is the outermost layer and contains tough fibrous connective tissue with many blood vessels and nerves. The **arachnoid mater** (or "arachnoid") is a thin, netlike membrane without blood vessels. The **pia mater** is a very thin layer, with many nerves and blood vessels (Figure 11.2).

The three layers of the meninges extend over the brain as well as the spinal cord, and in some regions of the brain the dura mater extends inward between the lobes of the brain, forming partitions.

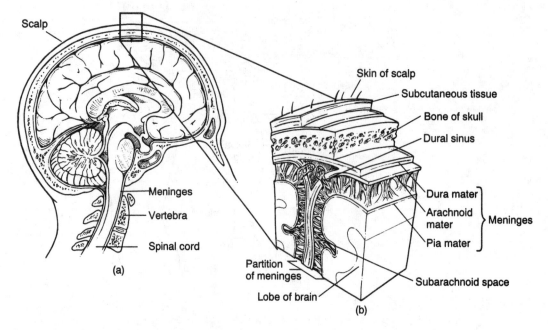

**FIGURE 11.2** *Structures enclosing the nervous system. (a) The relationship of the brain and spinal cord to the bones that enclose them (b) Details of the three layers of the meninges*

The space between the arachnoid mater and pia mater is called the **subarachnoid space**. It contains clear, watery fluid known as cerebrospinal fluid.

**Cerebrospinal fluid** is also found in the central canal of the spinal cord and the cavities of the brain. It is a lymphlike fluid that services the nutritional and gaseous needs of nerve cells of the CNS. A procedure called a "spinal tap" is used to obtain cerebrospinal fluid for analysis when nerve-related disease is suspected.

Thirty-one pairs of projections called **nerve roots** are located along the sides of the spinal cord. The roots closest to the dorsal aspect of the body are called **dorsal nerve roots**. They are the sites of cell bodies and axons of sensory nerves traveling to the spinal cord. The roots at the ventral aspect of the body are called **ventral nerve roots** (Figure 11.3). They contain the axons of motor neurons extending out from the spinal cord. Injury to the dorsal root results in loss of sensation from sense receptors (**anesthesia**), while injury to the ventral root makes one unable to respond to impulses (**paralysis**). Within the spinal cord, the dorsal and ventral roots arise from the **dorsal and anterior horns** of the spinal cord.

The spinal cord has two major functions in nerve coordination. It serves as a coordinating center for the reflex arc. It also serves as a connecting network between the peripheral nervous system and the brain. This connection is accomplished by means of axons extending from interneurons of the spinal cord and traveling upward in bundles called **nerve tracts**. These are called the **ascending tracts**. Other nerve tracts carry information down from the brain for trans-

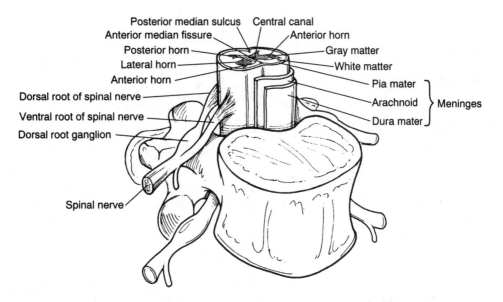

**FIGURE 11.3**   *The spinal cord in cross-section. The nerve tissue is covered by three meninges and surrounded by the bone of the vertebra. Note the dorsal and ventral nerve roots emerging from the spinal cord. (The dorsal roots are swollen with cell bodies of sensory nerves.) Within the cord, the roots arise from the posterior and anterior horns, respectively.*

mission to the muscles and glands. They are called **descending tracts**. The nerve tracts provide a two-way system of communication between the brain and the muscles and glands.

## The Brain

The brain is the organizing and processing center of the nervous system (Figure 11.4). It is the site of consciousness, sensation, memory, and coordination. The brain receives impulses from the spinal cord and 12 pairs of cranial nerves arising in the senses and other organs. It develops appropriate responses and sends forth these responses by motor neurons. The brain also initiates activities, such as memory, without environmental stimuli.

The brain is composed of left and right **hemispheres**. Its tissue is covered by the meninges as in the spinal cord and is serviced by lymphlike cerebrospinal fluid flowing through its cavities and in the subarachnoid space. The brain also has a large bed of capillaries for nutrient and gas exchange and for waste disposal. The brain consumes about 25 percent of all the oxygen used in the body and is extremely sensitive to oxygen or glucose deprivation.

The outer portion of the brain is gray ("gray matter") since it has the cell bodies of neurons, while the inner portion is white ("white matter") because it is composed primarily of the axons with myelin sheaths. The brain is divided into three major portions: the cerebrum, the cerebellum, and the brain stem.

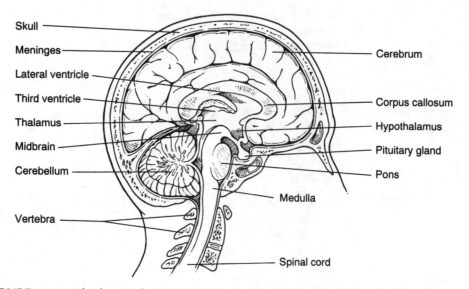

**FIGURE 11.4** *The human brain in position in the skull as seen from the lateral aspect. Various structures are shown in position relative to one another.*

## The Cerebrum

The **cerebrum** is the largest part of the brain, having nerve centers for sensory and motor activities. It is concerned with complex mental functions and consists of two large hemispheres connected by a bridge of nerve fibers called the **corpus callosum**. The surface of the cerebrum contains numerous **convolutions**, also called **gyri** (singular **gyrus**), and numerous grooves. A shallow groove is a **sulcus**, while a deep groove is a **fissure**.

The cerebral hemispheres accommodate over 10 billion cell bodies. Each hemisphere is divided into four lobes: the **frontal lobe**, at the anterior portion; the **parietal lobe**, posterior to the frontal lobe and separated from the frontal lobe by the **central sulcus**; the **temporal lobe**, located below the frontal lobe and separated from it by the **lateral sulcus**; and the **occipital lobe**, at the posterior portion of each hemisphere. Another portion of the brain called the **insula** is covered by portions of the frontal, parietal, and temporal lobes.

The cerebrum contains neurons for interpreting sensory impulses from the sense organs and initiating voluntary responses to the stimuli. It is the center for reasoning and memory, and it determines a person's intelligence and personality.

The primary **motor area** of the cerebrum is in the frontal lobe (Figure 11.5). Large pyramid-shaped cells exist here. Impulses from the cells cross over from one side of the brain to the other at the **corticospinal tract** to stimulate motor areas on the opposite sides of the body. A region of the frontal lobe called **Broca's area** also is concerned with motor function. **Sensory areas** are located in several lobes. Their impulses give rise to sensations, feelings, and emotions. The temporal lobes contain areas for hearing, while the occipital lobes contain areas for vision. The sense of smell is centered deep within the cerebrum.

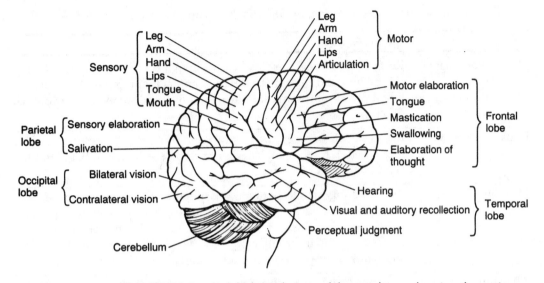

**FIGURE 11.5** *A right lateral view of the cerebrum showing the major areas and lobes, and the functions they regulate.*

Other regions of the cerebrum, especially areas of the frontal lobes, are associated with learning, reason, logic, foresight, and creativity. Regions of the parietal lobes help one to understand speech and express thoughts. Visual patterns are interpreted in the occipital lobe.

A series of interconnected cavities known as **ventricles** lie within the cerebral hemispheres. Cerebrospinal fluid fills the cavities and flows into the central canal of the spinal cord. Two large lateral ventricles extend into the cerebral hemispheres, and a third ventricle is found near the corpus callosum. The fourth ventricle is located in the brain stem (Figure 11.6).

### The Cerebellum

The **cerebellum** is a large mass of gray and white tissue lying adjacent to the medulla oblongata and serving as a coordinating center for motor activity. The cerebellum receives stimuli from the cerebrum and determines which muscles are to contract (Table 11.2). During walking, for example, the forebrain determines the muscles used and the strength and sequence of the contractions.

The cerebellum consists of two lateral hemispheres separated by a layer of dura mater. The cerebellum communicates with other parts of the central nervous system by three pairs of nerve tracts called **cerebellar peduncles.** As a reflex center for the coordination of skeletal muscle activity, the cerebellum helps to maintain posture.

### The Brain Stem

The nerve tissue connecting the cerebrum to the spinal cord contains a number of structures collectively known as the **brain stem**. Included in the brain stem is the diencephalon, the midbrain, the pons, and the medulla oblongata.

**TABLE 11.2**  *A Summary of Brain Parts and Their Functions*

| Structure | Specific Function |
| --- | --- |
| Medulla oblongata | Receives and integrates signals from spinal cord; sends signals to the cerebellum and thalamus; contains centers that regulate heart beat, blood pressure, respiratory rate, coughing, and some other involuntary functions |
| Pons | Relays signals between the medulla and more superior parts of the brain, between the hemispheres of the cerebellum, and between the cerebellum and cerebrum |
| Midbrain | Relays sensory signals between the spinal cord and the thalamus, and motor signals between the cerebral cortex and the pons and spinal cord; controls reflexive movements of the head and eyeballs in response to visual stimuli; controls reflexive movements of the head and trunk in response to auditory stimuli |
| Thalamus | Contains many nuclei through which it relays all sensory signals (except smell signals) to the cerebral cortex; relays motor signals from the cerebral cortex toward the spinal cord; relays signals to the cerebral cortex that maintain consciousness; processes some crude sensations |
| Hypothalamus | Receives sensory signals from internal organs by way of the thalamus and uses these signals to control actions of the autonomic nervous system and pituitary gland, thereby helping to maintain homeostasis; provides structural and functional connection between the nervous and endocrine systems by its relationship to the pituitary gland; in combination with the limbic system participates in physiological response to emotional experiences |
| Cerebellum | Receives sensory signals from the eyes; coordinates organs of balance and receptors in muscles, tendons, and joints |
| Cerebrum | Contains areas that receive and process sensory signals (somatic sensory area, visual area, auditory area) and that initiate motor signals for voluntary movements (somatic motor area and speech area); contains association areas where sensory signals are interpreted, memories are stored, and complex processing occurs; contains tracts of association fibers that relay signals between the cerebral cortex and other parts of the nervous system |
| Basal nuclei | Help to control muscle tone and thereby help to coordinate voluntary movements |
| Limbic system | Contains pleasure and punishment centers; plays a role in emotional feelings; hippocampus determines what memories will be stored |
| Reticular formation | Contains nuclei involved in wakefulness and sleep |

The **diencephalon** is the region above the midbrain between the hemispheres of the cerebrum. It encompasses the third ventricle and is organized into masses of gray matter called **nuclei**. One such nucleus, called the **thalamus**, is a relay center for sensory impulses extending to the cerebral cortex. Descending fibers communicate from the cortex with the thalamus, which appears to be associated with such sensations as pain and temperature.

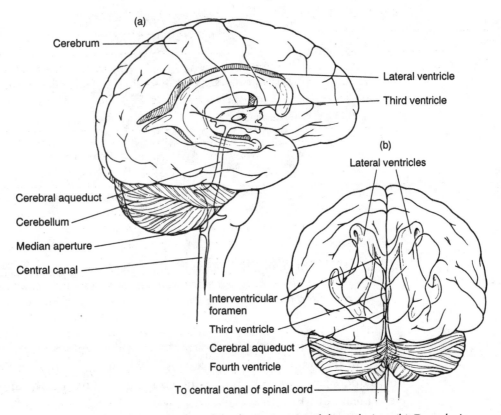

**FIGURE 11.6**  *The ventricles of the brain in (a) Left lateral view (b) Dorsal view*

Another nucleus of the diencephalon is the **hypothalamus**. The hypothalamus sends impulses to and receives them from the cerebrum and thalamus. Nerve cells of the hypothalamus produce hormones, including many stored by the pituitary gland. These hormones regulate the activity of a variety of visceral organs. Hunger, regulation of body weight and body temperature, and water balance are also associated with the hypothalamus.

A collection of structures ringing the edge of the brain stem comprise the **limbic system**. The limbic system is involved in emotional experiences. It is associated with feelings such as fear, anger, pleasure, and sorrow. Therefore, it has a substantial influence on a person's behavior.

Another portion of the brain stem, the **midbrain**, is located between the pons and diencephalon (Figure 11.7). Nerve fibers of the midbrain join the brain stem and spinal cord to the cerebrum, and nerve cells within the midbrain function as reflex centers. Corticospinal tracts connecting the cerebrum and spinal cord are found at the underside of the midbrain.

The brain stem also contains a rounded bulge called the **pons**, which separates the midbrain from the medulla oblongata. The pons consists mainly of nerve fibers that relay impulses from the medulla oblongata to the cerebrum and back to the medulla.

The brain stem also contains the **medulla oblongata**, the swollen tip of the spinal cord connecting with the remainder of the brain.

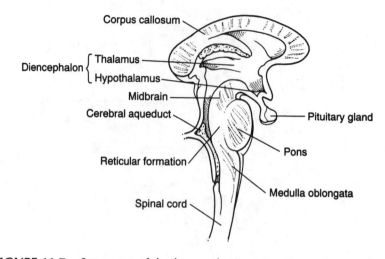

**FIGURE 11.7**   *Structures of the human brain stem. The major structures in descending order are the diencephalon, midbrain, pons, and medulla oblongata.*

The medulla oblongata is a passageway in which nerves extend to and from the brain to other organs. All ascending and descending nerve fibers pass through the medulla oblongata. The fourth ventricle is located within it.

Masses of gray matter, the nuclei, serve as control centers within the medulla oblongata. Among the activities controlled are the rate of heartbeat, contraction of smooth muscles in the walls of certain blood vessels, and regulation of breathing activities. Sneezing, coughing, vomiting, and swallowing are among the other activities regulated by the medulla.

The medulla also contains a network of nerve fibers called the **reticular formation**. The reticular formation extends into the pons and midbrain as well. Nerve fibers in the network are responsible for activating the cerebral cortex when sensory impulses are received. They arouse the cortex and prepare it to interpret sensory impulses or stimulate the process of thinking.

## THE PERIPHERAL NERVOUS SYSTEM

The brain and spinal cord are connected to every other part of the body and to the environment by a collection of nerves and cell bodies called the **peripheral nervous system**. The peripheral nervous system is composed of all the nervous tissue outside the brain and spinal cord. It is composed primarily of the peripheral nerves, the ganglia associated with them, and the sensory receptors. The nerve fibers of the peripheral nervous system may be afferent or efferent. Afferent nerve fibers conduct nerve impulses toward the nervous system, while efferent nerve fibers conduct them away. Nearly all peripheral nerves are mixed nerves containing both kinds of fibers. The afferent nerve fibers (sensory) arise in the senses. The efferent (motor nerves) arise in the central nervous system and include the somatic nerve fibers and autonomic nerve fibers. Somatic nerve fibers innervate skeletal muscles, while autonomic nerve fibers innenervate the smooth and cardiac muscle and the glands. The peripheral nervous system is subdivided into the sensory somatic system and the autonomic system.

## The Sensory Somatic System

The sensory somatic system carries nerve impulses from the senses to the central nervous system for interpretation. The system also carries impulses away from the central nervous system to the skeletal muscles and glands if a response is indicated. The system permits one to be aware of the external environment and to react to it. The awareness and the reactions occur on a voluntary basis.

The sensory somatic system is composed of 12 pairs of cranial nerves and 31 pairs of spinal nerves. The 12 pairs of **cranial nerves** arise from various locations in the brain and pass through the underside of the brain (Figure 11.8). The first pair of cranial nerves, the olfactory nerves, begins at the cerebrum, but the remaining 11 pairs originate in the brain stem. The cranial nerves then lead to various portions of the head, neck, and body trunk.

Most cranial nerves contain both sensory and motor axons, but certain cranial nerves contain only sensory axons. Those concerned with smell and vision are examples of the latter. Other cranial nerves involved primarily with glands and muscles contain motor axons.

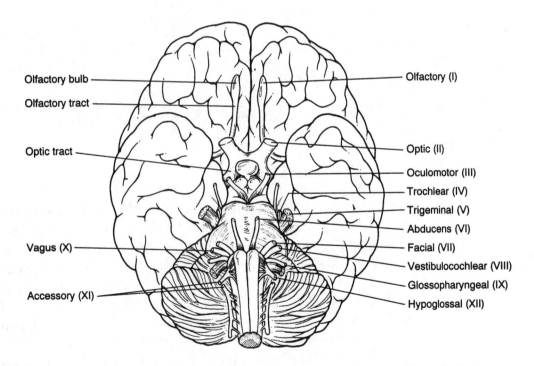

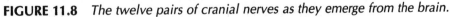

**FIGURE 11.8**   *The twelve pairs of cranial nerves as they emerge from the brain.*

**TABLE 11.3**  *The Cranial Nerves in Humans*

| Number | Name | Functions (s, sensory; m, motor) | Origin or Brain Terminus |
|---|---|---|---|
| I | Olfactory | (s) Smell | Cerebral hemispheres |
| II | Optic | (s) Vision | Thalamus |
| III | Oculomotor | (m) Eye movement | Midbrain |
| IV | Trochlear | (m) Eye movement | Midbrain |
| V | Trigeminal | (m) Swallowing movements<br>(s) Sensitivity of face and tongue | Midbrain and pons<br>Medulla oblongata |
| VI | Abducens | (m) Eye movement | Medulla oblongata |
| VII | Facial | (m) Facial movement | Medulla oblongata |
| VIII | Auditory or vestibulocochlear | (s) Hearing<br>(s) Balance | Medulla oblongata |
| IX | Glossopharyngeal | (s, m) Tongue and pharynx | Medulla oblongata |
| X | Vagus | (s, m) Heart, blood vessels, viscera | Medulla oblongata |
| XI | Spinal accessory | (m) Neck muscles and viscera | Medulla oblongata |
| XII | Hypoglossal | (m) Tongue muscles | Medulla oblongata |

**TABLE 11.4**  *A Comparison of Cranial Nerves and Spinal Nerves*

| Characteristic | Cranial Nerves | Spinal Nerves |
|---|---|---|
| Origin | Base of brain | Spinal cord |
| Distribution | Mainly to head and neck | Skin, skeletal muscles, joints, blood vessels, sweat glands, and mucosa except of head and neck |
| Structure | Some composed of sensory fibers only; some of both motor axons and sensory dendrites; motor fibers belong to voluntary or autonomic nervous systems | All composed of both sensory dendrites and motor axons; some somatic or voluntary, some autonomic |
| Function | Vision, hearing, sense of smell, sense of taste, eye movements | Sensations, movements, and sweat secretion |

Cranial nerves are designated by a number and a name. Table 11.3 indicates the names, numbers, and functions of the cranial nerves. Often the cell bodies of the nerves are located outside the brain in masses of cell bodies called **ganglia** (singular **ganglion**). This is true for sensory axons. The cell bodies of motor neurons are generally located within the gray matter of the brain.

Thirty one pairs of **spinal nerves** also are included in the sensory somatic system. The nerves communicate between the spinal cord and various parts of the body appendages and trunk. They are grouped as cervical nerves (8 pairs), thoracic nerves (12 pairs), lumbar nerves (5 pairs), sacral nerves (5 pairs) and coccygeal nerves (1 pair).

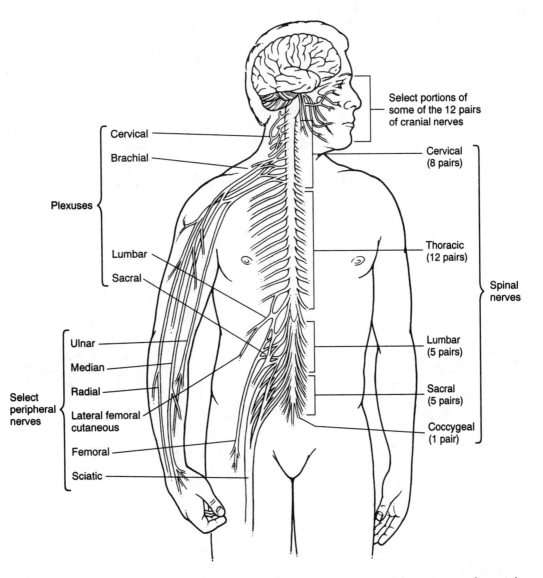

**FIGURE 11.9**  *The peripheral nervous system. The system consists of the 12 pairs of cranial nerves and the 31 pairs of spinal nerves. Note the plexuses in the cervical and brachial regions.*

The spinal nerves emerge from the spinal cord by the **dorsal and ventral roots** (also called **dorsal and ventral horns**). The dorsal root contains a ganglion in which the cell bodies of sensory neurons are contained. The ventral root has no ganglia because the cell bodies are in the gray matter of the spinal cord (Table 11.4).

Complex networks called **plexuses** are points at which the spinal nerves combine temporarily before passing to their respective points of termination. In a plexus, spinal nerves are recombined so that fibers originating from different spinal nerves extend to a body part together. The three major plexuses are the cervical plexus in the neck, the brachial plexus in the shoulder, and the lumbosacral plexus in the lumbar regions of the back (Figure 11.9).

# The Autonomic System

The **autonomic system** operates on an **involuntary basis**. It is a portion of the peripheral nervous system that functions without conscious control. The system coordinates functions of the visceral organs such as the cardiac muscle, smooth muscle, and visceral glands.

The autonomic system consists of two groups of motor neurons and a set of ganglia, lying between the neurons. The first group of motor neurons arise in the central nervous system and extend to a ganglion. They are called **preganglionic neurons** or preganglionic

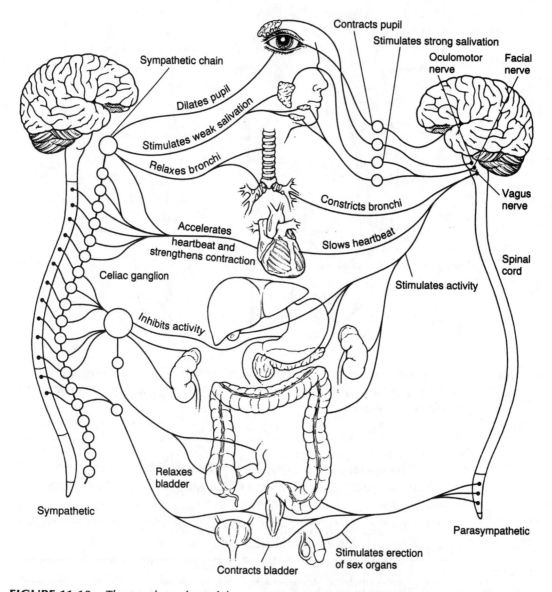

**FIGURE 11.10**   *The two branches of the autonomic nervous system. Nerves of the sympathetic system (left) prepare the body for an emergency. Note that the system consists of nerves before the sympathetic chain of ganglia (preganglionic nerves) and after the sympathetic chain (postganglionic nerves). Nerves of the parasympathetic system (right) return the body to normal. Many of these nerves arise from cranial nerves at the upper spinal cord.*

fibers. The second group of fibers extend from the ganglion to the body organs and are called **postganglionic neurons** or postganglionic fibers. They lie outside the CNS and are part of the PNS.

The autonomic system is subdivided into the sympathetic division and the parasympathetic division. The nerves of the **sympathetic division** duplicate the action of epinephrine (Chapter 13) by preparing the body for an emergency. Sympathetic impulses to muscles increase the heartbeat, constrict the arteries, dilate the pupils, and bring about other changes for dealing with a crisis (Figure 11.10). Impulses of the **parasympathetic division** return the body to normal after the crisis has passed. These impulses slow the heartbeat, dilate the arteries, constrict the pupils, and adjust other organs. In this sense they reestablish the body's homeostasis.

The different effects that sympathetic and parasympathetic divisions have on organs is partly due to the **neurotransmitters** secreted. The preganglionic fibers of the sympathetic and parasympathetic divisions secrete the neurotransmitter **acetylcholine**, while the postganglionic fibers of the sympathetic division secrete norepinephrine (or, noradrenaline), and the postganglionic fibers of the parasympathetic nervous system secrete acetylcholine (Table 11.5). Hence, the postganglionic fibers of the sympathetic division are called **adrenergic** fibers (from noradrenaline) while the postganglionic fibers of the parasympathetic division are called cholinergic fibers (from acetylcholine).

The two divisions of the autonomic nervous system are antagonistic, but not completely, since each division can activate certain muscles while inhibiting others. Most activity of the autonomic nervous system is involuntary, but the brain exerts some control by means of impulses arising from the medulla oblongata.

**TABLE 11.5**    *A Comparison of the Sympathetic and Parasympathetic Nervous Systems*

| Characteristic | Sympathetic System | Parasympathetic System |
|---|---|---|
| General effect | Prepares body to cope with stressful situations; mediates abnormal configuration of body functions | Restores body to resting state after stressful situation; actively maintains normal configuration of body functions |
| Extent of effect | Widespread throughout the body | Localized in tissues |
| Neurotransmitter released at synapse | Norepinephrine (usually) | Acetylcholine |
| Duration of effect | Lasting | Brief |
| Outflow from CNS | Thoracolumbar levels of spinal cord | Craniosacral levels (from brain and spinal cord) |
| Location of ganglia | Chain and collateral ganglia | Terminal ganglia |
| Number of post-ganglionic fibers | Many | Few |

# REVIEW QUESTIONS

**PART A—Completion: Add the word or words that correctly complete each of the following statements.**

1. The brain and spinal cord are components of the _____ .

2. Most of the substance of the brain and spinal cord is composed of _____ .

3. The peripheral nervous system is composed primarily of the axons and dendrites of sensory neurons and _____ .

4. The axons and dendrites extend to the muscles and glands and form _____ .

5. The spinal cord passes downward from the brain and extends through a bony tunnel found in the _____ .

6. The spinal cord is continuous with the _____ .

7. The inner material of the spinal cord is gray matter, while the outer portion is _____ .

8. The three membranes surrounding the spinal cord and protecting it are called _____ .

9. The outermost membrane surrounding the spinal cord is the _____ .

10. The inner membrane surrounding the spinal cord and having many nerves and blood vessels is the _____ .

11. Clear, watery fluid is found in the central canal of the spinal cord and is called _____ .

12. The ventral roots of the spinal cord contain the axons of _____ .

13. Injury to the ventral root of the spinal cord results in a condition called _____ .

14. The nerve tracts carrying information away from the brain through the spinal cord are known as _____ .

15. The spinal cord serves as a coordinating center for the _____ .

16. The brain receives impulses from the spinal cord as well as from 12 pairs of _____ .

17. The brain is composed of two major _____ .

18. Cerebrospinal fluid flows through the cavities of the brain as well as in the _____ .

19. The brain consumes about one quarter of the body's _____ .

20. The inner portion of the brain is composed of white matter while the outer portion is composed of _____ .

21. The numerous convolutions of the brain are called _____ .

22. The anterior portion of each hemisphere is occupied by a lobe called the _____ .

23. At the posterior portion of the brain hemispheres is a lobe called the _____ .

24. In the cerebrum of the brain, impulses cross over to opposite sides of the brain at the _____ .

25. The region of the frontal lobe concerned with motor function is _____ .

26. The sense of smell is located deep within the portion of the brain known as the _____ .

27. The interpretation of visual patterns occurs in the lobe of the brain known as the _____ .

28. The cavities of the brain are called the _____ .

29. The brain region lying adjacent to the medulla and serving as a coordinating center for motor activity is the _____ .

30. The cerebellum is composed of two lateral _____ .

31. Because it is a reflex center for coordinating muscle activity, the cerebellum helps maintain _____ .

32. The diencephalon is organized into masses of gray matter called _____ .

33. The nucleus of the diencephalon that relays sensory impulses into the cerebral cortex is the _____ .

34. The nucleus of the diencephalon that produces hormones stored in the pituitary gland is the _____ .

35. Emotional experiences such as fear, anger, pleasure, and sorrow are regulated in a ring of tissue at the edge of the brain stem called the _____ .

36. The rounded bulge in the brain stem separating the midbrain from the medulla is the _____ .

37. The swollen tip of the brain connecting the spinal cord to the remainder of the brain is the _____ .

38. A network of nerve fibers in the medulla that are responsible for activating the cerebral cortex is the _____ .

39. The sensory somatic system is composed of spinal nerves and _____ .

40. The human body has spinal nerves that number _____ .

41. The cranial nerve that is concerned with smell is the _____ .

42. The cranial nerve that regulates swallowing nerves and sensitivity of the face and tongue is the _____ .

43. The cranial nerve that has sensory and motor functions with respect to the heart, blood vessels, and organs of the viscera is the _____ .

44. The cell bodies of nerves are commonly located outside the brain in groups known as _____ .

45. The spinal nerves communicate impulses between various parts of the body and the _____ .

46. Bodies where spinal nerves combine temporarily before passing to the destination points are known as _____ .

47. The autonomic nervous system operates on a basis that is _____ .

48. The nerve fiber extending in the autonomic system from the ganglion to the body organs is called the _____ .

49. The nerves of the sympathetic division of the autonomic nervous system duplicate the action of the hormone _____ .

50. Once a crisis has passed the body is returned to normal by impulses of the portion of the autonomic nervous system known as the _____ .

**PART B—Multiple Choice: Circle the letter of the item that correctly completes each of the following statements.**

1. Cell bodies and axons of interneurons make up most of the substance of the
   (A) peripheral nervous system
   (B) sensory nervous system
   (C) central nervous system
   (D) autonomic nervous system

2. The cell bodies of neurons of the peripheral nervous system are located
   (A) in the body organs
   (B) at the body surface
   (C) in the sacral vertebrae
   (D) within or near the central nervous system

3. Most nerves contain
   (A) only cell bodies
   (B) only dendrites
   (C) sensory and motor axons and dendrites
   (D) association neurons only

4. All of the following are functions of the peripheral nervous system except
   (A) it interprets sensations and stimuli
   (B) it connects the body to the external environment
   (C) it carries response to the muscles and organs
   (D) it carries stimuli to the central nervous system

5. All the following apply to the spinal cord except
   (A) it is continuous with the brain
   (B) it terminates at the intervertebral disk separating the first and second lumbar vertebrae
   (C) the outside portion is gray
   (D) it is surrounded by meninges

6. All the following are meninges except
   (A) the pia mater
   (B) the corpus mater
   (C) the arachnoid mater
   (D) the dura mater

7. White matter of the nervous system is white because
   (A) cytoplasm is white
   (B) dendrites are white
   (C) the pia mater contains white pigments
   (D) myelin in the myelin sheaths is white

8. The cerebrospinal fluid may be found
   (A) within the dura mater
   (B) only in the myelin sheath
   (C) in the central canal of the spinal cord
   (D) in the cytoplasm of cells of the brain

9. The dorsal nerve roots are the sites of
   (A) gray matter of the brain
   (B) cell bodies and axons of sensory nerves
   (C) oligodendrocytes
   (D) attachment for the meninges

10. Destruction of the ventral nerve roots will result in
    (A) the inability to respond to stimuli
    (B) the inability to form cerebrospinal fluid
    (C) the ability to speak louder than usual
    (D) the ability to conduct more nerve impulses more efficiently

11. The descending tracts in the spinal cord
    (A) are composed solely of dendrites
    (B) have no cell bodies
    (C) carry impulses for transmission to muscles and glands
    (D) are extensions of the sensory organs at the body surface

12. The outer portion of the brain
    (A) is covered with cell bodies
    (B) has no meninges
    (C) exists in an oxygen-free environment
    (D) is composed of gray matter

13. All the following apply to the cerebrum except
    (A) it consists of two hemispheres
    (B) it has over ten billion cell bodies
    (C) it has numerous convolutions
    (D) it is one of the smaller parts of the brain

14. Each of the following is a lobe of the cerebrum except
    (A) parietal lobe
    (B) occipital lobe
    (C) thoracic lobe
    (D) temporal lobe

15. Areas for hearing are located in the cerebrum's
    (A) occipital lobe
    (B) lumbar lobe
    (C) cervical lobe
    (D) temporal lobe

16. The ventricles of the cerebrum carry the
    (A) axons of motor neurons
    (B) dendrites and cell bodies of interneurons
    (C) myelin sheaths of all neurons
    (D) cerebrospinal fluid

17. All the following apply to the cerebellum except
    (A) it is about the size of a large fist
    (B) it is a coordinating center for sensory activity
    (C) it has two lateral hemispheres
    (D) it lies adjacent to the medulla oblongata

18. One of the functions of the cerebellum is to
    (A) coordinate skeletal muscle activity
    (B) produce pituitary hormones
    (C) serve as a center for hearing
    (D) serve as a center for speech

19. The thalamus and hypothalamus are both located
    (A) in the cerebrum
    (B) in the diencephalon
    (C) next to the medulla oblongata
    (D) outside the brain

20. All the following functions are associated with the hypothalamus except
    (A) water balance
    (B) regulation of body temperature
    (C) emotional experiences
    (D) regulation of body weight

21. Nerve cells located in the midbrain function as
    (A) reflex centers
    (B) producers of hormones
    (C) hearing centers
    (D) sensor for pain and heat

22. The glossopharyngeal nerve has sensory and motor functions relating to the
    (A) ears and eyes
    (B) taste buds and eyes
    (C) tongue and pharynx
    (D) heart and blood vessels

23. All the following are activities regulated by
    (A) activation of the cerebral cortex
    (B) regulation of breathing activities
    (C) control of the rate of heartbeat
    (D) smooth muscle contraction in certain blood vessels.

24. The 12 pairs of cranial nerves and 31 pairs of spinal nerves make up the
    (A) central nervous system
    (B) spinal cord
    (C) sensory somatic system
    (D) autonomic nervous system

25. All the following apply to the autonomic nervous system except
    (A) it operates on an involuntary basis
    (B) it consists of sympathetic and parasympathetic divisions
    (C) it includes the cranial nerves
    (D) it is composed of preganglionic and postganglionic neurons.

*PART C—True/False:* **For each of the following statements, mark the letter "T" next to the statement if it is true. If the statement is false, change the <u>underlined</u> word to make the statement true.**

1. The peripheral nervous system is composed of the <u>axons</u> and dendrites of sensory neurons and motor neurons.

2. In the average adult, the spinal cord appears as a <u>gray</u> mass of tissue approximately 18 inches in length.

3. The spinal cord begins anatomically at the <u>obturator foramen</u> of the occipital bone.

4. The spinal cord is surrounded and protected by three membranes known as <u>sarcolemmas</u>.

5. The middle, thin, netlike covering of the spinal cord is the <u>dura mater</u>.

6. The watery fluid found within the spinal cord and brain is known as <u>plasma</u>.

7. Projections located along each side of the spinal cord are referred to as <u>nerve roots</u>.

8. Injury to the <u>ventral root</u> of the spinal cord leads to a loss of sensation and a condition called anesthesia.

9. The ascending and descending tracts provide a system of communication between the muscle and glands and the <u>spinal cord</u>.

10. The function of memory is associated most closely with the <u>spinal cord</u>.

11. The three major portions of the brain are the brain stem, the cerebellum, and the <u>pons</u>.

12. A shallow groove occurring within the brain tissue is correctly known as a <u>gyrus</u>.

13. The parietal lobe is located posterior to the frontal lobe and is separated from it by the <u>lateral sulcus</u>.

14. A person's intelligence, personality, and ability to initiate voluntary responses to stimuli are located in the brain portion called the <u>cerebrum</u>.

15. Centers for hearing are located in the cerebrum in the <u>occipital lobe</u>.

16. The fourth cranial nerve, known as the <u>abducens</u>, is responsible for eye movement.

17. The cavities within the brain carry cerebrospinal fluid and number <u>five</u>.

18. The cerebellum communicates with other parts of the central nervous system by three pairs of nerve tracts called <u>cerebellar peduncles</u>.

19. The thalamus and hypothalamus both are located within the <u>brain stem</u>.

20. The sensations of hunger, the regulation of body weight and temperature, and the water balance of the body are all associated with the <u>thalamus</u>.

21. Nerve fibers passing from the medulla oblongata to the cerebrum pass through a rounded bulge known as the <u>corpus corpora</u>.

22. The rate of heartbeat and the contraction of smooth muscle in the vessel walls are both regulated by impulses from the <u>medulla oblongata</u>.

23. Sensory and motor neurons are the principal components of the <u>peripheral nervous system</u>.

24. The <u>cranial nerves</u> are grouped as cervical, thoracic, lumbar, sacral, and coccygeal nerves.

25. The autonomic nervous system operates on a <u>voluntary basis</u>.

## PART A—Completion

1. central nervous system
2. interneurons
3. motor neurons
4. nerves
5. vertebrae
6. brain
7. white matter
8. meninges
9. dura mater
10. pia mater
11. cerebrospinal fluid
12. motor neurons
13. paralysis
14. descending tracts
15. reflex arc
16. cranial nerves
17. hemispheres
18. subarachnoid space
19. oxygen
20. gray matter
21. gyri
22. frontal lobe
23. occipital lobe
24. corticospinal tract
25. Broca's area
26. cerebrum
27. occipital lobe
28. ventricles
29. cerebellum
30. hemispheres
31. posture
32. nuclei
33. thalamus
34. hypothalamus
35. limbic system
36. pons
37. medulla oblongata
38. reticular formation
39. cranial nerves
40. 31 pairs
41. olfactory nerve
42. trigeminal nerve
43. vagus nerve
44. ganglia
45. spinal cord
46. plexuses
47. involuntary
48. postganglionic neuron
49. epinephrine
50. parasympathetic division

## PART B—Multiple Choice

| | | | | |
|---|---|---|---|---|
| 1. C | 6. B | 11. C | 16. D | 21. A |
| 2. D | 7. D | 12. D | 17. B | 22. C |
| 3. C | 8. C | 13. D | 18. A | 23. A |
| 4. A | 9. B | 14. C | 19. B | 24. C |
| 5. C | 10. A | 15. D | 20. C | 25. C |

## PART C—True/False

1. true
2. white
3. foramen magnum
4. meninges
5. arachnoid
6. cerebrospinal fluid
7. true
8. dorsal root
9. brain
10. brain
11. cerebrum
12. sulcus
13. central sulcus
14. true
15. temporal lobe
16. trochlear
17. four
18. true
19. diencephalon
20. true
21. pons
22. true
23. true
24. spinal nerves
25. involuntary

# CHAPTER 12

# THE SPECIAL SENSES

The special senses of the body include the major senses of sight, smell, hearing, taste, and several others, such as touch and equilibrium. All the senses have highly specialized receptors enabling them to respond to the appropriate stimuli. Those receptors that detect chemical stimuli are called **chemoreceptors**, while those detecting light are **photoreceptors**, and those detecting mechanical changes are **mechanoreceptors**. When the receptors are located at the body surface, they are known as **exteroreceptors**, while if they are within the body muscles, joints, or bones, they are called **proprioreceptors** (Table 12.1). The organs of the special senses are intimately associated with the nervous system, both structurally and functionally, and they depend upon the nervous system for conscious interpretation of the environmental changes they detect.

**TABLE 12.1**  *A Summary of the Special Senses*

| Sense Organ | Specific Receptors | Locations | Nature of Stimulus | Major Stimulus | Anatomic Location |
|---|---|---|---|---|---|
| Olfactory membrane | Olfactory cell | Exteroreceptors | Chemoreceptors | Chemicals in solution | Superior nasal cavity |
| Taste bud | Gustatory cell | Exteroreceptors | Chemoreceptors | Chemicals in solution | Dorsum of tongue, pharynx |
| Eye | Rods and cones | Exteroreceptors | Photoreceptors | Light intensity | Eye |
| Ear | | | | | |
| Cochlea | Organ of corti (hair cells) | Exteroreceptors | Mechanoreceptors | Vibration | Inner ear |
| Vestibular apparatus | Maculae and cristae (hair cells) | Proprioreceptors | Mechanoreceptors | Deflection | Inner ear |

## THE EYE

The eye is the organ of sight in the body. It gathers light from the environment and forms an image on nerve cells of the retina. The image is then transformed into nerve impulses, which are interpreted by the brain.

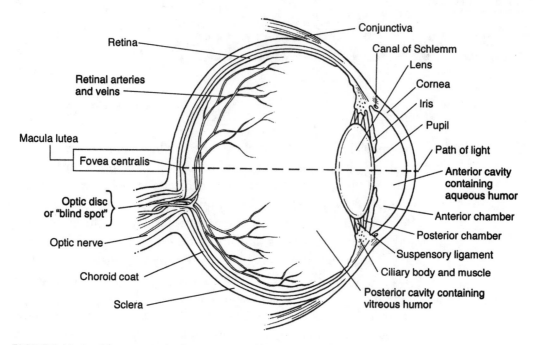

**FIGURE 12.1**   *The anatomical structures of the eye as seen in a longitudinal section of this organ.*

## Anatomy

The eye is a fluid-filled, somewhat movable sphere serviced by the anterior end of the **optic nerve**, a cranial nerve extending from and to the brain. Visual images initiated by receptors in the retina of the eye travel over the optic nerve, then pass through the optic tract, finally reaching the visual cortex, where they are interpreted.

Anatomically, the "eye" is synonymous with the **eyeball**. The eyeball is a bit longer than wide, with its anterior portion extending out from the sphere. The wall of the eye is composed of three layers, or coats: an outer, tough, fibrous coat consisting of the cornea and sclera; a middle, highly vascular coat containing the choroid layer, iris, and ciliary bodies; and an inner coat, the retina, which contains the receptors of sight (Figure 12.1).

The eye contains two fluid-filled cavities—the **anterior cavity** and the **posterior cavity**. The anterior cavity is subdivided into the **anterior chamber**, which lies between the iris and the lens. Both chambers are filled with a watery fluid called **aqueous humor**. The posterior cavity, which lies between the lens and the retina, contains a jellylike substance called **vitreous humor**.

The **pupil** of the eye is an opening in the iris of the eye. Two layers of smooth muscle compose the **iris**: a sphincter muscle layer, which constricts the pupil and makes it smaller, and a dilator muscle layer, which makes the pupil larger. The iris contains pigments that give color to the eye. The "white" of the eye is the visible portion of the **sclera** of the eye.

**TABLE 12.2**  *Eye Structures and Their Functions*

| Structure | Function |
| --- | --- |
| Cornea | Refracts light; important in focusing light onto the retina |
| Sclera | Maintains shape of eye and protects eye; also serves as site of muscle attachment |
| Iris | Controls amount of light passing through the pupil |
| Cilliary body | Changes shape of lens (accommodation) and secretes aqueous humor |
| Choroid | Absorbs light; contains blood vessels for eye structures |
| Retina | Absorbs light and stores vitamin A; receives light and forms image for transmission to brain |
| Lens | Refracts light; important in accommodation |
| Anterior cavity | Maintains shape of eye and refracts light through its aqueous humor |
| Posterior cavity | Maintains shape of eye and refracts light through its vitreous humor |
| Aqueous humor | Fills anterior cavity, helping to maintain shape of eye; refracts light; maintains intraocular pressure |
| Vitreous humor | Fills posterior cavity and maintains intraocular pressure; lends shape to eye and keeps retina firmly pressed against choroid; refracts light |

Behind the pupil is the **lens**, a transparent, biconvex disk of fibrous protein material in concentric layers. The lens is firmly attached to a structure called the **ciliary body** by a ligament called the **suspensory ligament**. Most of the ciliary body consists of the **ciliary muscle**. The ciliary body joins the iris at its periphery, and the remainder of the iris extends inward between the cornea and lens.

The innermost layer of the eyeball is the **retina**. The retina extends anteriorly as far as the posterior aspect of the ciliary body. Two layers comprise the retina: an outer pigmented layer, which adheres to the choroid layer; and an inner layer of nerve tissue, the retina proper (Table 12.2).

The inner, nerve layer of the retina consists of three layers of neurons: first, a layer of **receptor neurons** numbering about 100 million **rod cells** and 700 million **cone cells**, so-called because of their shape; second, a layer of **bipolar neurons**, the nerve cells that receive impulses initiated by the rod and cone cells; and, third, a layer of **ganglionic neurons** attached directly to the optic nerve.

The accessory structures of the eye include the eyebrows, eyelids, eyelashes, conjunctiva, and lacrimal apparatus. The **eyebrows** and **eyelashes** offer protection against the entry of foreign particles to the pupil, while the **eyelids** protect the anterior portion of the eye.

The eyelids are covered on their deep surfaces by the **conjunctiva**, the mucous membrane also folded over part of the eyeball. The conjunctiva covers the eyeball and lines the inner eyelids. The **lacrimal apparatus** contains the lacrimal glands that produce tears to bathe the eyeball and keep it moist.

# Physiology of Vision

The physiology of vision relies upon the two different types of cells in the retina: the rod-shaped cells (rods) and cone-shaped cells (cones). The **rods** permit vision when there is poor light. They form outlines or silhouettes of objects and are primarily concerned with **twilight vision**. They detect movement in the environment and use a visual pigment called **rhodopsin**.

The **cones** are most accurate where there is sufficient light to permit close, detailed vision. These receptors are most concerned with **daylight vision**. The cones enable one to see detail and are responsible for color vision. Cones are most concentrated in the **fovea centralis**, a small dent near the center of the retina. Seeing clearly means catching an image on the fovea centralis. Moving further away from the fovea centralis there are fewer cones, but the number of rods increases. Rods are at their greatest number at the outer edge of the retina. For this reason, the detection of movement and twilight vision is concerned chiefly with the outer periphery of the retina.

When light energy stimulates the rods and cones in the retina, chemical reactions rapidly take place in these visual receptors. These events give rise to nerve impulses, which are transported over the optic nerve and optic tract to the visual cortex of the brain where they are interpreted. No visual receptors are found at a region of the retina called the **optic disk**, and therefore, this portion of the retina is called the **blind spot**. The image reaching the retina occurs in the inverted position due to the optical properties of the lens of the eye, but the image is made upright by the cortex of the occipital lobe of the brain.

The pathway of light to the eye begins at the clear cornea, which lies at the surface of the eye. Light then passes through the pupil, which varies in size according to the distance of the object being viewed: the pupil is smaller when the object is close and larger when the object is farther away. Light rays then pass through the aqueous humor to the lens, the principal structure of light focusing. The lens is elastic and focuses light rays on the retina. This process of light focusing based on lens elasticity is called **accommodation**. When the object is distant, the lens becomes flattened; when the object is near, the lens becomes more rounded (convex). The lens changes its shape chiefly by the activity of the ciliary muscle acting on the suspensory ligament. During accommodation for near vision, for example, the ciliary muscle contracts and releases tension of the suspensory ligament, thereby allowing the lens to become more convex due to its natural elasticity and tendency to assume the shape of a sphere.

The lens, cornea, aqueous humor, and vitreous humor all represent **refracting media**, which are media that focus light rays and cause them to converge on the fovea centralis of the retina where the image is formed. The area served by an eye is called the **external visual field**. The external visual field usually overlaps that of the other eye, and this overlap is responsible for a three dimensional image. The extrinsic muscles of the eye bring about the movements of the eyeball that help one see a single image. Persons having **strabismus** (cross-eyes) have a condition in which the eyes do not converge together, and the person sees two images instead of one.

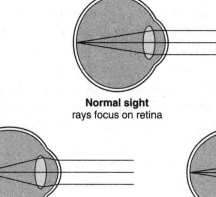

**Normal sight**
rays focus on retina

**Nearsightedness**
rays focus in front of retina

**Concave lens**
corrects nearsightedness

**Farsightedness**
rays focus behind retina

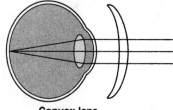

**Convex lens**
corrects farsightedness

**Astigmatism**
rays do not focus

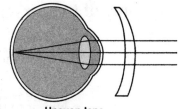

**Uneven lens**
corrects astigmatism

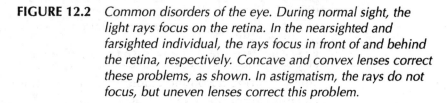

**FIGURE 12.2**  *Common disorders of the eye. During normal sight, the light rays focus on the retina. In the nearsighted and farsighted individual, the rays focus in front of and behind the retina, respectively. Concave and convex lenses correct these problems, as shown. In astigmatism, the rays do not focus, but uneven lenses correct this problem.*

## Visual Disorders

Two common disorders of the eye are myopia and hyperopia. In **myopia** (or **nearsightedness**), the image forms in front of the retina (Figure 12.2). This condition is due to elongation of the eyeball, or it may be caused by a lens that does not adjust sufficiently during accommodation. Glasses with biconcave lenses are used to focus the image on the retina. In **hyperopia** (**farsightedness**), the image forms behind the retina and is blurred because the eyeball is too short or because the lens is too flat to permit nearby vision. This defect often happens as the lens loses elasticity with age. Glasses with biconvex lenses are used to focus the image on the retina.

Another disorder of the eye is called **astigmatism**, caused by a lens or cornea that is curved irregularly. This irregular curvature results in a light refraction so that the rays fall on different areas of the retina, thereby producing a blurry image. Astigmatism is the inability to separate two closely placed points. The condition is correctly by using cylinder-shaped lenses.

Another visual defect is **colorblindness** resulting from the inability of cones to react to certain colors of the spectrum. For example, a person may be colorblind to red and green colors. In this case, red and green cannot be distinguished because of the lack of cones sensitive to red and green. Color blindness is usually a sex-linked genetic trait carried by females and expressed in males.

## THE EAR

The ear is the organ of hearing in the human body. Its purpose is to gather sound waves from the environment and transmit them to nerves in the inner ear. Here the sound waves are transformed into nerve impulses for transmission to the brain. The major region of hearing in the brain is the cortex of the temporal lobe of the cerebrum.

## Anatomy of the Ear

The ear consists of three major portions: the external ear, the middle ear, and the internal ear. The **external ear** contains the outer ear structure and is called the **pinna**, or auricle (Figure 12.3). Within the pinna lies the **external auditory canal** through which sound vibrations pass. The entry to the external auditory canal is called the **external auditory meatus**. At the proximal end of the external auditory canal is a membranous structure called the **tympanic membrane**, or **eardrum**.

The **middle ear** contains the middle ear bones, also called **ossicles**. They are the **malleus**, or hammer; the **incus**, or anvil; and the **stapes**, or stirrup. The stapes is connected at its proximal end to a membrane called the **oval window**, which leads to the inner ear.

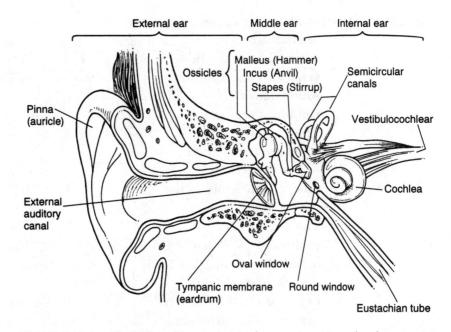

**FIGURE 12.3**    *The ear and its anatomical structures. Note the three main portions of the ear and the associated structures.*

A long, slender tube called the **Eustachian tube** leads from the pharynx to the middle ear. This canal is used to help maintain air pressure on both sides of the eardrum. When a person travels to a higher altitude, dense air is trapped in the middle ear. This dense air passes down the Eustachian tube in the pharynx, and a "pop" is felt. When a person travels to a lower altitude, low density air is trapped in the middle ear, and dense air travels up the Eustachian tube to equalize the pressure on both sides of the tympanic membrane.

The internal ear contains a large snail-like structure called the **cochlea**. Within the cochlea, there is a fluid called **perilymph**. Vibrations of the perilymph induce the nerve impulses that produce sound (Table 12.2) as discussed below.

## Physiology of Hearing

Hearing is the perception of sound vibrations made by an object and transformed into sound waves. The medium of vibration in hearing is air vibrations, which set up sound waves having pitch, intensity, and timbre. **Pitch** (frequency) is the number of vibrations per unit of time, often expressed as cycles per second. The **intensity** (strength) of sound waves varies with the height or amplitude of the sound wave, and it is usually expressed in decibels. The **timbre** (quality) of sound waves depends upon a single tone's extra constituents called overtones. Overtones vary with the object producing the sound.

The sense of hearing involves mechanical actions as sound waves are transformed to mechanical impulses. The sound waves enter the external auditory meatus and pass through the external auditory canal and impinge on the tympanic membrane. The energy of the sound waves causes the tympanic membrane to vibrate, and the vibrations are transferred directly to the three middle ear ossicles. The malleus, incus, and stapes vibrate in sequence as the sound waves are transmitted.

The innermost ossicle, the stapes, is connected to the **oval window**, a membrane at the opening of the cochlea. As the oval window vibrates, it sends vibrations to the perilymph in the cochlea. The perilymph vibrations are transmitted through the roof of the cochlea to the **organ of Corti**. The organ of Corti contains the dendrites of neurons that form the cochlear division of the acoustic nerves. These dendrites are arranged around **hair cells** of the organ of Corti. As the hair cells vibrate, they move against membranes, which stimulate impulses to arise in the dendrites. The impulses then travel over the cochlear branch of the acoustic nerve to the brain's temporal lobe, where interpretation of the sound waves is made.

# OTHER SENSES

Seeing and hearing are but two of the many senses present in the human body. Several of the other senses include taste, smell, various types of touch sensation, and the sense of balance.

## Taste

The sense of taste is called the **gustatory sense**. It is a chemical sense, in which the substances tasted are dissolved in fluid. Then the molecules can be detected by the taste buds on the tongue (Figure 12.4).

The **taste buds** are located on the upper surface of the tongue within tiny elevations called **papillae**. Other papillae are found on the soft palate, the walls of the pharynx, and in surrounding regions, but they are of minor importance compared to those found on the tongue.

The four basic or primary tastes are **sweet, sour, bitter**, and **salty**. The posterior portion of the tongue is sensitive to molecules stimulating the bitter taste, while those of sour taste stimulate the anterior lateral portions of the tongue, and those of salty taste stimulate the posterior lateral portions of the tongue. Salty and sweet tastes are detected on the extreme anterior portion of the tongue.

Molecules enter taste pores of the papillae and stimulate the specialized gustatory (taste) cells of the taste buds. These cells send impulses over sensory nerve fibers to a branch of the **facial nerve** or

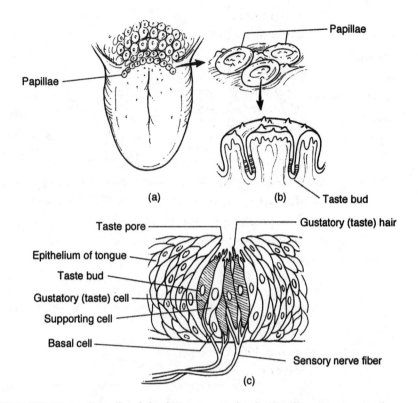

**FIGURE 12.4**   *Details of the human taste buds. Papillae (a) contain the taste buds (b) at their base. An individual taste cell within the taste bud (c) has numerous supporting and connective tissue, and each taste cell is associated with a sensory nerve fiber that delivers impulses to the brain for interpretation.*

the **glossopharyngeal nerve** and on to the brain. Impulses from these nerves pass through the medulla where they synapse with neurons leading to the thalamus. Neurons from the thalamus carry impulses to the temporal lobe of the cerebral cortex, where the taste stimuli are interpreted.

## Smell

The sense of smell is referred to as the **olfactory sense**. It is a chemical sense that requires contact between the nerve receptors and molecules of the substance sensed.

During smell, fine particles of substances enter the nose and stimulate special olfactory cells in the mucous membrane of the nose's uppermost portion. When stimulated, the receptors form impulses that leave the nose region over branches of the **olfactory nerve** (Figure 12.5). This nerve enters the skull through the cribiform plate of the ethmoid bone and passes through the olfactory bulb in the olfactory tract, which leads to the frontal and temporal lobes of the cerebrum. Interpretation of the stimuli is made here.

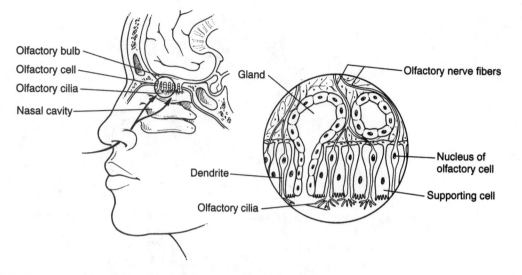

**FIGURE 12.5**    *The sense of smell. Molecules strike the olfactory cilia of the olfactory cells in the upper portion of the nose. These cells are nerve cells whose impulses are sent to the brain.*

The body can detect an infinite variety of smells. The mechanism of olfaction is unclear, but it is known that olfactory cells can be fatigued, and the awareness of an odor diminishes.

## Touch and Related Senses

The sense of touch and related senses such as pain, pressure, and vibration require receptors in the skin, muscles, joints, and visceral organs. Several types of receptors function in touch and related sensations: **Free nerve endings** in the skin, for example detect pain. Other receptors called **Merkel's disks** are touch receptors in the skin. Still other receptors called **Meisner's corpuscles** detect light and touch, and **Pacinian corpuscles** are receptors that detect pressure and vibrations in the skin. Sensations are relayed via impulses to the brain for interpretation (Figure 12.6).

## Equilibrium

The sense of equilibrium is derived from activity within the inner ear (where hearing also occurs). The inner ear contains a series of passageways and canals located within the temporal bone and known as a labyrinth. The labyrinth of the inner ear is divided into two portions: a **membranous labyrinth** and a **bony labyrinth**. The membranous labyrinth is enclosed within the bony labyrinth. The bony labyrinth is a maze of chambers hollowed out of the temporal bone and where the cochlea, vestibule, and semicircular canals are located. It is filled with perilymph, which bathes the membranous

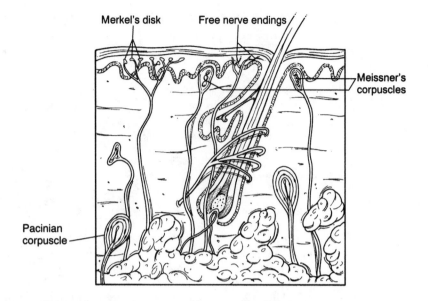

**FIGURE 12.6**  *A selection of touch senses in the human skin.*

labyrinth and is similar to cerebrospinal fluid. The membranous labyrinth contains endolymph, which is similar to intercellular fluid.

Within the bony labyrinth are three looped structures called **semicircular canals**. The semicircular canals contain endolymph and are connected to the cochlea at a region called the **vestibule**. Within the vestibule are two outgrowths called **utricle** and the **saccule**. A tiny canal joins the utricle and saccule together. The utricle and semicircular canals are concerned with the sense of equilibrium.

Each of the semicircular canals lies at right angles to the others, and each connects with the utricle. At their point of connection with the utricle, each canal has an enlarged portion called an **ampulla** (Figure 12.7). Within the ampulla lie a cluster of sensory hair cells. When the position of the head changes, movement of the endolymph in the semicircular canals stimulates the hair cells, and these stimulations generate impulses in the local nerve fibers. The nerve fibers carry the impulses along a branch of the **acoustic nerve** to the brain. The brain sends motor impulses to the muscle cells that make the adjustments to allow the body to remain in its state of balance, or dynamic equilibrium.

Smaller degrees of movement, such as those involved in maintaining **posture**, or static equilibrium, arise from a slightly different mechanism. Within the saccule and the utricle, there are tiny structures called **maculae**. Each macula is composed of hair cells and a membrane containing small bits of calcium carbonate called **otoliths** (ear stones). When the head changes position slightly, pressure on the membrane causes the otoliths to shift their position and exert a force on the hair cells. The hair cells then initiate impulses that travel over the acoustic nerve to the brain. The brain then makes the adjustments via motor impulses that maintain the equilibrium necessary for posture.

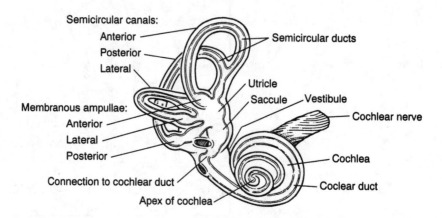

**FIGURE 12.7** *Details of the anatomy of the bony labyrinth where the sense of equilibrium originates. The bony labyrinth is a space in the temporal bone containing the semicircular canals, vestibule, and cochlea. Movement of the endolymph in the semicircular canals stimulates hair cells in the ampullae, and the stimulations are sent to the brain, which dispatches impulses to the muscles to adjust body movements.*

# REVIEW QUESTIONS

*PART A—Completion:* **Add the word or words that correctly complete each of the following statements.**

1. The purpose of the eye is to gather light from the environment and form an image on nerve cells of the _____ .

2. The nerve that carries impulses from the eye to the brain is the _____ .

3. The inner coat of the wall of the eye is composed of the _____ .

4. The outer wall of the eye consists of the cornea and the _____ .

5. The thin, watery fluid in the anterior chamber of the eye is _____ .

6. The jellylike substance that fills the vitreous chamber of the eye is called _____ .

7. The posterior chamber of the eye lies between the iris and the _____ .

8. The iris is composed of two layers of _____ .

9. The pupil of the eye is an opening in the portion of the eye known as the _____ .

10. The transparent biconcave lens is found behind the _____ .

11. The suspensory ligament attaches the lens to the structure called the _____ .

12. The nervous layer of the retina consists of three layers of _____ .

13. The eyelids are covered on their surfaces by the mucous membrane called the _____ .

14. Tears that bathe the eyeball and keep it moist are produced by the _____ .

15. Twilight vision is concerned with those retinal cells known as _____ .

16. Daylight vision and close, detailed vision are permitted by those retinal cells known as _____ .

17. The place where most cones are concentrated is the _____ .

18. The retinal cells in their greatest number at the outer edge of the retina are _____ .

19. The optic disc contains no visual receptors and is therefore called the _____ .

20. The lobe of the brain where visual patterns are interpreted is the _____ .

21. The process of light focusing due to the elasticity of the lens is called the _____ .

22. The change of lens shape to focus objects at various distances is under the control of a muscle called the _____ .

23. Persons having a condition in which the eyes do not converge together suffer from _____ .

24. The correct term for nearsightedness is _____ .

25. Nearsightedness can be corrected by utilizing glasses having lenses that are _____ .

26. The condition of farsightedness is correctly known as _____ .

27. Farsightedness can be corrected with glasses having lenses that are _____ .

28. Irregular curvature of the lens or cornea results in a disorder known as _____ .

29. The sex-linked genetic trait in which a person cannot detect certain colors in called _____ .

30. The major lobe of the brain in which hearing takes place is the _____ .

31. The technical name for the eardrum is the _____ .

32. The middle ear bones, which transmit sound to the inner ear, are known as the malleus, incus, and _____ .

33. The long, slender tube leading from the pharynx to the middle ear is the _____ .

34. The snail-like structure of the internal ear is called the _____ .

35. Three qualities of sound waves are intensity, pitch, and _____ .

36. Sound waves are transmitted by the middle ear bones to the membranous covering of the cochlea called the _____ .

37. At the roof of the cochlea, there exists the dendrites of neurons in a structure called the _____ .

38. Vibrations are transmitted to the brain for interpretation over the cochlear branch of the nerve called the _____ .

39. The sense of taste is technically known as the _____ .

40. Taste buds are located on the upper surface of the tongue within tiny elevations called _____ .

41. The four primary tastes are sweet, sour, salty, and _____ .

42. The sensations of salt and sweet are detected on the portion of the tongue that is _____ .

43. Impulses of taste are transmitted to the brain over the facial nerve or the _____ .

44. The sense of smell is called the _____ .

45. The interpretations of smell are made in the temporal lobe of the brain and in the _____ .

46. Touch receptors in the skin are known as _____ .

47. Receptors that detect pressure and vibrations in the skin are called the _____ .

48. The sense of equilibrium is associated with the canals and passageways found in the _____ .

49. Each of the semicircular canals used in equilibrium connects with an outgrowth called the _____ .

50. Maintaining posture depends on impulses arising in the semicircular canal and using small bits of calcium carbonate called _____ .

**PART B—Multiple Choice: Circle the letter of the item that correctly completes each of the following statements.**

1. The outer layer of the wall of the eye is composed of the
   (A) iris and ciliary bodies
   (B) retina
   (C) cornea and sclera
   (D) rod cells and cone cells

2. The vitreous humor of the eye may be found between the
   (A) iris and lens
   (B) pupil and iris
   (C) retina and fovea
   (D) lens and retina

3. The iris is composed of
   (A) pigmented blood vessels
   (B) two layers of smooth muscle
   (C) nerve cells
   (D) connective tissue cells called astrocytes

4. The color of a person's eye depends upon which pigments are contained in the
   (A) iris
   (B) retina
   (C) sclera
   (D) cornea

5. All the following are components of the retina of the eye except
   (A) bipolar neurons
   (B) ganglionic neurons
   (C) cone cells
   (D) ciliary muscles

6. The conjunctiva is a mucous membrane that
   (A) covers the sclera and extends to the retina
   (B) covers the eyeball and lines the eyelids
   (C) covers the rod cells but not the cone cells
   (D) is found only in the posterior chamber of the eye

7. Tears are produced by a set of glands next to the eyeball and known as the
   (A) ethmoid glands
   (B) lacrimal glands
   (C) submandibular glands
   (D) suspensory glands

8. The rod-shaped cells of the retina permit vision
   (A) where there is a large amount of light
   (B) in color
   (C) where there is poor light
   (D) only at the fovea

9. The highest concentration of cone-shaped cells is at the
   (A) ciliary body
   (B) iris
   (C) fovea centralis
   (D) temporal lobe

10. The highest concentration of rod-shaped cells is at the
    (A) suspensory ligament
    (B) outer edge of the retina
    (C) occipital lobe
    (D) blind spot

11. An inverted image is converted into an upright image at the
    (A) temporal lobe of the brain
    (B) blind spot
    (C) occipital lobe of the brain
    (D) optic disk

12. The process of accommodation is due to the
    (A) changing shape of the lens
    (B) replacement of cone cells by rod cells
    (C) formation of images in front of the retina
    (D) development of color blindness

13. Persons having the condition called strabismus have a condition in which
    (A) they cannot see red or green color
    (B) light rays focus behind the retina
    (C) the optic disk is damaged
    (D) the eyes do not converge together

14. A person who suffers from nearsightedness can be assisted by glasses having
    (A) flat lenses
    (B) biconvex lenses
    (C) biconcave lenses
    (D) lenses that bring images together

15. A person suffering from farsightedness can be helped by wearing glasses that have
    (A) biconvex lenses
    (B) biconcave lenses
    (C) flat lenses
    (D) lenses that bring images together

16. In a person who is farsighted
    (A) the image forms in front of the retina
    (B) no image forms
    (C) the image forms behind the retina
    (D) the eyes are cross-eyed

17. In most individuals, color blindness is a
    (A) result of injury to the eye
    (B) genetic trait
    (C) result of astigmatism
    (D) due to poorly fitted glasses

18. All the following are parts of the outer ear except the
    (A) tympanic membrane
    (B) pinna
    (C) external auditory canal
    (D) ear ossicles

19. The Eustachian tube leads from the
    (A) middle ear to the inner ear
    (B) pinna to the cochlea
    (C) cochlea to the semicircular canals
    (D) pharynx to the inner ear

20. All the following are characteristics of sound waves except
    (A) intensity
    (B) timbre
    (C) refraction
    (D) pitch

21. Sound vibrations are conducted from the stapes to the peri-lymph of the cochlea by the
    (A) oval window
    (B) organ of Corti
    (C) bipolar neurons
    (D) pinna

22. Molecules that stimulate the sour taste are detected at the
    (A) posterior portion of the tongue
    (B) areas below the tongue
    (C) anterior portion of the tongue
    (D) nowhere on the tongue

23. The taste buds send their impulses to the brain for interpretation utilizing the
    (A) facial or glossopharyngeal nerves
    (B) vagus or accessory nerves
    (C) olfactory and optic nerves
    (D) trochlear and trigeminal nerves

24. The olfactory nerve is primarily concerned with carrying impulses from the
    (A) tongue
    (B) ear
    (C) nose
    (D) skin

25. The utricle, saccule, and vestibule are all concerned primarily with the sense of
    (A) hearing
    (B) seeing
    (C) equilibrium
    (D) pressure

**PART C—True/False: For each of the following statements, mark the letter "T" next to the statement if it is true. If the statement is false, change the underlined word to make the statement true.**

1. The nerve that transmits impulses from the eye to the brain is the <u>olfactory nerve</u>.

2. The <u>inner</u> layer of the eyeball is a highly vascular layer consisting of the choroid layer, iris, and ciliary bodies.

3. The two components of the outer layer of the eyeball are the cornea and the <u>retina</u>.

4. The posterior chamber of the eye is found between the iris and the <u>retina</u>.

5. Two layers of <u>striated</u> muscle compose the iris.

6. The lens is a transparent <u>biconcave</u> disc of fibrous protein material occurring in concentric layers.

7. A ligament called the <u>suspensory ligament</u> attaches the lens to the ciliary body.

8. The neurons that receive impulses initiated by rod and cone cells are <u>unipolar</u> neurons.

9. The mucous membrane folded over part of the eyeball and lining the eyelid is the <u>conjunctiva</u>.

10. Tears are produced in the eye to bathe the eyeball by the <u>nasal</u> apparatus.

11. The <u>rod</u> cells permit close, detailed vision and are concerned with color vision.

12. Cone cells are highly concentrated in the <u>fovea centralis</u>, a small dent near the posterior portion of the center of the retina.

13. The number of rod cells <u>decreases</u> as the distance from the fovea centralis increases.

14. The principal structure for focusing light rays on the retina is the <u>iris</u>.

15. The <u>intrinsic</u> muscles of the eye are responsible for eyeball movements that help one see a single, three dimensional image.

16. The condition called astigmatism is due to an irregular curvature of the lens or <u>retina</u>.

17. The major site for hearing in the brain is the temporal lobe of the <u>cerebellum</u>.

18. The three middle ear bones that transmit sound waves are known as the malleus, <u>ethmoid</u>, and stapes.

19. The Eustachian tube leading to the <u>inner</u> ear helps maintain air pressure on both sides of the ear drum.

20. The intensity of sound waves is usually expressed in <u>decibels</u>.

21. Perilymph vibrations in the inner ear are transmitted to dendrites located in the <u>organ of Corti</u> in the roof of the cochlea.

22. Most papillae that contain the taste buds are found on the <u>soft palate</u> of the mouth.

23. Taste stimuli pass through the <u>cerebellum</u> of the brain on their way to the temporal lobe of the cerebrum where the stimuli are interpreted.

24. The variety of smells that the body can detect is <u>four</u>.

25. The semicircular canals of the inner ear are concerned primarily with the sense of <u>equilibrium</u>.

**Answers**

## PART A—Completion

1. retina
2. optic nerve
3. retina
4. sclera
5. aqueous humor
6. vitreous humor
7. lens
8. smooth muscle
9. iris
10. pupil
11. ciliary body
12. neurons
13. conjunctiva
14. lacrimal apparatus
15. rods
16. cones
17. fovea centralis
18. rods
19. blind spot
20. occipital lobe
21. accommodation
22. ciliary muscle
23. strabismus
24. myopia
25. biconcave
26. hyperopia
27. biconvex
28. astigmatism
29. color blindness
30. temporal lobe
31. tympanic membrane
32. stapes
33. Eustachian tube
34. cochlea
35. timbre
36. oval window
37. organ of Corti
38. acoustic nerve
39. gustatory sense
40. papillae
41. bitter
42. anterior
43. glossopharyngeal nerve
44. olfactory sense
45. frontal lobe
46. Merkel's disks
47. Pacinian corpuscles
48. inner ear
49. utricle
50. otoliths

## PART B—Multiple Choice

| | | | | |
|---|---|---|---|---|
| 1. C | 6. B | 11. C | 16. C | 21. A |
| 2. D | 7. B | 12. A | 17. B | 22. C |
| 3. B | 8. C | 13. D | 18. D | 23. A |
| 4. A | 9. C | 14. C | 19. D | 24. C |
| 5. D | 10. B | 15. A | 20. C | 25. C |

## Part C—True/False

1. optic nerve
2. middle
3. sclera
4. lens
5. smooth
6. biconvex
7. true
8. bipolar
9. true
10. lacrimal
11. cone
12. true
13. increases
14. lens
15. extrinsic
16. cornea
17. cerebrum
18. incus
19. middle
20. true
21. true
22. tongue
23. thalamus
24. infinite
25. true

# CHAPTER 13

# THE ENDOCRINE SYSTEM

The human endocrine system is composed of several **endocrine glands** that secrete hormones into the blood. The blood transports the hormones to their target cells where they bring about biochemical and physiological changes in the metabolic activities of cells. For example, hormones stimulate growth and development; they promote water retention; they raise or lower the level of blood glucose; they promote sodium retention; and they induce the development of male sex characteristics. Hormones bind to specific cell membrane receptors or to intracellular receptors within the cells.

Endocrine glands produce three types of hormones: steroid hormones, protein hormones, and amine hormones. **Steroid hormones** are lipids synthesized from cholesterol (Table 13.1). They are complex rings of carbon and hydrogen atoms that include cortisol, cortisone, estrogen, progesterone, and testosterone.

**Protein** (or, peptide) **hormones** are composed of amino acids bound by peptide bonds. Most body hormones are proteins including insulin from the pancreas, calcitonin from the thyroid gland, and hormones from the pituitary gland. **Amine hormones** have amino groups associated with the molecule. Although the hormones are derived from amino acids, there are no peptide bonds in the molecule. Examples of amine hormones are thyroxine from the thyroid gland and epinephrine from the adrenal gland.

Researchers have found that steroid hormones dissolve in phospholipids and readily pass through cell membranes. In the cell cytoplasm, the hormones combine with proteins, and the complex stimulates the activity of specific genes, which encode certain types of messenger RNA molecules. This activity leads to the synthesis of certain proteins that bring about alterations in the cell.

**TABLE 13.1**   *Two Main Types of Hormones*

| Type of Hormone | Examples |
| --- | --- |
| Steroid | Estrogens, testosterone, aldosterone, cortisol |
| Nonsteroid | |
|   Amines | Norepinephrine, epinephrine |
|   Peptides | ADH, oxytocin |
|   Proteins | Insulin, somatotropin, prolactin |
|   Glycoproteins | FSH, LH, TSH |

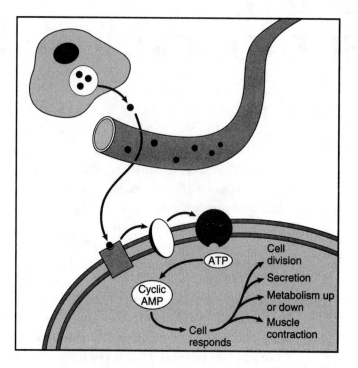

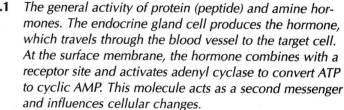

**FIGURE 13.1** *The general activity of protein (peptide) and amine hormones. The endocrine gland cell produces the hormone, which travels through the blood vessel to the target cell. At the surface membrane, the hormone combines with a receptor site and activates adenyl cyclase to convert ATP to cyclic AMP. This molecule acts as a second messenger and influences cellular changes.*

Many of the protein (peptide) and amine hormones act as **first messenger** molecules and attach to receptor sites at the cell membranes of target cells. From these receptor sites, they increase the activity of certain enzymes in the cell membranes. The principal enzyme affected is called **adenyl cyclase** (Figure 13.1). Adenyl cyclase then converts ATP molecules in the cell to molecules of cyclic adenosine monophosphate (cAMP). This molecule, known as the **second messenger**, disperses throughout the cell and influences cellular changes such as increased protein synthesis, altered membrane permeability, and enzyme activations.

Certain endocrine glands exhibit an autocrine or a paracrine effect. The **autocrine** effect is one in which the hormone secreted by the endocrine gland acts on cells of the gland that produced it. The **paracrine** effect is one in which the hormone produced by an endocrine gland acts on cells that are not their normal target cells.

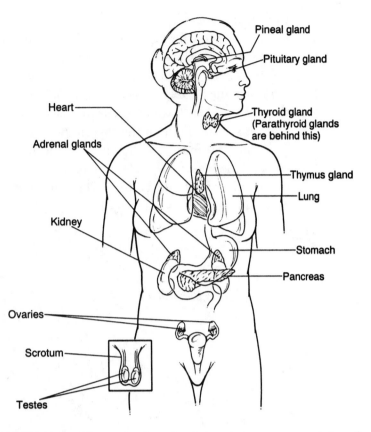

**FIGURE 13.2** *The endocrine glands in relation to some of the other major body organs.*

## PITUITARY GLAND

The pituitary gland is a pea-sized structure lying in the inferior aspect of the brain in a saddle of the sphenoid bone called the sella turcica (Figure 13.2). The gland has an anterior lobe called the **adenohypophysis**, and a posterior lobe called the **neurohypophysis**. The neurohypophysis is not really an endocrine gland, but rather a reservoir for hormones produced by the hypothalamus, which lies above it. A stalk called the infundibulum connects the pituitary gland to the floor of the hypothalamus.

The anterior lobe (adenohypophysis) produces several important hormones of the body. The first hormone, **human growth hormone (HGH),** is also known as **somatotropin**, or somatotropic hormone (STH). HGH accelerates body growth by stimulating the cellular uptake of amino acids and proteins by encouraging protein synthesis and by mobilizing fats. The hormone is a protein composed of 191 amino acids. Its release is inhibited by high levels of HGH in the blood.

Inadequate secretion of HGH in childhood results in pituitary **dwarfism**, while oversecretion during childhood leads to **gigantism**. Oversecretion of HGH in adults results in **acromegaly**, a condition in which a person's appearance changes because of thickening bones and growth of soft tissues. These physiological changes occur primarily in the face, hands, and feet.

The second hormone of the anterior pituitary is **thyroid-stimulating hormone (TSH)**. This hormone affects the development of the thyroid gland and stimulates its uptake of iodine. It also controls the synthesis and release of the thyroid hormones.

A third hormone of the anterior pituitary gland is **adrenocorticotropic hormone (ACTH)**. The target tissue of ACTH is the cortex of the adrenal gland. Here, ACTH influences growth of the tissue and stimulates the secretion of glucocorticoids.

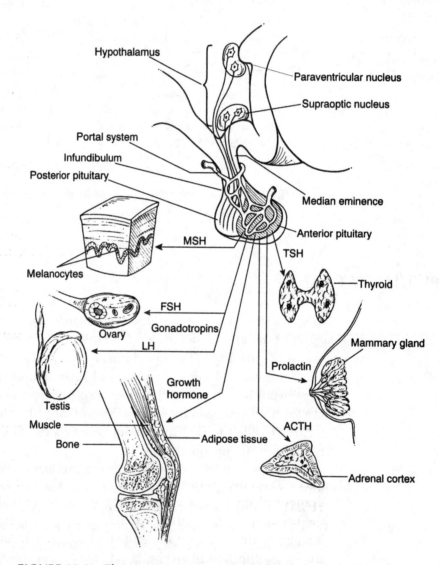

**FIGURE 13.3**   *The anterior pituitary gland seen from the left lateral aspect. The hormones and their target organs are illustrated. Note the relationship of the gland to the hypothalamus and its blood supply. The hormones of the posterior pituitary gland are not shown.*

**TABLE 13.2**   *Hormones of the Human Pituitary Gland*

| Hormone | Target Organ(s) | Principal Effects |
|---|---|---|
| *Anterior lobe:* | | |
| Follicle-stimulating hormone (FSH) | Ovarian follicle and seminiferous tubules of testis | Stimulates growth of follicles and tubules |
| Luteinizing hormone (LH) or Interstitial-cell-stimulating hormone (ICSH) | Corpus luteum and interstitial cells of testis | Stimulates formation of corpus luteum from follicle; stimulates production of progesterone by the ovaries and testosterone by the testes |
| Prolactin (or) Lactogenic hormone (LTH) | Female mammary gland | Stimulates milk production after parturition |
| Somatotropic hormone (STH) | Body tissues | Increases synthesis of protein |
| Growth hormone (HGH) | Body tissues | Increases synthesis of protein |
| Adrenocorticotropic hormone (ACTH) | Cortex of adrenal gland, skin, liver, mammary gland | Stimulates production of corticosteroids; increases metabolic rate, glycogen deposition in liver, darkening of skin, milk production |
| Thyroid-stimulating hormone (TSH) | Thyroid gland | Stimulate production of thyroxine |
| Melanocyte-stimulating hormone (MSH) | Skin melanocytes | Controls skin pigmentation |
| *Posterior lobe:* | | |
| *Vasopressin (or) antidiuretic hormone (ADH) | Smooth muscle, especially of arterioles and kidney tubules | Constricts blood vessels, thus raising blood pressure; stimulates reabsorption of water by kidney |
| *Oxytocin | Uterus and ducts of mammary gland | Stimulates contraction of uterus if primed by ovarian hormones; enables milk-letdown reflex by mammary gland |

* Produced by the hypothalamus

The fourth hormone of the anterior pituitary is **prolactin**, also known as the **lactogenic hormone (LTH)**. The hormone acts on the mammary glands, where it stimulates the production of milk in the lactating female.

The next hormone is called **follicle-stimulating hormone (FSH)**. FSH acts on the ovaries and testes. In the female, it stimulates development of the follicle in the ovary, while in the male it stimulates the production of sperm cells. FSH is one of the **gonadotropins**, the hormones that regulate the gonads (sex organs). Another gonadotropin of the anterior lobe is **luteinizing hormone (LH)**. In the female, LH stimulates maturation of follicle cells and promotes ovulation. It also stimulates the corpus luteum to secrete estrogens and progesterone (Chapter 23). In the male, LH stimulates the cells of the testes to produce testosterone (Figure 13.3).

The posterior lobe of the pituitary gland (neurohypophysis) stores two important hormones originating in the **hypothalamus**, the region of the brain above the pituitary. The pituitary gland is connected to the hypothalamus by a stalk of tissue known as the **infundibulum**. A nerve bundle connects the posterior pituitary with the tissue of the hypothalamus and runs through the infundibulum. Neurosecretory cells of the hypothalamus synthesize the two neurohormones transported down the nerve bundle to the posterior pituitary gland for temporary storage and release on demand in response to nervous stimuli from neurons of the hypothalamus.

The first hormone released from the posterior pituitary is **antidiuretic hormone (ADH)**, also known as **vasopressin**. ADH acts on the kidney tubules where it facilitates the elevation of blood pressure by regulating the reabsorption of water. Hyposecretion results in **diabetes insipidus**, characterized by excessive urine production and excessive thirst.

The second hormone of the posterior pituitary is **oxytocin** (Table 13.2). The target tissue of oxytocin is the smooth muscle of the uterus and mammary glands where it stimulates uterine contractions and milk secretion, respectively. Secretion is regulated in part by suckling.

## THYROID GLAND

The thyroid gland is located in the neck tissues near the trachea. It consists of two lateral lobes interconnected by an isthmus. The apex of each lobe is lateral to the lower third of the thyroid cartilage of the larynx, and the base is lateral to the upper portion of the trachea. The functional units are cluster of cells called follicles (Figure 13.4).

The thyroid gland secretes at least three hormones: thyroxine, triiodothyronine, and calcitonin. **Thyroxine** and **triiodothyronine** increase the rate of protein synthesis in body cells and the rate of energy release of these cells. They also regulate the individual's rate of growth, and they stimulate the maturity of the nervous system. The synthesis of the two hormones is regulated by TSH from the anterior pituitary gland.

The thyroid hormones accelerate the rate of cellular metabolism in all parts of the body. The hormones stimulate the activity of enzymes associated with the metabolism of glucose and thereby increase the basal metabolic rate, as well as the amount of oxygen consumed in the body cells and the amount of heat produced. The hormones also regulate the increase in the number of receptors in blood vessels and thus play a role in maintaining blood pressure.

In order to produce thyroxin and triiodothyronine, **iodine** atoms must be available in the diet. If unavailable, the thyroid glands swell with fluid, and an enlargement of the neck occurs. This enlargement is called **goiter**. Adding iodine to the diet relieves the goiter.

Insufficient secretion of thyroxin in infants and children results in **cretinism**. Symptoms of cretinism include stunted growth, thickened facial features, abnormal bone growth, mental retardation, and general lethargy. The condition is treated with hormone therapy. An excess of thyroxin and triiodothyronine leads to **Graves' disease** (exophthalmic goiter). Often this condition is due to a tumor in the thyroid gland. Symptoms include loss of weight, rapid pulse, increased appetite, and increased metabolic rate. Bulging eyes (exophthalamus) is sometimes seen. Insufficient secretion of thyroxin in adults results in **myxedema**. Symptoms of this disease include weight gain, slow pulse, decreased metabolic rate, lack of energy, and general weakness.

The third hormone of the thyroid gland is **calcitonin** (Table 13.3). Calcitonin regulates the level of calcium in the blood and is antagonistic to the hormone from the parathyroid gland. Calcitonin lowers the blood calcium and increases the level of calcium in the bone, while the parathyroid hormone does the reverse.

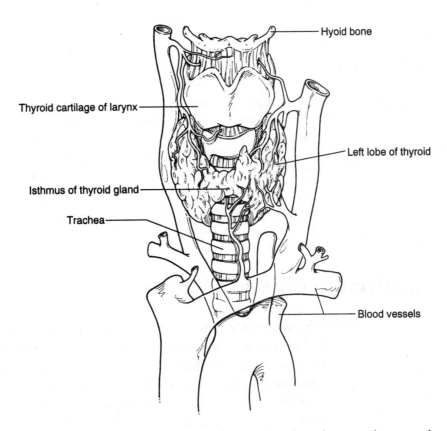

**FIGURE 13.4**   *The thyroid gland seen in place from the ventral aspect of the body. Two lateral lobes of the gland are connected by an isthmus.*

**TABLE 13.3**  *Major Endocrine Glands and Their Secretions*

| Gland | Location | Products | Target Organs or Function |
|---|---|---|---|
| Pituitary gland | Below the brain | See Table 10.2 | |
| Pineal gland | Upper surface of the brain | Melatonin | Reproductive organs, especially ovaries |
| Thyroid gland | In front of and below voice box (larynx) | Thyroxine Triiodothyronine | Body tissue cells |
| | | Calcitonin | Rapidly increases calcium deposition in bones |
| Parathyroid glands | Embedded in posterior surface of thyroid gland | Parathormone | Slowly increases number and size of osteoclasts and so increases calcium concentration in the blood |
| Thymus gland | Thorax tissue | Lymphocyte-maturing hormone | Lymphocytes |
| Pancreas (islets of Langerhans) | In fold of stomach | | |
| beta cells | | Insulin | Facilitates cellular glucose uptake, especially in the liver |
| alpha cells | | Glucagon | Facilitates breakdown of glycogen in liver and release of glucose into blood |
| Adrenal gland | Upper surface of the kidney | | |
| medulla | Inner core of gland with nervous origin | Epinephrine Norepinephrine | Functions in fight-or-flight reaction (emergency response) |
| cortex | Outer part of gland | i. glucocorticoids | Regulate sugar and protein metabolism |
| | | ii. mineralocorticoids | Regulate sodium and mineral balance |
| | | iii. sex hormones | Affect secondary sex hormones |

# PARATHYROID GLANDS

The parathyroid glands are four tiny masses of gland tissue located on the posterior surface of the thyroid gland, each about the size of an apple seed. The glands secrete a hormone called **parathyroid hormone**, also known as **parathormone**. Parathyroid hormone regulates the level of calcium in the blood by increasing the calcium level (Figure 13.5). It acts antagonistically to calcitonin from the thyroid gland and decreases the level of calcium in the bones by stimulating the activity of osteoclasts. It also influences the reabsorption of calcium in the kidney tubules and the intestinal mucosa.

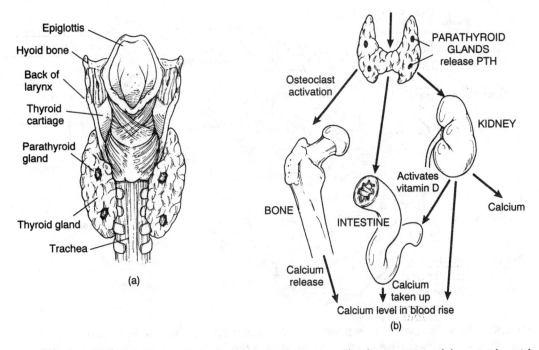

**FIGURE 13.5**   *The parathyroid gland and its hormone. (a) The four masses of the parathyroid gland in place on the dorsal surface of the thyroid gland. (b) When the blood level of calcium is low, the parathyroid gland releases parathyroid hormone (PTH). The hormone increases blood calcium levels by its action on the bone, kidney tubules, and intestine.*

The discovery of the activity of parathyroid glands was related to their removal during routine thyroid surgery. Physicians noted in some cases that uncontrolled muscle spasms occurred in some patients and found that calcium metabolism was upset. Locating the four parathyroid glands and understanding their activity resolved the problem. Parathyroid hormone affects the activation of vitamin D, which then regulates intestinal absorption of calcium. The conversion of vitamin D to its active form is accomplished in the kidneys under the control of the hormone. Conditions related to oversecretion of the thyroid hormone generally arise from a tumor in the gland. Softened and deformed bones are characteristic signs.

## PANCREAS

The pancreas is a large flattened glandular organ near the stomach in the abdominal cavity. It lies in the fold of the stomach (Figure 13.6) and has both digestive and endocrine functions. Its digestive functions consist of producing enzymes for the digestive process; its endocrine function consists of producing two hormones: insulin and glucagon.

**Insulin** is a protein hormone composed of 51 amino acids in two chains. It acts throughout the body and promotes the movement of glucose molecules into body cells, thereby decreasing the level of

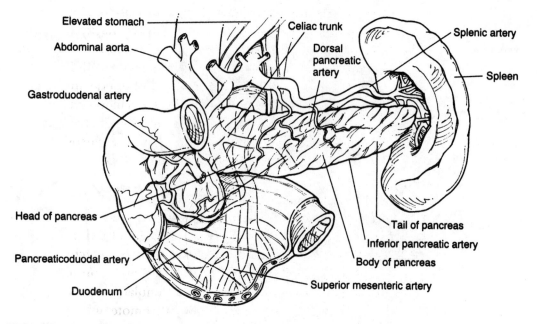

**FIGURE 13.6** *The pancreas in position in the abdominal cavity as seen from the ventral aspect. The stomach is elevated to the left to show the pancreas underneath.*

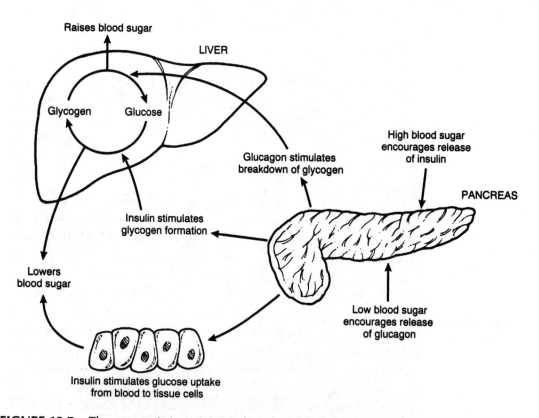

**FIGURE 13.7** *The antagonistic activities of insulin and glucagon, two hormones of the pancreas. When blood sugar is high, insulin affects the tissue cells and liver to remove glucose from the blood. Low blood sugar stimulates glucagon release, and glucagon affects the liver to increase the blood glucose level.*

blood glucose. Insulin is produced by **beta cells** of the **islets of Langerhans** of the pancreas. When these cells are inactive, insulin is lacking in the body, and a condition called insulin-dependent **diabetes mellitus** results. Insufficient glucose enters the cells for metabolism, and this insufficiency results in a lack of energy in the body and a tired feeling. The kidney expels the excess blood glucose into the urine. Also, since there is a high concentration of glucose in the urine, the kidneys expel much water to dilute the glucose. Therefore, the diabetic urinates frequently, which results in excessive thirst.

The second hormone of the pancreas is **glucagon** (Figure 13.7). Glucagon is produced by the **alpha cells** of the pancreas. Glucagon stimulates the breakdown of glycogen (glycogenolysis) in the liver. This breakdown results in molecules of glucose, which are then released to the blood. Thus, glucagon brings about an increase in the level of blood glucose, while insulin stimulates a decrease. The hormone also acts through cyclic AMP to promote gluconeogensis, the formation of glucose from amino acid and fatty acid molecules. The effect is to remove amino acids from the blood.

## ADRENAL GLANDS

The adrenal glands are located on the superior borders of the kidneys, one gland per kidney. There are two major portions of each gland: the **medulla**, which is the inner tissue of the adrenal gland, and the **cortex**, which is the outer tissue of the gland. The medulla secretes hormones that complement the action of the sympathetic nervous system, while the cortex secretes hormones that help regulate the mineral balance, energy balance, and reproductive functions of the body.

The hormones of the adrenal cortex are referred to as mineralocorticoids and glucocorticoids. The important **mineralocorticoids** include **aldosterone** (Figure 13.8). In the blood and body fluid, mineralocorticoids regulate the concentration of electrolytes, especially sodium and potassium. The secretion of mineralocorticoids is regulated by the electrolyte concentration in the blood. The **glucocorticoids**, by comparison, affect the metabolism of carbohydrates, proteins, and fats. They also promote constriction of the blood vessels and serve as antiinflammatory compounds. The important glucocorticoids include **cortisol** (cortisone). The secretion of glucocorticoids is regulated by adrenocorticotropic hormone (ACTH) from the anterior pituitary gland by a negative feedback mechanism.

The cortex of the adrenal glands also produces **steroid hormones** that influence the sexual characteristics. These hormones supplement the sex hormones from the gonads.

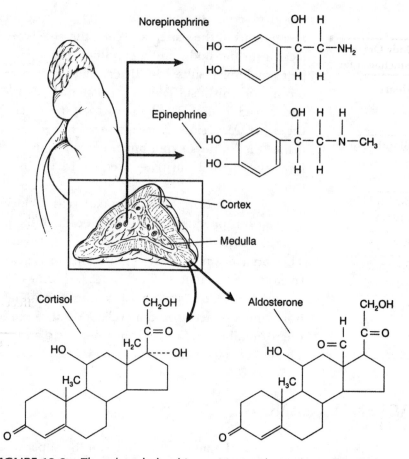

**FIGURE 13.8**   *The adrenal gland in position at the surface of the kidney, the two main regions of the adrenal gland, and representative hormones from each region. Cortisol and aldosterone are steroid hormones with the typical steroid configuration of atoms.*

The medulla of the adrenal glands secretes a series of amine hormones known as **catecholamines**. The hormones include **epinephrine** and **norepinephrine** (Table 13.4). Both hormones function in preparing the body for greater physical performance, the "fight or flight" response. In this regard, both hormones work with the sympathetic nervous system.

Various diseases arise from inadequate or overadequate secretions of adrenal cortex hormones. **Addison's disease**, resulting from inadequate secretions, is accompanied by sodium and potassium imbalance, a bronze skin tone, dehydration, hypotension, and general weakness. Oversecretion results in **Cushing's syndrome**, accompanied by a puffy face, hypertension, and general muscle weakness.

**TABLE 13.4**   *A Comparison of Epinephrine and Norepinephrine*

| Body Organ or Function Affected | Epinephrine | Norepinephrine |
|---|---|---|
| Heart | Rate increases<br>Force of contraction increases | Rate increases<br>Little or no effect on force of contraction |
| Blood vessels | Vessels in skeletal muscle vasodilate, thereby decreasing resistance to blood flow | Vessels in skeletal muscles vasoconstrict, thereby increasing resistance to blood flow |
| Systemic blood pressure | Some increase due to increased cardiac output | Great increase due to vasoconstriction |
| Airways | Dilation | Less effect |
| Reticular formation of brain | Activated | Little effect |
| Liver | Promotes change of glycogen to glucose—increasing blood sugar | Little effect on blood sugar |
| Metabolic rate | Increases | Little or no effect |

## OTHER ENDOCRINE GLANDS

Other glands of the endocrine system include the ovaries and the testes. In the **ovaries**, the cell clusters (follicles) secrete estrogens (female hormones), which influence development of the secondary sex characteristics of the female. The **corpus luteum** (see Chapter 23) secretes progesterone, and the **placenta** secretes estrogens and progesterone and other hormones. In the male, the **testes** secrete testosterone, as well as other androgens (male hormones) that influence secondary male characteristics (Chapter 22).

Another endocrine gland is the **pineal gland**, a small, cone-shaped gland located in the midbrain on the roof of the third ventricle. The pineal gland is attached to the thalamus, and it secretes a hormone called melatonin. **Melatonin** is believed to regulate the secretions of other hormones, and it may influence mating behavior as well as the day–night cycle.

The **thymus gland** is located posterior to the sternum in the upper thorax (Figure 13.9). This gland secretes **thymosins**, a series of hormones used for the maturation and development of T-lymphocytes. T-lymphocytes are important cells of the body's immune system. The thymus gland is large in the fetus and newborn, but it decreases in size with age.

Several **digestive glands** are located within the lining of the stomach and small intestine. These digestive glands secrete gastrin, secretin, and other hormones functioning in the digestive processes.

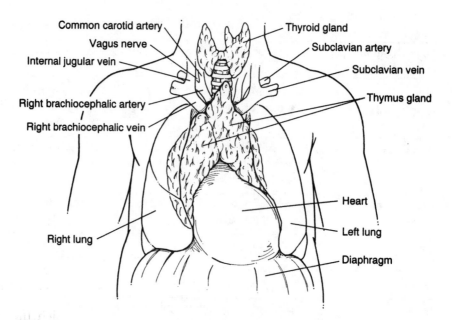

**FIGURE 13.9** *The thymus gland shown in position in the thoracic cavity of a very young child. The gland decreases in size with age and is barely visible by the teenage years.*

Another endocrine gland that we shall consider is the collection of tissue cells of the body. The cells of various organs including the liver, kidney, heart, and lungs, secrete hormones called **prostaglandins**. Prostaglandins are lipid hormones produced in extremely small quantities. They produce a variety of effects on body tissues, such as smooth muscle tissues, other endocrine gland tissues, and reproductive tissues.

The final gland we shall consider includes cells of the **kidney**. These cells produce a hormone called **erythropoetin**. Erythropoetin functions in the maturation of red blood cells taking place in the bone marrow. Erythropoetin has only been known to exist since the 1980s and its method of activity has not yet been elucidated.

# REVIEW QUESTIONS

*PART A—Completion:* **Add the word or words that correctly complete each of the following statements.**

1. The chemical products of the endocrine glands are _____ .

2. Steroid hormones are lipid molecules synthesized from _____ .

3. Another name for the anterior lobe of the pituitary gland is _____ .

4. Oversecretion of the human growth hormone during childhood results in _____ .

5. The thyroid gland is located in the tissues of the _____ .

6. In order to produce thyroxin, the diet must contain atoms of _____ .

7. Insufficient secretion of thyroxin in infants and children can result in a disease called _____ .

8. An excess of thyroxin in a human can result in a condition known as _____ .

9. The parathyroid glands are located on the posterior surface of the _____ .

10. The hormone secreted by the parathyroid glands regulates the body's level of _____ .

11. The largest glandular organ of the abdominal cavity is the _____ .

12. Diabetes mellitus is a condition in which the cells receive an insufficient supply of _____ .

13. The hormone that regulates the passage of glucose in the cells is called _____ .

14. In addition to insulin, the pancreas also produces the hormone _____ .

15. Insulin is produced by cells of the pancreas known as _____ .

16. The adrenal glands may be found on the superior borders of the _____ .

17. The outer tissue of the adrenal gland is known as the _____ .

18. Aldosterone is an example of adrenal hormones called _____ .

19. The metabolism of carbohydrates, protein, and fats is regulated by hormones of the adrenal glands called _____ .

20. The activity of the adrenal gland is regulated by a pituitary hormone abbreviated as _____ .

21. An insufficient secretion of hormones from the adrenal cortex may result in the disease called _____ .

22. The placenta of a woman secretes the hormone _____ .

23. The cone-shaped endocrine gland located in the midbrain is the _____ .

24. The thymus gland plays an important role in the development of white blood cells called _____ .

25. Numerous cells in various organs of the body secrete lipid hormones called _____ .

26. The hormone erythropoetin functions during the maturation of _____ .

27. All hormones are transported in the body by the _____ .

28. Estrogen, progesterone, and cortisol are examples of hormones composed of molecules of _____ .

29. When hormones interact with cells they affect the activity of an enzyme called _____ .

30. The pituitary gland lies in the saddle of a sphenoid bone called the _____ .

31. The hormones released by the posterior pituitary gland are produced by a portion of the brain called the _____ .

32. The hormone TSH is secreted by the pituitary gland and has an effect on the _____ .

33. The hormone that acts on the mammary glands and stimulates the production of milk is called _____ .

34. One hormone of the posterior pituitary gland acts on the kidney tubules and is called _____ .

35. The hormone whose target tissue is the uterus where it induces contractions is called _____ .

36. In the absence of dietary iodine, the thyroid gland swells and produces a condition called _____ .

37. The calcium-regulating hormone of the thyroid gland is called _____ .

38. Another name for the parathyroid hormone is _____ .

39. The hormone insulin is composed exclusively of _____ .

40. The function of the hormone glucagon is to stimulate the breakdown of the carbohydrate _____ .

41. The inner tissue of the adrenal gland is called the _____ .

42. The concentration of electrolytes in the body is regulated by adrenal hormones called _____ .

43. An important hormone of the adrenal medulla which functions in the "fight or flight" response is called _____ .

44. Oversecretion of hormones from the adrenal cortex can result in _____ .

45. Hormones from the ovaries influence the secondary sex characteristics of the _____ .

46. The hormone that is believed to regulate mating behavior and the day–night cycle is called _____ .

47. Female sex hormones are known as _____ .

48. The location of the pineal gland is within the _____ .

49. The maturation and development of T-lymphocytes is regulated by hormones called _____ .

50. The hormones gastrin and secretin function in the process of _____ .

**PART B—Multiple Choice: Circle the letter of the item that correctly completes each of the following statements.**

1. The products of the body's endocrine glands are
   (A) hormones
   (B) enzymes
   (C) minerals
   (D) ions

2. Hormones may consist of all the following except
   (A) steroid molecules
   (B) protein molecules
   (C) carbohydrate molecules
   (D) amine molecules

3. The pituitary gland lies in the
   (A) abdominal cavity
   (B) inferior aspect of the brain
   (C) along the femoral artery
   (D) in the tissues of the neck

4. All the following hormones are produced by the anterior lobe of the pituitary gland except
   (A) HGH
   (B) TSH
   (C) prolactin
   (D) insulin

5. The oversecretion of HGH in adults may result in the condition called
   (A) diabetes mellitus
   (B) acromegaly
   (C) Addison's disease
   (D) Cushing's disease

6. The neurohypophysis is another name for the
   (A) posterior lobe of the pituitary gland
   (B) medulla of the kidney
   (C) follicle that secretes estrogens
   (D) placenta

7. In the female body, the luteinizing hormone
   (A) stimulates TSH production
   (B) regulates mineral metabolism in the body
   (C) stimulates uterine contractions
   (D) acts to promote ovulation

8. The target tissue of ACTH is the
   (A) thymus gland
   (B) medulla of the adrenal gland
   (C) cortex of the adrenal gland
   (D) beta cells of the pancreas

9. The hormone responsible for contractions of the uterus is known as
   (A) insulin
   (B) UCH
   (C) glucagon
   (D) oxytocin

10. The thyroid gland is located
    (A) within the brain
    (B) in the lower abdominal cavity
    (C) near the trachea
    (D) behind the spleen

11. In order for the thyroid gland to produce thyroxin
    (A) iodine must be available
    (B) carbohydrate molecules must be available
    (C) calcium levels must be low
    (D) iron levels must be low

12. Symptoms of cretinism include
    (A) excessive urination and thirst
    (B) electrolyte imbalance in the body
    (C) stunted growth and thickened facial features
    (D) depressed calcium absorption in the digestive tract

13. Graves' disease can result from an excess of
    (A) calcium in the blood
    (B) thyroxin in the blood
    (C) catecholamines in the respiratory passageways
    (D) glucagon in the pancreas

14. Calcitonin and the parathyroid hormone are both concerned with the level of
    (A) pituitary hormones in the blood
    (B) glucose in the blood
    (C) thymosins in the blood
    (D) calcium in the blood

15. The symptoms of diabetes mellitus may include all the following except
    (A) frequent urination
    (B) excessive thirst
    (C) high glucose content of the urine
    (D) abnormal mineral absorption

16. The hormone antagonistic to insulin is
    (A) FSH
    (B) glucagon
    (C) vasopressin
    (D) estrogen

17. The parathyroid glands are located close to the
    (A) pancreas
    (B) lower abdominal cavity
    (C) brain
    (D) trachea

18. In patients with diabetes mellitus
    (A) insufficient glucose enters the cells
    (B) the medulla of the adrenal gland is damaged
    (C) progesterone is not produced by the placenta
    (D) the hypothalamus is nonfunctional

19. The endocrine gland located in the abdominal cavity is the
    (A) pineal gland
    (B) thymus gland
    (C) pancreas
    (D) thyroid gland

20. The two major portions of the adrenal gland are the
    (A) medulla and cortex
    (B) exocrine and endocrine
    (C) renal and subrenal
    (D) posterior and anterior portions

21. The hormones of the adrenal glands complement the action of the
    (A) sensory nervous system
    (B) central nervous system
    (C) sympathetic nervous system
    (D) external nervous system

22. The concentration of sodium and potassium ions in the blood and body fluid is regulated by hormones known as
    (A) glucocorticoids
    (B) androgens
    (C) adrenergic hormones
    (D) mineralocorticoids

23. Hormones that influence the secondary sex characteristics may be produced by both the
    (A) pancreas and pineal gland
    (B) thyroid and parathyroid glands
    (C) thymus and pituitary glands
    (D) adrenal glands and reproductive organs

24. The secretion of melatonin is related to the
    (A) pancreas
    (B) thymus gland
    (C) pineal gland
    (D) pituitary gland

25. The proper functioning of the body's immune system depends in part on the activity of the
    (A) thyroid gland
    (B) thymus gland
    (C) parathyroid gland
    (D) adrenal gland

**PART C—True/False:** For each of the following statements, mark the letter "T" next to the statement if it is true. If the statement is false, change the underlined word to make the statement true.

1. The hormones gastrin and secretin are located within the linings of the stomach and underline.

2. The cone-shaped gland located in the brain is known as the pineal gland.

3. The development of T-lymphocytes is regulated by hormones known as mineralocorticoids.

4. Secondary male characteristics are influenced by hormones known as androgens.

5. The beta cells of the pancreas are responsible for the production of glucagon.

6. The parathyroid hormone acts in a manner that is antagonistic to the activity of calcitonin.

7. The adrenal gland lies in the fold of the stomach.

8. Both thyroxin and triiodothyronine increase the rate of carbohydrate synthesis in the body.

9. Contractions of the uterus may be stimulated by the hormone vasopressin.

10. The hormone ACTH is produced by the adrenal gland, and it regulates the activity of the adrenal cortex.

11. The follicle stimulating hormone is a product of the <u>posterior</u> pituitary gland that acts on the ovaries and testes.

12. The <u>anterior</u> pituitary gland receives and stores hormones from the hypothalamus.

13. Swelling of the thyroid glands due to a lack of iodine is referred to as <u>goiter</u>.

14. Insufficient secretion of thyroxin in adults may result in a condition called <u>myxedema</u>.

15. The composition of the hormone insulin is <u>carbohydrate</u>.

16. The hormone aldosterone is an example of the <u>glucocorticoids</u>.

17. Two important catecholamine hormones are epinephrine and <u>insulin</u>.

18. An inadequate secretion of hormones from the adrenal cortex can result in <u>Cushing's</u> disease.

19. Many of the tissue cells of the body produce lipid hormones called <u>prostaglandins</u>.

20. The maturation of red blood cells is controlled by the hormone <u>melatonin</u>.

21. The pituitary gland lies in a saddle of the <u>ethmoid</u> bone.

22. Another name for human growth hormone is <u>somatotropin</u>.

23. The hormone prolactin stimulates the production of <u>urine</u> in the human body.

24. The thymus gland is located in the tissue of the <u>leg</u>.

25. Glucagon and insulin are both hormones produced by the <u>pancreas</u>.

## PART A—Completion

1. hormones
2. sterols
3. adenohypophysis
4. gigantism
5. neck
6. iodine
7. cretinism
8. Graves' disease
9. thyroid gland
10. calcium
11. pancreas
12. glucose
13. insulin
14. glucagon
15. beta cells
16. kidney
17. cortex
18. mineralocorticoids
19. glucocorticoids
20. ACTH
21. Addison's disease
22. progesterone
23. pineal gland
24. T-lymphocytes
25. prostaglandins
26. red blood cells
27. blood
28. steroid
29. adenyl cyclase
30. sella turcica
31. hypothalamus
32. thyroid gland
33. prolactin
34. antidiuretic hormone
35. oxytocin
36. goiter
37. calcitonin
38. parathormone
39. protein
40. glycogen
41. medulla
42. mineralocorticoids
43. epinephrine
44. Cushing's syndrome
45. female
46. melatonin
47. estrogens
48. brain
49. thymosins
50. digestion

## PART B—Multiple Choice

| | | | | |
|---|---|---|---|---|
| 1. A | 6. A | 11. A | 16. B | 21. C |
| 2. C | 7. D | 12. C | 17. D | 22. D |
| 3. B | 8. C | 13. B | 18. A | 23. D |
| 4. D | 9. D | 14. D | 19. C | 24. C |
| 5. B | 10. C | 15. D | 20. A | 25. B |

## PART C—True/False

1. small intestine
2. true
3. thymosins
4. true
5. alpha
6. true
7. pancreas
8. protein
9. oxytocin
10. pituitary gland
11. anterior
12. posterior
13. true
14. true
15. protein
16. mineralocorticoids
17. norepinephrine
18. Addison's
19. true
20. erythropoetin
21. sphenoid
22. true
23. milk
24. thorax
25. true

# CHAPTER 14

# THE BLOOD

The blood is one of the body's connective tissues. It is the medium for transporting oxygen from the lungs to the cells, and it transports a portion of carbon dioxide waste from the cells to the lungs. Components of the blood protect the body from disease by recognizing and engulfing microorganisms and foreign molecules in the bloodstream. Other components of the blood transport metabolic wastes from the cells to the kidneys, nutrients from the digestive system to the cells, and hormones throughout the body.

The blood is composed of red blood cells, white blood cells, and fragments of cells called platelets (Figure 14.1). These three blood components are referred to as the **formed elements**. They are

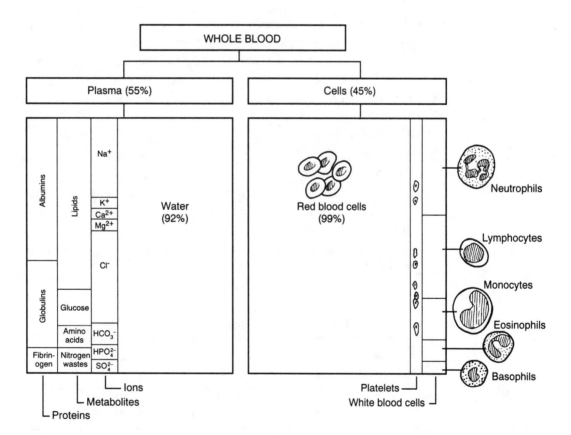

**FIGURE 14.1** *The composition of human blood. The two major components of whole blood are plasma and formed elements. The plasma contains water and numerous dissolved materials, including proteins, metabolites (nutrients and waste products), and ions. The great majority of formed elements are red blood cells.*

suspended in a pale, somewhat yellow fluid known as **plasma**. The blood of an average human takes up about 8 percent of the body weight. The blood is more viscous than water and normally has a pH of about 7.35 to 7.45.

## PLASMA

The plasma is the fluid portion of the blood. It contains about 92 percent water, 7 percent protein, and 1 percent ions such as sodium, calcium, bicarbonate, chloride, and potassium. It also contains metabolic waste products produced by the cells as well as hormones, nutrients, and dissolved gases. When the blood-clotting proteins are removed from the plasma, the remainder is known as **serum**. Serum is commonly used in immunological studies, and it is employed as a source of antibodies for immune therapy.

## Plasma Proteins

The proteins in plasma are of three basic types: albumins, globulins, and fibrinogen (Table 14.1). **Albumin** proteins contribute to the viscosity of the blood and are partly responsible for maintaining a consistent pH in the blood. They are also involved in maintaining the osmotic pressure of the blood, and they transport fatty acids and hormones.

**Globulins** make up about 40 percent of the plasma proteins. One group of globulins called **gamma globulins** are antibody molecules produced in the immune system. These molecules unite specifically with the substances that stimulated their formation (antigens), and they represent a primary mechanism of the body's defense. Other globulin proteins, known as alpha and beta globulins, bind to hormones, vitamins, and other substances in the bloodstream and help transport them.

**TABLE 14.1**   *The Major Components of Blood*

| Component | Examples |
| --- | --- |
| Water | |
| Salts | Sodium, potassium, calcium, magnesium, chloride, bicarbonate |
| Plasma Proteins | Albumin, globulins, fibrinogen |
| Blood Cells (Formed Elements) | White cells, red cells, platelets |
| Substances Transported by Blood | Sugars, amino acids, fatty acids, glycerol hormones, nitrogenous wastes, carbon dioxide, oxygen |

Approximately 7 percent of the protein in the plasma is a liver product called **fibrinogen**. Together with other proteins, fibrinogen is involved in the process of blood clotting.

The proteins of the plasma generally remain in the bloodstream because they cannot easily pass through the walls of blood capillaries. By remaining in the bloodstream, they encourage the osmosis of water molecules from the tissue fluids into the bloodstream (Chapter 21).

# RED BLOOD CELLS

Red blood cells are also known as **erythrocytes**. Their primary purpose in the body is to transport oxygen, a function accomplished by the pigment **hemoglobin** contained in their cytoplasm. Technically speaking, red blood cells are not cells, because they have little internal organization and no nucleus or organelles. They are sacs filled with hemoglobin, and for this reason, they are often called **corpuscles**. However, we shall refer to them as cells, because they have been historically considered in this manner.

## Morphology, Number, and Production

There are approximately 5.4 million red blood cells per cubic millimeter (or microliter) of blood in an adult male. A female has approximately 4.8 million per cubic millimeter (or microliter) of blood (Figure 14.2). Each red blood cell is a flexible, biconcave disk (thinner at the center than at the edge). Red blood cells tend to form stacks, a formation known as the **Rouleaux formation**.

Red blood cells shrink in solutions that contain excessive salt or other material. The water flows out of the red blood cell, in the direction of the higher salt concentration, by the process of osmosis. This leads to **crenation** of the red blood cells. When placed in solution with lower than normal concentration of salt or other solute, the red blood cells swell. This is because osmosis encourages water to pass into the cells in the direction of higher salt concentration. The red blood cells burst and release their hemoglobin in a process called **hemolysis**.

Red blood cells are produced in the red marrow of bones such as the vertebrae, ribs, and body of the sternum. The process of red blood cell formation, called **erythropoiesis**, begins with cells called **hemocytoblasts** (or **stem cells**). The production process is complex, and the cells pass through multiple stages before finally emerging as red blood cells. In the process, hemoglobin accumulates in the cytoplasm, and the nucleus, organelles, and other cellular features disappear. Mature red blood cells enter the capillaries from the bone marrow by squeezing through the capillary walls in a process

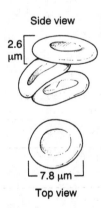

Side view

2.6 μm

7.8 μm

Top view

**FIGURE 14.2**
*Red blood cells (erythrocytes) and their structure.*

called **diapedesis**. Red blood cell production is regulated by a hormone called **erythropoietin**. Erythropoietin is secreted by numerous body cells, primarily by those of the kidneys.

# Hemoglobin

The red pigment **hemoglobin** is an oxygen-binding substance composed of four polypeptide chains. Two polypeptide chains are called **alpha chains**, and two are called **beta chains**. Each chain is composed of approximately 150 amino acid molecules.

Each polypeptide chain is attached to a **heme group**. This group contains iron ions (Figure 14.3). Oxygen molecules bind loosely with the heme portion of the hemoglobin molecule to form **oxyhemoglobin**. Since a single hemoglobin has four heme groups, it can transport four oxygen molecules. The binding is weak, and the flow of oxygen into the red blood cells at the lungs takes place by diffusion. Hemoglobin also transports a small amount of carbon dioxide (but the major portion is transported through the plasma dissolved as bicarbonate ions as noted in Chapter 21). Hemoglobin combined with carbon dioxide is called **carboxyhemoglobin**.

Carbon monoxide is a poisonous gas whose molecules combine readily with hemoglobin and bind tightly to the molecule. By taking up the space reserved for oxygen, carbon monoxide molecules reduce the amount of available hemoglobin and encourage death by oxygen starvation.

# Destruction of Red Blood Cells

Red blood cells circulate in the bloodstream for approximately 120 days. Older cells and damaged cells are engulfed by phagocytes (macrophages) in the spleen, liver, and bone marrow, and the red blood cells are broken down. The polypeptide chains are degraded to release their amino acids, which can be reused for new hemoglobin molecules; the iron portion of the hemoglobin is separated and brought to the bone marrow for new hemoglobin formation, and the excess iron is stored in the liver (Chapter 18); the remainder of the heme portion of the hemoglobin is changed to a greenish pigment called **biliverdin**, which is converted to a yellowish-orange pigment called **bilirubin**. Bilirubin is transported from the spleen to the liver and is excreted in the bile (Chapter 18). When the bile reaches the intestine, bacteria convert some bilirubin to **urobilinogen**, which gives color to the feces. Urobilinogen is also transported back to the liver and then to the general circulation. Eventually it reaches the kidney, where it gives color to the urine.

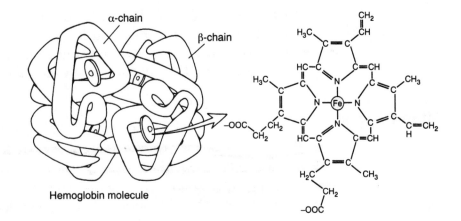

**FIGURE 14.3**  *The hemoglobin molecule of red blood cells. Two alpha chains and two beta chains of polypeptides make up the bulk of the molecule. A heme group is attached to each of the four polypeptide molecules. Note the iron ion (Fe) at the center of the heme group. This is where the oxygen molecule binds for transport.*

# Anemia

A reduction or lack of red blood cells is referred to as anemia. The anemia can be **iron-deficiency anemia** if a deficiency of iron exists in the diet. Without iron, the body fails to synthesize hemoglobin, and the ability to transport oxygen to the cells is reduced. Patients feel exhausted because ATP production is slowed in the oxygen-poor environment.

Another form of anemia is **pernicious anemia**. This condition is due to a lack of **vitamin $B_{12}$** or it may be due to a deficiency of a glycoprotein called **intrinsic factor**, which is essential for vitamin $B_{12}$ absorption. Vitamin $B_{12}$ and intrinsic factor are both required for red blood cell maturation. Without these substances, the membranes of immature red blood cells rupture easily and cannot withstand the chemical environment of the bloodstream. The result is fewer-than-normal red blood cells, and consequently, a reduced oxygen-carrying capacity.

Two other forms of anemia are aplastic anemia and sickle cell anemia. **Aplastic anemia** develops when the production of red blood cells is hindered. Such things as drugs, toxic poisons, and X-rays may be the cause. **Sickle cell anemia** results from hemoglobin containing an incorrect amino acid in the beta polypeptide chain. This mistake derives from a defect in the gene that encodes the polypeptide. The abnormal hemoglobin molecule cross-links with other hemoglobin molecules, and long crystals develop. The crystals deform the cell and cause it to assume a shrunken shape resembling a sickle. The deformed cells rupture easily or are trapped in tiny capillaries, causing blockages.

A final type of anemia is **Cooley's anemia**. This is an inherited disorder in which the body cannot synthesize one or more of the polypeptide chains of hemoglobin. Without hemoglobin, oxygen is poorly transported in the body, and energy metabolism suffers.

## Blood Groups

The surface of human red blood cells contains one, two, both, or no protein molecules known as **antigens** (Figure 14.4). The antigens are designated A and B. Depending on which antigens are present on the red blood cells, a person may have type A blood (A antigen only), type B blood (B antigen only), type AB blood (A and B antigen), or type O blood (no antigens). The antigens have no apparent significance in the physiology of the body.

In addition to the red blood cell antigens, a person also possesses blood group **antibodies** in the serum. A person of group A has anti-B antibodies in the serum; a person of blood type B has anti-A antibodies; a person of blood type AB has no anti-A or anti-B antibodies; and a person of blood type O has both anti-A and anti-B antibodies. Like the blood group antigens, these antibodies have no apparent physiological significance.

In emergency cases where a **blood transfusion** must be made, bloods can be transferred as long as one takes into consideration the

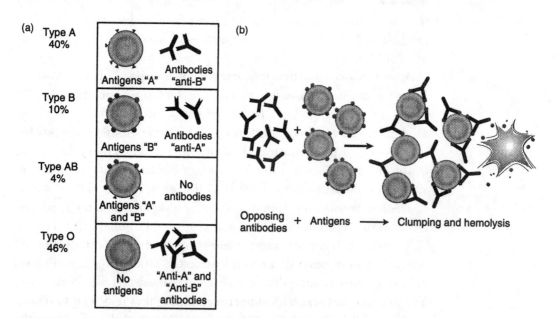

**FIGURE 14.4**  *Blood groups and blood typing. (a) The four blood groups are shown with the types of antigens and antibodies found in each group. The blood "type" is the same as the blood group antigens found on the red blood cell. (b) When mixing blood during blood transfusions, it is extremely important that the same antigens and antibodies are not brought together in the recipient's bloodstream. If they are combined, a reaction will take place, as displayed in the figure. The clumping and destruction of red blood cells (hemolysis) may present a fatal experience.*

antigens present in the donor's blood and the antibodies present in the recipient's serum. To avoid a serious transfusion reaction, the same antigens and antibodies must not be brought together. For example, if a donor is Type A and the recipient is Type AB, the transfusion can be made, because the donor will be contributing A antigens and the recipient does not have anti-A antibodies in the serum. However, if a donor had Type AB blood and the recipient had Type B, the transfusion must not be made, because the Type AB donor has A antigens on the red blood cell and the recipient has anti-A antibodies on the serum. If the two bloods were brought together, the red blood cells would clump, and a possibly lethal transfusion reaction would take place. A person having Type O blood is said to be the universal donor, because there are no A or B antigens on the red blood cells, and donations can be made to persons of other blood types. A person having Type AB blood is said to be the universal recipient, because this person has no anti-A or anti-B antibodies in the serum, and this person can receive bloods of the other three types.

Another antigen of significance is the **Rh antigen**. Approximately 85–90 percent of the American population has this antigen on the red blood cell and is said to be **Rh-positive**. Approximately 10–15 percent of Americans lack this factor, and the blood type is said to be **Rh-negative**. Thus, a person may be blood Type A+ if the person has the A antigen and the Rh antigen; or the person may be B– if he or she has the B antigen but lacks the Rh antigen. The Rh antigen, like the others, has no apparent significance in the normal physiology of the body.

The Rh factor is important in the condition known as **erythroblastosis fetalis**, or **hemolytic disease of the newborn**. This condition develops when an Rh-positive male (e.g., A+) fathers a child with an Rh-negative female (e.g., AB–). In this case, there is a 3:1 possibility that the child will have a blood type that is Rh-positive (e.g., A+ or O+). During the birth process, some of the child's blood cells enter the mother's bloodstream and stimulate her immune system to produce anti-Rh antibodies. These antibodies will have no affect on the child, but they will remain in the mother's bloodstream. If the woman has another child in the future and if the child's blood is Rh-positive, (e.g., A+), then the Rh antibodies will enter the bloodstream of the second child by passing across the placenta. There the antibodies will react with the Rh antigens on the surface of the red blood cells, and the reaction will result in excessive clumping that may lead to death.

To avoid the possibility of hemolytic disease of the newborn, the Rh-negative woman is given an injection of Rh antibodies (RhoGAM) upon the birth of her first child. The anti-Rh antibodies unite with the Rh antigens in her bloodstream and neutralize them. This action prevents the Rh antibodies from stimulating her immune system, and no anti-Rh antibodies are produced. When the second child is developing, no antibodies will be present to clump its blood.

The woman must receive another injection of RhoGAM after the birth of her second child to avoid the production of anti-Rh antibodies that could harm the third child.

# WHITE BLOOD CELLS

White blood cells are also called **leukocytes** (Table 14.2). Their primary function in the body is to defend the tissues against infection and substances foreign to the body. A normal adult has approximately 7000 white blood cells per cubic millimeter of blood.

White blood cells of various types develop by a complex process in the body's red bone marrow. All white blood cells enter the circulation by diapedesis, and some complete their maturation elsewhere. The white blood cells live for several hours or several months, depending on their type, and many white blood cells leave the circulation by diapedesis to mingle among the tissue cells.

## Types of White Blood Cells

There are two major groups of white blood cells: the granulocytes and the agranulocytes. **Granulocytes** have granules in their cytoplasm and include neutrophils, eosinophils, and basophils. **Agranulocytes** have no granules in their cytoplasm and include monocytes and the lymphocytes.

**Neutrophils** comprise about 60 percent of the total white blood cell count in the body. Their granules stain with neutral dyes and have a purplish color. The nucleus of the neutrophil usually has between two and five lobes, and the cell is often called a **polymorphonuclear** white blood cell. The neutrophil's principal function is phagocytosis, and neutrophils accumulate rapidly at an infection site.

**TABLE 14.2**  *Microscopic Appearance of Leukocytes after Wright's Staining*

| Leukocytes | Appearance |
| --- | --- |
| Granulocytes | |
|   Neutrophils | Fine, light blue cytoplasmic granules; 3- to 5-lobed nucleus |
|   Eosinophils | Bright red cytoplasmic granules; 2-lobed nucleus |
|   Basophils | Large, dark purplish blue cytoplasmic granules; irregular nucleus; often S-shaped |
| Agranulocytes | |
|   Lymphocytes | Thin layer of nongranular, blue cytoplasm; large, bright purple nucleus |
|   Monocytes | Thick layer of nongranular cytoplasm; large horseshoe- or kidney-shaped, purple nuclei |

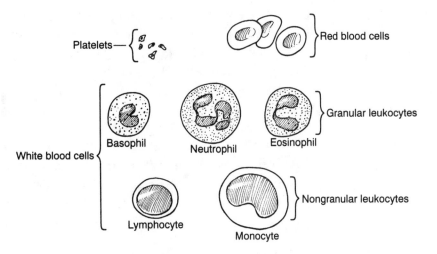

**FIGURE 14.5**   *A display of white blood cells compared to one another and to red blood cells and platelets. Note the distinctive granulation and nuclei present in the various types of white blood cells.*

**Basophils** and **eosinophils** both have granules in their cytoplasm: the basophil's granules stain with basic dyes and appear blue, while the eosinophil's granules stain with acidic dyes and appear red (Figure 14.5). Each type of cell accounts for approximately 1 percent of the total white blood cell count. Both cells are believed to function in allergy reactions and during inflammation. In addition, basophils may have a function in the clotting mechanism.

**Lymphocytes** have no granules in their cytoplasm. They account for about 30 percent of all the white blood cells and occur in two types: B-lymphocytes and T-lymphocytes. **B-lymphocytes** are stimulated by the antigens of microorganisms during the immune response, and they revert to plasma cells. The plasma cells then produce antibody molecules, which enter the bloodstream and interact with the microorganisms that stimulated their production. The interaction generally leads to destruction of the microorganism. B-lymphocytes are found primarily in the body's lymph nodes.

**T-lymphocytes** are also found in the lymph nodes. Before reaching the lymph nodes, the young cells mature in the thymus gland. On stimulation by antigens, the T-lymphocytes leave the lymph nodes and proceed to the infection site where they interact cell-to-cell with microorganisms and destroy them. B-lymphocytes and T-lymphocytes are the key cells of the immune system (Chapter 16), and they provide body defense through this system.

**Monocytes** make up between 6 and 8 percent of the white blood cells (Table 14.3). They have a very large nucleus, which is indented along one margin. Monocytes squeeze through the capillary walls by diapedesis and enter the tissue environment where they perform phagocytosis on microorganisms. In the tissues, the monocytes change into large phagocytic cells called **macrophages**. Macrophages begin the immune response by engulfing microorganisms and delivering their antigens to the lymphocytes in the lymph nodes, where the immune system is located.

Examining the white blood cell population of the body can provide valuable insight into disease. For example, an elevated white blood cell count usually indicates a microbial infection. In addition, noting whether particular white blood cells are in high numbers can also be valuable. These data are obtained by a **differential white blood cell count**. A general reduction of white blood cells is called **leukopenia**, while an above-average population of white blood cells is called **leukemia**.

## PLATELETS

Platelets (also called thrombocytes) are blood elements produced in the bone marrow. Technically, **platelets** are not cells because they consist of fragments of cytoplasm enclosed by membranes. They form from large cells called **megakaryocytes**, derived from hemocytobasts. Bits of megakaryocyte cytoplasm pinch off within membranes and are released into the circulation. Approximately 300,000 platelets exist per cubic millimeter of blood.

Platelets function in two important processes: they form platelet plugs, and they are involved in the blood-clotting mechanism. **Platelet plugs** form at the severed part of a blood vessel when platelets react with collagen fibers in the wall of the blood vessels (Figure 14.6). The platelets stick to the fibers, and form a mass that patches the hole in the vessel. This reaction happens within seconds of injury to the vessel.

## Blood Clotting

Blood-clot formation occurs in larger injuries sustained in blood vessels. The process of clot formation results in a mass of protein fibers, trapped blood cells, and platelets that repair the injury sustained. The elements involved in clotting are referred to as **clotting factors**.

When the tissue or blood vessel is injured, the blood clotting mechanism is activated by either an intrinsic pathway or extrinsic pathway. The intrinsic pathway involves factors found only in the blood, while the extrinsic pathway is initiated by factors outside the bloodstream.

In the **intrinsic pathway**, a clotting factor called **platelet factor** is released by blood platelets and by endothelial cells lining the blood vessels. The factor interacts with calcium ions and numerous other clotting factors to yield **platelet-derived thromboplastin**. Thromboplastin forms a prothrombin activator, a lipoprotein that activates a globular liver protein known as **prothrombin**. In the reaction, prothrombin is converted to an active form called **thrombin**. Calcium ions are essential for this conversion.

**TABLE 14.3**   *The Characteristics of Formed Elements of the Blood*

| Cells | Number | Function | Role in Disease |
|---|---|---|---|
| Red blood cells (Erythrocytes) | Male: 5.4 million/mm³ Female: 4.8 million/mm³ | Oxygen transport; carbon dioxide transport | Too few: anemia Too many: polycythemia |
| Platelets (Thrombocytes) | About 300,000/mm³ | Essential for clotting | Too few: clotting malfunctions; bleeding; easy bruising |
| White blood cells (Leukocytes) | About 7000/mm³ total | | |
| Neutrophils | About 60% of WBC | Phagocytosis | Too many; may be due to bacterial infection, inflammation, leukemia |
| Eosinophils | About 1% of WBC | Some role in allergic response | Too many may result from allergic reaction, parasitic infections |
| Basophils | About 1% of WBC | Possible role in allergic response | |
| Lymphocytes | About 30% of WBC | Produce antibodies; destroy foreign cells | Atypical lymphocytes present in infectious mononucleosis; too many may result in leukemia |
| Monocytes | 6 to 8% of WBC | Phagocytosis; differentiate in tissues to form macrophages | May increase in monocytic leukemia, tuberculosis, fungal infections |

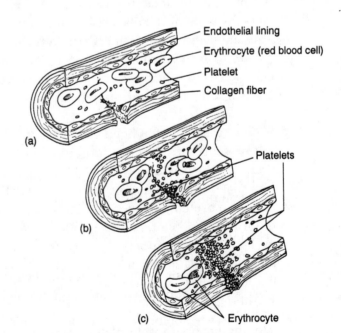

**FIGURE 14.6**   *Formation of a platelet plug. (a) A break occurs in the wall of a blood vessel. (b) Platelets adhere to each other and to the collagen fibers of the blood vessel wall. (c) The platelet plug helps control the loss of blood at the injury site*

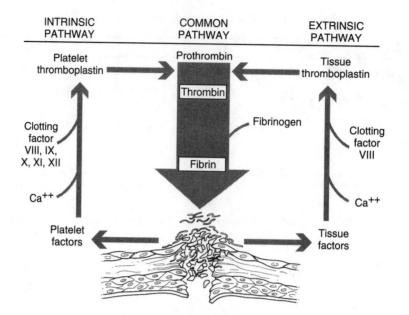

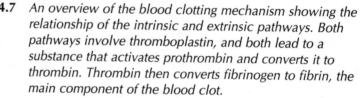

**FIGURE 14.7** *An overview of the blood clotting mechanism showing the relationship of the intrinsic and extrinsic pathways. Both pathways involve thromboplastin, and both lead to a substance that activates prothrombin and converts it to thrombin. Thrombin then converts fibrinogen to fibrin, the main component of the blood clot.*

Prothrombin can also be activated by the **extrinsic pathway**. In this process, tissue factors from damaged blood vessels or on cell surfaces throughout the body react with clotting factor VII and calcium ions and activate the factor. Factor VII then activates other clotting factors to form **tissue-derived thromboplastin**. Together with calcium ions and other factors, the thromboplastin forms a prothrombin activator. The prothrombin activator converts prothrombin to thrombin.

Once thrombin has been produced, it functions as an enzyme. In the presence of calcium, thrombin activates the liver protein **fibrinogen**, which is dissolved in the plasma. The activation converts fibrinogen to **fibrin**, an insoluble protein (Figure 14.7). Threads of the fibrin protein then accumulate with platelets and red blood cells to form a blood clot. Plasma gels at the site, and the clot soon loses its liquid and contracts. During the contraction, the clot pulls together the damaged ends of the blood vessel or tissue, and the clot hardens. Dense fibrin threads soon form, and the clot is complete.

Although blood clotting is essential to good health, there are situations in which blood clots can be damaging to the body. For example, a blood clot may form in blood vessels roughened by the accumulation of plaques consisting of cholesterol molecules. The cholesterol accumulates with other lipids on the inner wall of vessels and leads to a condition called **atherosclerosis**.

Plaques of cholesterol also stimulate the production of blood clots associated with a condition called **thrombosis**. A blood clot so-formed is called a **thrombus**. It may impede the flow of blood in the coronary arteries (Chapter 15) and result in the condition called coronary thrombosis. Should the clot move to another part of the body, the clot is called an embolus. This condition is called **embolism**.

# REVIEW QUESTIONS

**PART A—Completion:** Add the word or words that correctly complete each of the following statements.

1. The red blood cells, white blood cells, and platelets are blood components known as _____ .  *formed elements*

2. The pH of the blood is usually about _____ .  *7 - 8.*

3. The blood is responsible for the transport of substances that lend chemical coordination to the body and are known as _____ .  *plasma protein*

4. The fluid portion of the blood is the _____ .

5. When the blood-clotting proteins are removed from the plasma, the plasma is then known as _____ .

6. Among the major ions transported by the blood are calcium, potassium, chloride, bicarbonate, and _____ .

7. Those plasma proteins that contribute to the viscosity of the blood and are partly responsible for maintaining the constant pH of the blood are known as _____ .

8. The antibodies belong to a major group of plasma proteins called _____ .

9. Approximately 7 percent of the plasma protein consists of a liver product known as _____ .

10. The plasma proteins encourage the movement of water molecules from the tissue fluids into the bloodstream, and this movement of fluids is known as _____ .

11. The major pigment that transports oxygen within red blood cells is _____ .

12. Red blood cells are commonly known by their alternative name _____ .

13. The number of red blood cells per cubic millimeter in an adult male is approximately _____ .

14. The number of red blood cells per cubic millimeter in an adult female is _____ .

15. The shape of a red blood cell is that of a _____ .

16. When a red blood cell shrinks in solutions that contain excessive salt, the process is called _____ .

17. When red blood cells swell and burst in solutions that contain low concentrations of salt, the process is _____ .

18. Red blood cells are formed in the bone marrow through a process known as _____ .

19. When mature, the red blood cells enter the body capillaries from the bone marrow by squeezing through the walls of the capillaries in a process that is called _____ .

20. The two polypeptide chains of the hemoglobin molecule are known as _____ .

21. The iron-containing group of the hemoglobin molecule that binds to oxygen molecules is known as the _____ .

22. When oxygen is bound to the hemoglobin molecule, the hemoglobin molecule is known as _____ .

23. When carbon dioxide is bound to the hemoglobin molecule, the hemoglobin molecule is known as _____ .

24. The main portion of carbon dioxide is transported through the plasma dissolved as _____ .

25. Red blood cells circulate in the human bloodstream for approximately _____ .

26. After breakdown, the hemoglobin pigment of red blood cells is eventually converted to a bile pigment known as _____ .

27. The production of red blood cells is regulated by a hormone known as _____ .

28. A lack of vitamin $B_{12}$ may result in a condition known as _____ .

29. Such things as X-rays and drugs may prevent the production of red blood cells, a condition called _____ .

30. A genetic defect that encodes hemoglobin may lead to a deformity of the red blood cells and a disease known as _____ .

31. When the body fails to synthesize one of more of the polypeptide chains of hemoglobin, the condition that results is called _____ .

32. An alternative name for white blood cells is _____ .

33. A normal adult has a white blood cell count per cubic millimeter that numbers about _____ .

34. The white blood cells develop within the _____ .

35. Neutrophils and basophils are types of white blood cells known as _____ .

36. Lymphocytes and monocytes have no granules in their cytoplasm and are therefore known as _____ .

37. The principal function of the neutrophil is _____ .

38. The granules of the eosinophil stain with acidic dyes and appear _____ .

39. The percentage of the total white blood count that is basophils is approximately _____ .

40. The lymphocytes are the important cells of the body system known as the _____ .

41. When the monocytes enter the tissue environment, they change into large, phagocytic cells called _____ .

42. Antibodies are produced by white blood cells known as _____ .

43. A general reduction of white blood cells in the body is referred to as _____ .

44. Blood platelets are produced in the bone marrow by large cells called _____ .

45. A mass of platelets and collagen fibers that patch a hole in a blood vessel is known as a _____ .

46. In the intrinsic pathway, an important factor that initiates the chemical pathway is called Factor _____ .

47. In the extrinsic pathway, substances from damaged blood vessels release an activating substance called _____ .

48. The substance thrombin is responsible for activating the conversion of fibrinogen to _____ .

49. The accumulation of cholesterol substances within the inner wall of a blood vessel encourages a condition called _____ .

50. A condition in which a blood clot moves from one part of the body to another is known as a _____ .

**PART B—Multiple Choice:** Circle the letter of the item that correctly completes each of the following statements.

1. All the following are functions of the blood except
   (A) it transports oxygen from the lungs to the cells
   (B) it protects the body from disease
   (C) it transports nutrients from the digestive system to the body cells
   (D) it coordinates all the activities of the body cells

2. All the following are characteristics of the blood except
   (A) it contains about 92 percent water
   (B) its pH varies between 6.2 and 6.7
   (C) it is more viscous than water
   (D) it contains about 7 percent protein

3. Albumins, globulins, and fibrinogen are three of the
   (A) hormones found in the blood
   (B) salts found in the blood
   (C) gases transported by the blood
   (D) proteins present in the blood

4. The antibody molecules produced in the immune system belong to a group of proteins called
   (A) gamma globulins
   (B) albumin proteins
   (C) clotting proteins
   (D) alpha and beta proteins

5. Which of the following characteristics applies to the red blood cell?
   (A) it is alternatively referred to as a leukocyte
   (B) it has the shape of a biconcave disk
   (C) it has a distinctive nucleus and organelles
   (D) it is incapable of transporting oxygen to the cells

6. When red blood cells are placed in solution that contains no salt, the red blood cells
   (A) tend to shrink
   (B) tend to swell
   (C) undergo crenation
   (D) lose their nucleus and cytoplasm

7. The squeezing of red blood cells from the bone marrow into the capillaries is a process known as
   (A) hemolysis
   (B) phagocytosis
   (C) diapedesis
   (D) thrombosis

8. All the following are found in hemoglobin molecules except
   (A) two polypeptide chains
   (B) iron atoms
   (C) carbohydrate molecules
   (D) heme groups

9. The major portion of carbon dioxide molecules is transported in the blood
   (A) attached to the red blood cells
   (B) attached to the white blood cells
   (C) dissolved in plasma as bicarbonate ions
   (D) attached to the surface of platelets.

10. Old and damaged red blood cells are broken down in the
    (A) pancreas, kidney, and small intestine
    (B) liver, spleen, and bone marrow
    (C) thyroid, thymus, and pituitary gland
    (D) small intestine, brain, and vena cava

11. Bilirubin is a bile pigment formed from the breakdown of
    (A) hemoglobin
    (B) white blood cells
    (C) platelets
    (D) fibrinogen proteins

12. Pernicious anemia is related to a
    (A) lack of iron in the body
    (B) genetically inherited defect
    (C) lack of vitamin $B_{12}$ in the body
    (D) insufficiency of calcium in the body

13. Sickle cell anemia is derived from a(n)
    (A) attack of white blood cells on red blood cells
    (B) deficiency of iodine in the body
    (C) spontaneous clotting of the blood
    (D) defect traced to the genes of the body

14. Which of the following applies to the neutrophils of the body?
    (A) they are types of white blood cells
    (B) they have no granules in their cytoplasm
    (C) their primary function is in the production of antibodies
    (D) their granules stain blue with basic dyes

15. The primary cells of the body's immune system are the
    (A) basophils and eosinophils
    (B) polymorphonuclear white blood cell
    (C) B-lymphocytes and T-lymphocytes
    (D) monocytes

16. Approximately 6 to 8 percent of the white blood cells consist of
    phagocytic cells known as
    (A) lymphocytes
    (B) neutrophils
    (C) platelets
    (D) monocytes

17. An overpopulation of white blood cells is a characteristic of a
    form of cancer known as
    (A) melanoma
    (B) leukemia
    (C) parasitemia
    (D) leukopenia

18. The primary function of the white blood cells in the body is in
    (A) transport of nutrients
    (B) body defense
    (C) excretion of waste products
    (D) body movement

19. The clotting protein called prothrombin is manufactured in the
    (A) liver
    (B) bone marrow
    (C) bloodstream
    (D) kidney

20. Which of the following is a reaction in the extrinsic pathway of blood clotting?
    (A) Factor XII converts to Factor V
    (B) prothrombin is converted to thrombin
    (C) thromboplastin is converted to prothrombin
    (D) fibrin is converted to fibrinogen

21. The accumulation of cholesterol along the inner walls of blood vessels can lead to a condition called
    (A) embolism
    (B) thrombosis
    (C) atherosclerosis
    (D) plug formation

22. A person who has blood type A may donate blood to a person who has
    (A) blood type O or blood type AB
    (B) blood type A only
    (C) blood type A or blood type AB
    (D) blood type O only

23. A person who has blood type B may receive blood from a person who has
    (A) blood type A only
    (B) blood type B or blood type AB
    (C) blood type B or blood type O
    (D) blood type O or blood type AB

24. Hemolytic disease of the newborn may develop when the
    (A) male is Rh-positive and female is Rh-negative
    (B) male is Rh-negative and female is Rh-positive
    (C) both male and female are Rh-positive
    (D) both male and female are Rh-negative

25. To prevent hemolytic disease of the newborn from occurring in succeeding pregnancies, shortly after the birth of a child a woman is given an injection of
(A) Rh antigen
(B) anti-Rh antibodies
(C) penicillin
(D) A and B antigens

**PART C—True/False:** For each of the following statements, mark the letter "T" next to the statement if it is true. If the statement is false, change the underlined word to make the statement true.

1. The formed elements of the blood are suspended in a pale, somewhat yellow fluid known as serum.

2. Those proteins that contribute to the viscosity of the blood and help maintain a consistent pH in the blood are known as globulin proteins.

3. The gamma globulins are well known as antigen molecules, which are produced in the immune system.

4. Red blood cells have no nucleus or organelles, but instead they are filled with the red pigment hemoglobin.

5. The red blood cell appears as a biconvex disk.

6. When red blood cells are suspended in solutions that contain excessive amounts of salt, the cells swell in a process called crenation.

7. When red blood cells are placed in a solution that contains no salt, they tend to burst in a process called hemolysis.

8. Red blood cell formation goes on in the red bone marrow by a process called lymphopoiesis.

9. The hemoglobin molecule consists of four chains of polypeptides.

10. When carbon monoxide enters red blood cells it binds weakly to the hemoglobin molecule.

11. Red blood cells circulate in the bloodstream for approximately 320 days.

12. Bilirubin is a pigment produced from the hemoglobin of red blood cells and excreted by the liver to the bile.

13. The hormone erythropoietin regulates and stimulates the development of red blood cells in the bone marrow.

14. Failure of the body to synthesize one or more polypeptide chains of hemoglobin results in a condition known as pernicious anemia.

15. Eosinophils, basophils, and neutrophils are all types of white blood cells known as <u>agranulocytes</u>.

16. Because the nucleus of a <u>lymphocyte</u> occurs in two to five lobes, the cell is known as a polymorphonuclear cell.

17. Basophils are among the <u>most</u> numerous of all the white blood cells in the circulation.

18. T-lymphocytes and B-lymphocytes are the important cells of the body's <u>endocrine</u> system.

19. The important phagocytes of the body include neutrophils, macrophages, and <u>lymphocytes</u>.

20. Leukopenia is a <u>higher</u> than normal count of white blood cells in the body.

21. In the intrinsic pathway of blood clotting, <u>sodium</u> ions are required for at least one of the chemical conversions.

22. A blood clot that moves from one body location to another is known as a <u>thrombus</u>.

23. A person whose blood type is AB may donate blood to a person whose blood type is <u>O</u>.

24. A person having blood type O is known as the universal <u>recipient</u>.

25. When a male who is Rh-positive and a female who is Rh-negative have a child, there is a <u>100 percent</u> probability that the child's blood type will be Rh-positive.

**Answers**

## PART A—Completion

1. formed elements
2. 7.35 to 7.45
3. hormones
4. plasma
5. serum
6. sodium
7. albumin proteins
8. globulin proteins
9. fibrinogen
10. osmosis
11. hemoglobin
12. erythrocytes
13. 5.8 million
14. 4.8 million
15. biconcave disk
16. crenation
17. hemolysis
18. erythropoiesis
19. diapedesis
20. alpha and beta chains
21. heme group
22. oxyhemoglobin
23. carboxyhemoglobin
24. bicarbonate ions
25. 120 days
26. bilirubin
27. erythropoietin
28. pernicious anemia
29. aplastic anemia
30. sickle cell anemia
31. Cooley's anemia
32. leukocytes
33. 7000
34. red bone marrow
35. granulocytes
36. agranulocytes
37. phagocytosis
38. red
39. one percent
40. immune system
41. macrophages
42. lymphocytes
43. leukopenia
44. megakaryocytes
45. platelet plug
46. XII
47. thromboplastin
48. fibrin
49. atherosclerosis
50. embolism

## Part B—Multiple Choice

| | | | | |
|---|---|---|---|---|
| 1. D | 6. B | 11. A | 16. D | 21. C |
| 2. B | 7. C | 12. C | 17. B | 22. C |
| 3. D | 8. C | 13. D | 18. B | 23. C |
| 4. A | 9. C | 14. A | 19. A | 24. A |
| 5. B | 10. B | 15. C | 20. B | 25. B |

## Part C—True/False

1. plasma
2. albumin
3. antibody
4. true
5. biconcave
6. shrink
7. true
8. erythropoiesis
9. two
10. tightly
11. 120
12. true
13. true
14. Cooley's
15. granulocytes
16. neutrophil
17. least
18. immune
19. monocytes
20. lower
21. calcium
22. embolus
23. AB
24. donor
25. 75 percent

# CHAPTER 15

# THE CARDIOVASCULAR SYSTEM

The cardiovascular system is responsible for delivering the nutrients and oxygen to the tissue cells and removing their metabolic waste products. The system also brings hormones to their target cells. It is composed of a pump (the heart) and a set of tubes that carry blood (the blood vessels). Virtually all parts of the body are serviced by the cardiovascular system.

## THE HEART

The heart is the pump of the cardiovascular system. It consists of two main pumping chambers, the **ventricles**, and two receiving chambers called the **atria** (singular **atrium**). The heart propels blood through the arteries, capillaries, and veins of the cardiovascular system and supplies blood to all the body cells. It is roughly the size of a clenched fist, is hollow and cone-shaped, and, typically, weighs less than a pound.

The heart is found in the mediastinum of the thorax, approximately between the second and fifth ribs. It lies anterior to the vertebral column and posterior to the sternum and is flanked by the lungs, which overlap it. The heart tips slightly to the left and assumes an oblique position in the thoracic cavity (Figure 15.1).

Two saclike membranes called the **pericardium** enclose the heart. The outer membrane is known as the **parietal pericardium**, while the inner membrane is the **visceral pericardium**, or **epicardium**. The pericardial cavity is the fluid-filled space between the parietal pericardium and visceral pericardium (epicardium). The epicardium is often filled with fat, especially in the older years. Inflammation of the pericardium is called **pericarditis**. The epicardium is considered the outer layer of the heart.

The main constituent of the heart is the second, or middle layer, called the **myocardium**. The myocardium is composed of cardiac muscle cells, crisscrossed and arranged in bundles. Within the myocardium, the muscle cells are connected to one another by tissue fibers arranged in a criss-crossing connecting network to compose a fibrous skeleton. The network of connective tissue fibers is

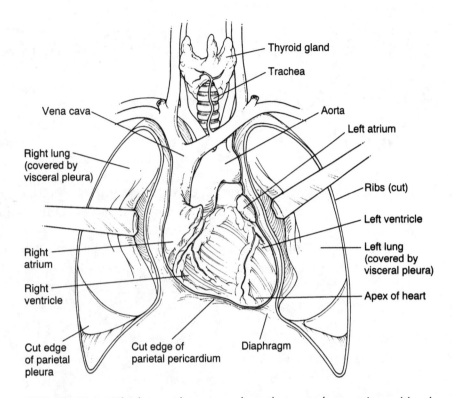

**FIGURE 15.1**   *The human heart seen from the ventral aspect in position in the thoracic cavity. Note the position of the adjacent organs.*

dense and reinforces the myocardium internally. Ropelike rings of fibrous tissue provide additional support around the valves and at the point where the large vessels arise from the heart.

The third layer of the heart is the inner layer called the **endocardium**. The endocardium is a layer of endothelium lying atop a layer of thin connective tissue. The endocardium lines the heart chambers and makes up most of the tissue of the heart valves. Inflammation of the heart valves is called **endocarditis.**

## Chambers and Vessels of the Heart

There are four chambers in the heart. The two superior chambers are called **atria** while the two inferior chambers are called **ventricles**.

The heart chambers are separated longitudinally by a wall-like mass of tissue called the **cardiac septum**. Between the atria, the septum is called the **interatrial septum**; between the ventricles it is called **interventricular septum** (Figure 15.2).

The atria are receiving chambers of the heart. Blood returning in veins from the body tissues enter the atria and is held there until the ventricles empty out. Each atrium has a flat, wrinkled extension called the **auricle**, which fills with blood when the atrium is full. The auricle increases the volume of the atrium.

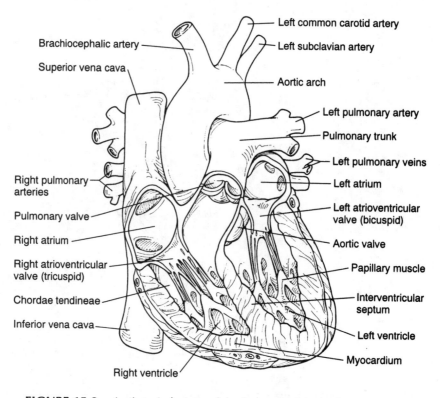

**FIGURE 15.2**   *An internal view of the heart displaying the major structures of this organ. The details of the valves can be seen, and the important vessels are shown.*

Three veins empty into the right atrium: the **superior vena cava,** which returns blood from the head and neck; the **inferior vena cava,** which returns blood from the inferior regions of the body; and the **coronary sinus,** which receives blood from the heart muscle and returns it to the right atrium. The left atrium receives blood from the lungs by means of the pulmonary veins.

The **ventricles** are the pumping chambers of the heart lying inferior to the atria. The right ventricle pumps blood to the lungs, while the left ventricle pumps blood to the body's organs, tissues, and cells.

## Blood Pathway through the Heart

There are two main cardiovascular circuits in the body. The first circuit, called the **pulmonary circuit,** extends from the heart to the lungs and back to the heart. The second circuit, the **systemic circuit,** extends from the heart to the body's cells and then back to the heart (Figure 15.3).

The pulmonary circuit begins on the right side of the heart. Blood from the body organs enters the right atrium through the

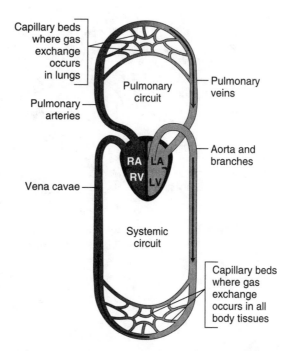

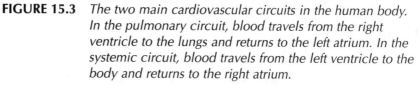

**FIGURE 15.3**  *The two main cardiovascular circuits in the human body. In the pulmonary circuit, blood travels from the right ventricle to the lungs and returns to the left atrium. In the systemic circuit, blood travels from the left ventricle to the body and returns to the right atrium.*

vena cavae. This blood is poor in oxygen and rich in carbon dioxide. The blood passes through a valve from the right atrium into the right ventricle. The right ventricle then pumps blood to the lungs through the **pulmonary arteries**. At the lungs, carbon dioxide leaves the blood and enters the lung spaces, while oxygen passes out of the lung spaces into the blood. The oxygen-rich blood then returns through the **pulmonary veins** to the left side of the heart, thus completing the pulmonary circuit.

At the left side of the heart the systemic circuit begins. The blood enters the left atrium, then passes through a valve into the left ventricle. The left ventricle pumps the oxygen-rich blood into the **aorta**, the largest artery of the body. From the aorta, the blood is distributed to other **arteries** of the systemic circuit. Blood flows to the head, chest, abdominal region, and all other parts of the body through the arteries.

At the body cells, the blood releases its oxygen and takes on carbon dioxide. The carbon dioxide-rich blood (also, oxygen-poor) then returns to the heart through the **veins** of the cardiovascular system and the vena cava. When it reaches the heart, the blood enters the right atrium, thus completing the systemic circuit.

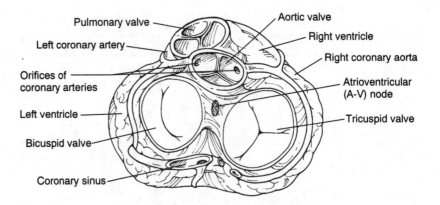

**FIGURE 15.4**   *A cutaway superior view of the heart after removal of the atria. Note the structure of the bicuspid (two-flaps) and tricuspid (three-flaps) valves. The aortic and pulmonary valves are semilunar valves. The coronary artery, vein, and sinus are visible.*

## Heart Valves

A set of heart valves ensures that blood flows through the heart in one direction and does not back up. There are four heart valves involved in this activity. Two valves are called **atrioventricular valves**, and two valves are **semilunar valves** (Figure 15.4)

One atrioventricular valve is located on the right side of the heart, between the right atrium and right ventricle. This valve is called the **tricuspid valve**, because it has three valves, or cusps. The second atrioventricular valve is found on the left side of the heart, between the left atrium and the left ventricle. It is called the **bicuspid valve** because is has only two flaps; the valve is also called the **mitral valve**. The atrioventricular valves ensure that blood flows from the atria to the ventricles and does not back up into the atria when the ventricles contract.

White cords of collagen anchor the heart valve to papillary muscles of the ventricle wall. The cords of tissue, called **chordae tendineae**, prevent the valve flaps from moving backward into the atria. When the chordae tendineae are damaged or when damage occurs in the valves themselves, the valves tend to flap backward. In the mitral valve this condition is called **mitral valve prolapse**.

The two **semilunar valves** are found within the arteries extending from the ventricles. The **right semilunar valve** (pulmonary valve) exists in the pulmonary artery, which extends from the right ventricle toward the lungs. The valve prevents blood from flowing backward into the ventricle when the ventricle relaxes. The **left semilunar valve** (aortic valve) is found in the aorta. It prevents blood from flowing backward into the left ventricle (Table 15.1).

**TABLE 15.1**  *A Summary of the Heart Valves*

| Valve | Location | Function |
| --- | --- | --- |
| Tricuspid valve | Between right atrium and right ventricle | Prevents blood from moving from right ventricle back into right atrium during ventricular contraction |
| Pulmonary semilunar valves | Between right ventricle and pulmonary artery | Prevents blood from moving back from pulmonary artery into right ventricle during ventricular relaxation |
| Bicuspid (mitral valve) | Between left atrium and left ventricle | Prevents blood from moving from left ventricle during ventricular contraction |
| Aortic semilunar valve | Between left ventricle and aorta | Prevents blood from moving back from aorta into ventricle during ventricular relaxation |

## The Coronary Circulation

Because the heart is composed of living tissue, it must be supplied with blood. The arteries that supply this blood are the **coronary arteries**; the vessels that drain the cardiac muscle are **coronary veins**. Coronary arteries supply oxygen-rich blood to the muscle of the heart and join with the coronary veins, which collect the oxygen-poor blood and return it to the **coronary sinus** (Figure 15.5). The sinus delivers blood into the right atrium.

Prolonged blockage of the coronary arteries by blood clots is called **coronary thrombosis**. This condition can result in the death of the heart muscle cells. When the cells die, they often form a blockage called an **infarct**. Such a condition is called **myocardial infarction**. Often it is associated with a condition called **heart attack**.

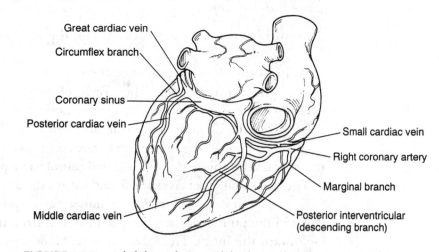

**FIGURE 15.5**  *A left lateral view of the heart showing some of the coronary arteries and veins. The coronary sinus lies close to the vena cavae and empties into the right atrium with these veins.*

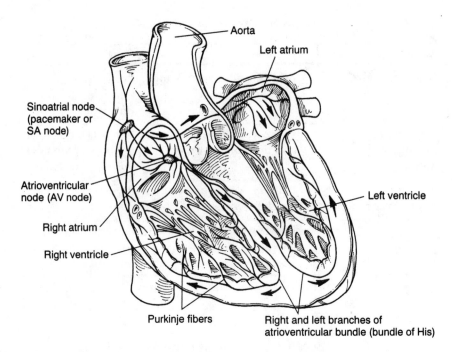

**FIGURE 15.6**    *The conduction system of the heart. Impulses are generated at the SA node and pass laterally to other parts of the heart through Purkinje fibers. Impulses from the SA node also stimulate the AV node, and impulses pass from this node to the bundle of His and other Purkinje fibers.*

## Cardiac Muscle

**Cardiac muscle** is similar to skeletal muscle in its physiology and biochemistry. One structural difference is notable: Skeletal muscle cells tend to be long and cylindrical, while cardiac muscle cells are shorter, broader, and more branched and interconnected. The junctions between cardiac muscle cells occur at connections called **intercalated disks**. Intercalated disks contain desmosomes and gap junctions, which form a tighter bond than the junctions at skeletal muscle cells. Cardiac muscle cells therefore function as more integrated units than skeletal muscle cells.

Cardiac muscle cells derive their energy in metabolism in the same way as skeletal muscle cells. However, cardiac muscle cells generally require more energy, and therefore there is more metabolic activity going on in their cells.

The contractions of cardiac muscle cells are generally independent of impulses received from the nervous system. The heart initiates and distributes nerve impulses to contract its cells through a system of specialized tissues. The first component of the system consists of cells of the **sinoatrial (SA) node**. This mass of nervelike cells is located in the wall of the right atrium. Impulses generated at the sinoatrial node ultimately reach to all parts of the heart. The SA node sets the pace for heart contractions and is commonly known as the **pacemaker** (Figure 15.6). The SA node depolarizes without other nerve intervention at a rate approximating 70 to 80 times per minute. The depolarization wave initiated in the SA node sets the pace for the sinus rhythm.

Impulses are distributed to the heart tissues from the SA node through a series of nerve fibers called **Purkinje fibers** and by cell-to-cell transmission of impulses by the ventricular muscle cells. Impulses spread via the fibers to the second major node of the heart called the **atrioventricular node (AV)**. The atrioventricular node is located in the interatrial septum, the septum between the two atria. Impulses reaching the AV node are amplified by and then exit to a large group of Purkinje fibers known as the **bundle of His**. The fibers then divide to left and right bundles and penetrate to the myocardium and all other regions of the heart, where they provide nerve impulses for cardiac nerve contraction. Nerve impulses within the heart follow the typical depolarization and repolarization found in all nerve impulses in nerve cells (Chapter 10).

Impulse transmission through the heart's conduction system is detected in an **electrocardiogram**. In a typical electrocardiogram record, three recognizable waves are seen in each cardiac cycle. The first wave, the **P wave**, is an upward wave indicating depolarization of the atria and the spread of an impulse from the SA node through the atria, causing them to contract. The second wave, the **QRS wave**, has downward, large upward, and downward waves. It represents contractions of the ventricles. Then comes a dome-shaped deflection called the **T wave**. This represents repolarization of the ventricles (Figure 15.7).

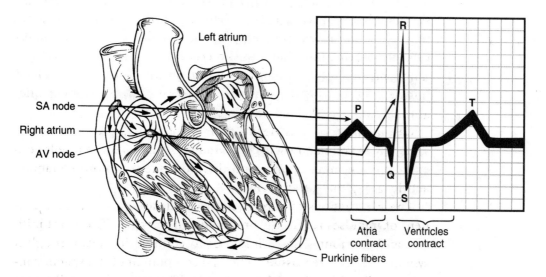

**FIGURE 15.7**  *The cardiac cycle as seen in the waves of an electrocardiogram. Impulses from the SA node cause the atria to contract and are reflected in the P wave. After a momentary pause, the impulses from the AV node cause the ventricles to contract. These contractions are shown in the QRS wave. The T wave represents the time that the muscles are recovering from the contraction.*

Although external stimulation is not required for heart activity, nervous control of the heart can be exerted by fibers of the autonomic nervous system. The rate of heartbeat and ventricular contraction is increased by impulses from the sympathetic nervous system and decreased by the parasympathetic nervous system (Chapter 11).

Although the heart's rhythm is usually regular, irregular heart rhythms may occur. These are called **arrhythmias**. When the heart is contracting rapidly or irregularly, it is said to be fibrillating; the condition is called **fibrillation**. A strong electric shock can be used to defibrillate the heart.

## The Heart Cycle

The alternating contractions and relaxations of the heart chambers is referred to as the **cardiac cycle**. The term **systole** refers to heart contractions; the term **diastole** refers to the relaxation periods of the heart. Therefore, the heart cycle consists of systole and diastole.

While systole is occurring in the ventricles (blood is being pumped out), the atria are in a condition of diastole and are filling with blood. When the blood pressure in the atria exceeds that in the ventricles, the blood forces itself against the atrioventricular valves and flows into the ventricles. Now the blood pressure in the ventricles rises and, during contraction, the blood flows out of the ventricles into either the pulmonary artery or the aorta. The amount of blood pumped out of a ventricle during each systole is called the **stroke volume**. The amount of blood pumped out of a ventricle per minute is known as the **cardiac output**.

The heart beats approximately 70 to 75 times each minute, and a cardiac cycle consumes slightly less than one second. Since the stroke volume in an adult averages about 70 ml of blood, the average cardiac output is about 5250 ml of blood per minute. The flow of blood is controlled by pressure changes, and blood will flow through any opening that is available.

When the atrioventricular valves shut, the heart emits a sound, often described as a "**lub**." This is the first heart sound. Next, blood flows through the pulmonary arteries and aorta, and the semilunar valves shut. This sound is referred to as a "**dub**." The heart sounds are therefore described as "lub-dub." Unusual heart sounds, called **murmurs**, indicate a problem at the valves. Heart sounds can be heard with a stethoscope.

## THE BLOOD VESSELS

Blood vessels form a network of tubes carrying blood between the heart and body cells and tissues. These tubes are an essential element of the cardiovascular system because they provide the environment in which the blood cells can perform their function.

Those vessels carrying blood away from the heart are **arteries**. Arteries divide into smaller vessels called **arterioles**, which branch out to various parts of the body. Arterioles enter organs and tissues where they branch into microscopic vessels called **capillaries**. Blood in the capillaries services the tissues. The capillaries then emerge from the cellular environment and form small veins called **venules**. Venules merge with other venules to form larger tubes, the **veins**, which convey blood back toward the heart (Table 15.2).

All the blood vessels except the capillaries have three distinct layers of cells. These layers are called **tunics**. They surround the tubular opening of the blood vessel called the lumen. The inner tunic lining the vessel lumen is the tunica (tunica interna). This layer is composed of a thin layer of endothelium, with simple squamous epithelium lying on a connective tissue basement membrane. The middle layer is the tunica media. It consists primarily of smooth muscle cells and fibers of elastic connective tissue. The outer layer of the blood vessel is the tunic adventitia, also called the tunic externa. The principal tissue of this layer is collagen fibers loosely woven. The fibers protect the blood vessels and provide anchorage to the surrounding structures.

**TABLE 15.2**   *A Comparison of the Body's Blood Vessels*

| Vessel | Structure | Function |
|---|---|---|
| Artery | Thick, strong wall with three layers—endothelial lining, middle layer of smooth muscle and elastic tissue, and outer layer of connective tissue | Carries high-pressure blood from heart to arterioles |
| Arteriole | Thinner wall than artery, but with three layers; smaller arterioles have endothelial lining, some smooth muscle tissue, and small amount of connective tissue | Connects artery to capillary; helps to control blood flow into capillary by undergoing vasoconstriction or vasodilation |
| Capillary | Single layer of squamous epithelium | Provides semipermeable membrane through which nutrients, gases, and wastes are exchanged between blood and tissue cells; connects arteriole to venule |
| Venule | Thinner wall, less smooth muscle and elastic tissue than arteriole | Connects capillary to vein |
| Vein | Thinner wall than artery, but with similar layers; middle layer more poorly developed; some with flaplike valves | Carries low-pressure blood from venule to heart; valves prevent back flow of blood; serves as blood reservoir |

## Arteries

The arteries are composed of a hollow core called the **lumen**, surrounded by three layers of tissue. The innermost layer of tissue consists of simple squamous epithelium and is referred to as the **endothelium**. The inner layer also contains elastic tissue. The middle layer is composed of smooth muscle as well as elastic tissue. The outer layer consists primarily of elastic fibers and collagen.

Arteries have the ability to expand and accommodate blood pulsing through them when the heart contracts. This expansion is due to the elastic connective tissue. When the heart relaxes, the elastic tissue recoils and forces the blood forward.

Another characteristic of the arteries owing to its smooth muscle is the ability to contract. The narrowing of the artery lumen, called **vasoconstriction**, can be induced by nerve impulses from the sympathetic nervous system. An increase in lumen size, called **vasodilation**, occurs in the absence of impulses from this system.

## Arterioles, Capillaries, and Venules

The walls of large **arterioles** are similar to those of arteries, but as the arterioles decrease in size, they consist primarily of endothelium with a small amount of smooth muscle. Like arteries, arterioles undergo vasoconstriction and vasodilation.

The microscopic capillaries connect arterioles to venules. Capillaries can be found near practically every body cell, especially those having high metabolic activity (these cells require more oxygen than other cells). The exchange of nutrients and gases occurs between the blood and the cells in the capillaries as explained by Starling's law of fluid movement (Chapter 21). Capillary walls have only a single layer of endothelium, and substances pass through them easily. Many capillaries contain **sphincters** at the entry to the capillary. Sphincters are circular muscles that open and close to regulate the flow of blood through the capillary.

Venules form when several capillaries combine with one another. The venules collect blood and deliver it to the veins. The walls of small venules are similar to those of small arterioles, but large venules have the same structure as veins.

## Veins

Veins are similar to arteries in structure, except that the outer layer in veins is thicker and the middle layer is thinner. Blood flowing through a vein lacks pressure, so the blood flows smoothly as compared to blood in the arteries, which spurts. Since the blood flows with less pressure, the vein walls need not be as strong as artery wall.

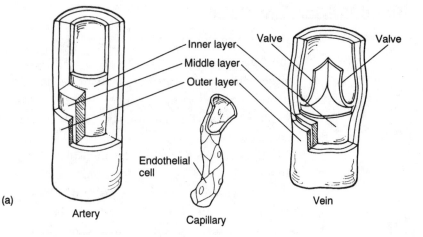

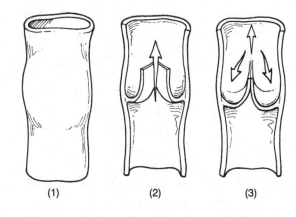

**FIGURE 15.8**   *Blood vessels and valves. (a) The structure of the artery, vein, and capillary compared. Three tissue layers are found in arteries and veins, but only a single layer of endothelial cells makes up the capillary. The walls of the artery are thicker than the venous walls to resist the pressure of the blood. (b) Arteries have no valves, but valves are found in veins. In the vein the area at a valve (1) is dilated. The valve flaps open when blood flows through (2), then the flaps come together (3) closing the cavity and preventing the backflow of blood.*

To assist the blood flow, many veins have infoldings of the inner tissue layer to form **valves**. Valves prevent the backflow of venous blood, especially in the limbs (Figure 15.8). When the valves weaken, blood leaks backward, and the pressure of the blood dilates the wall of the vein. Dilated veins are called **varicose veins**. Aging, prolonged standing, and pregnancy contribute to varicose veins.

The veins serve as blood reservoirs because they contain much of the body's blood. Approximately 60 percent of the blood is carried in the veins and venules.

## Blood Pressure and Pulse

The flow of blood through the body's circulation influences the **blood pressure**, which is the pressure blood exerts on the walls of vessels. Blood pressure is caused by cardiac output (higher cardiac output, higher blood pressure) as well as resistance to blood flow. Resistance arises from blood viscosity, blood vessel length, and blood vessel radius.

The blood pressure can be measured by an instrument called a **sphygmomanometer**. A sphygmomanometer contains a column of mercury, which rises when a pulse of blood is detected. When the heart pumps blood from the left ventricle, the pressure on the walls of nearby arteries is sufficient to raise the column of mercury to a height of 120 millimeters (mm). This is called the **systolic pressure** (the pressure occurring during systole). During diastole the blood continues to exert pressure on the wall of the blood vessel. This pressure, the **diastolic pressure**, is sufficient to raise the column of mercury to 80 mm. Therefore, the average blood pressure for an individual may be expressed as systolic pressure "over" diastolic pressure, or "120/80." Various conditions can increase or decrease the blood pressure. For instance, narrowing of the arteries can increase blood pressure significantly.

As the left ventricle of the heart contracts and relaxes, it pushes blood through the arteries and creates the **pulse**. Pulse is stronger near the heart and weaker as blood becomes more distant from the heart. It is common to take the pulse in the radial artery of the wrist. The carotid artery and the popliteal artery can also be used to measure the pulse. The pulse is the same as the heart rate and averages about 70–75 beats per minute. A rapid pulse reflects a rapid heart rate and a condition called **tachycardia**. A slow pulse reflects a slow heart rate and a condition called **bradycardia**.

The volume of blood coursing through the blood circulation is approximately five liters. A decrease in this volume, such as by excessive bleeding, can result in reduced blood pressure. Reduced heart rate can also result in lower blood pressure because the volume of blood being put out by the heart is less.

## Regulation of Blood Flow

The homeostasis of the body requires that blood flow be regulated and blood pressure be maintained within a certain range. High blood pressure can damage tissues of the heart, brain, kidneys, and other organs, while low blood pressure can cause inadequate amounts of oxygen and nutrients to reach body cells.

The blood flow can be regulated by regulatory centers in the brain and other regions, and by various chemicals throughout the body. The **vasomotor center** is a cluster of sympathetic neurons in the medulla of the brain. Impulses from this center control the diameter of blood vessels and the blood pressure by means of impulses to the smooth muscles of the arteriole walls. This control helps maintain the blood pressure throughout the body. Decreasing the number of sympathetic impulses results in vasodilation and lowered blood pressure.

Another form of regulation is exerted by **baroreceptors**. Baroreceptors are clusters of neurons found in the major arteries of the neck and chest, such as the aorta and carotid arteries. Baroreceptors send impulses to increase or decrease the activity of cardiac muscle and thereby regulate blood flow. Baroreceptors also send impulses to the vasomotor center and indirectly control blood flow by means of these neurons.

Certain neurons are sensitive to the chemicals found in the blood. These neurons, called **chemoreceptors**, are clustered in two masses of tissue where the common carotid arteries branch into the external and internal carotid arteries. The clusters are called **carotid bodies**. Other chemoreceptors are found in **aortic bodies**, small masses of tissue located near the aorta. Chemoreceptors react to abnormal levels of oxygen, carbon dioxide, and hydrogen ions in the blood. They relay impulses to the vasomotor center, which regulates blood flow.

Blood flow can also be regulated by neuron centers in the **cerebral cortex**. During periods of intense anger, for example, the cortex neurons send impulses to the hypothalamus and on to the vasomotor center. During times of great depression, the blood pressure can be lowered by the same pathway.

Certain chemicals can affect blood pressure by inducing vasoconstriction. Two examples of chemicals that act in this way are **epinephrine** and **norepinephrine**. Both are hormones of the adrenal glands, and both can increase the rate of heart activity, thereby increasing the rate of blood flow. Another hormone, **antidiuretic hormone** (Chapter 13), is a product of the hypothalamus that causes vasoconstriction by regulating the amount of water in the blood.

## Shock

**Shock** is a condition that develops when the cardiovascular system fails to deliver sufficient oxygen and nutrients to the body cells. During shock, the membranes of cells become dysfunctional, and cellular metabolism is abnormal. Cellular death often occurs. Such things as hemorrhage, dehydration, and excess vomiting or sweating can bring about shock. Shock can be hypovolemic (decreased volume from loss of blood) or obstructive (mechanical obstruction to blood flow such as a thrombus) or septic (toxin damage from microorganisms).

Some symptoms of shock include low blood pressure and clammy, cool, pale skin due to constriction of the blood vessels in the skin. Another characteristic symptom is sweating. The person becomes disoriented and often faints due to lack of oxygen in the brain tissues. Tachycardia is often observed. The pulse is weak and rapid because of the cardiac output. Blood transfusions can relieve the symptoms of shock.

## Circulation Patterns

As noted previously, blood circulation follows pulmonary and systemic circuits. The major artery of the systemic circuit is the aorta, while the major veins are the inferior and superior vena cavae. Other arteries of the body arise from the aorta as it extends from the left ventricle and turns to form an arch. At this point, **coronary arteries** have emerged to supply the heart muscle, and major arteries to the neck and head have also emerged.

The arching **aorta** then descends the body along the spinal cord and gives rise to numerous other major **arteries**, which carry blood to the major organs of the body. Figure 15.9 illustrates the location of many of these important arteries. Among the significant ones are the carotid arteries of the neck; the subclavian arteries of the shoulder; the axillary, brachial, radial, and ulnar arteries of the upper limb; the intercostal arteries of the thoracic wall; the phrenic arteries of the diaphragm; the celiac artery, which gives rise to the left gastric, splenic, and hepatic arteries; the superior mesenteric artery, which extends to the small intestine; the renal arteries to the kidneys; the common iliac arteries to the lower limbs; and the external iliac artery, which gives rise to the femoral, popliteal, and tibial arteries.

After the blood has serviced the cells, the **capillaries** lead to venules, veins, and eventually to the vena cavae. The two vena cavae unite just before the right atrium, and the blood flows into the atrium by a single, major **vena cava**. Figure 15.10 illustrates the major veins of the venous system. Among the important ones are the jugular veins in the neck; the axillary, brachial, radial, and ulnar veins, which extend from the upper limb; the azygous and hemiazygous veins from the thoracic muscles; the splenic, inferior mesenteric, and superior mesenteric veins from abdominal organs; and the common iliac veins from the lower limbs.

The **pulmonary circuit** takes blood from the right ventricle to the tissues of the lungs and returns it to the left atrium. Two pulmonary arteries extend to the two lungs, and after gas exchange has taken place, the capillaries unite to form the pulmonary veins. Pulmonary veins emerge from the right and left lung and unite to form the single, pulmonary vein. The pulmonary artery is the only artery that carries oxygen-poor blood, while the pulmonary vein is the only vein to carry oxygen-rich blood.

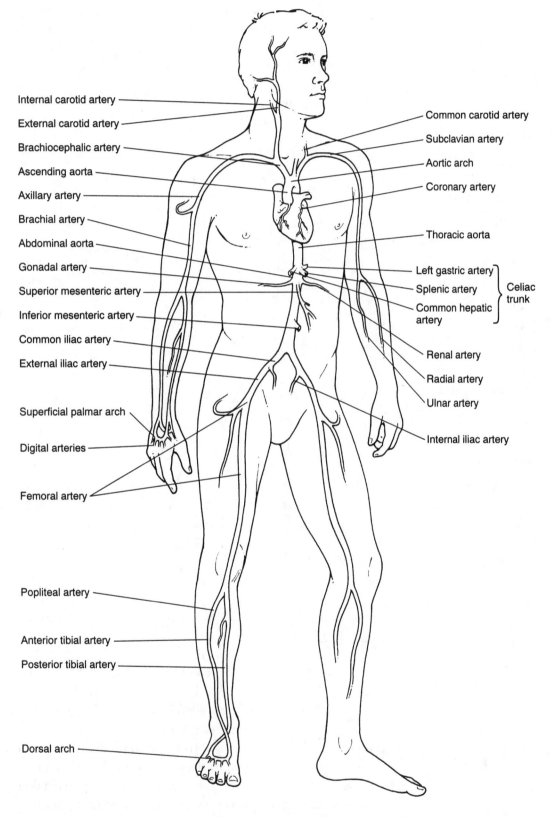

Internal carotid artery

External carotid artery

Brachiocephalic artery

Ascending aorta

Axillary artery

Brachial artery

Abdominal aorta

Gonadal artery

Superior mesenteric artery

Inferior mesenteric artery

Common iliac artery

External iliac artery

Superficial palmar arch

Digital arteries

Femoral artery

Popliteal artery

Anterior tibial artery

Posterior tibial artery

Dorsal arch

Common carotid artery

Subclavian artery

Aortic arch

Coronary artery

Thoracic aorta

Left gastric artery

Splenic artery

Common hepatic artery

Celiac trunk

Renal artery

Radial artery

Ulnar artery

Internal iliac artery

**FIGURE 15.9**   *The major arteries of the human body, excluding the pulmonary arteries.*

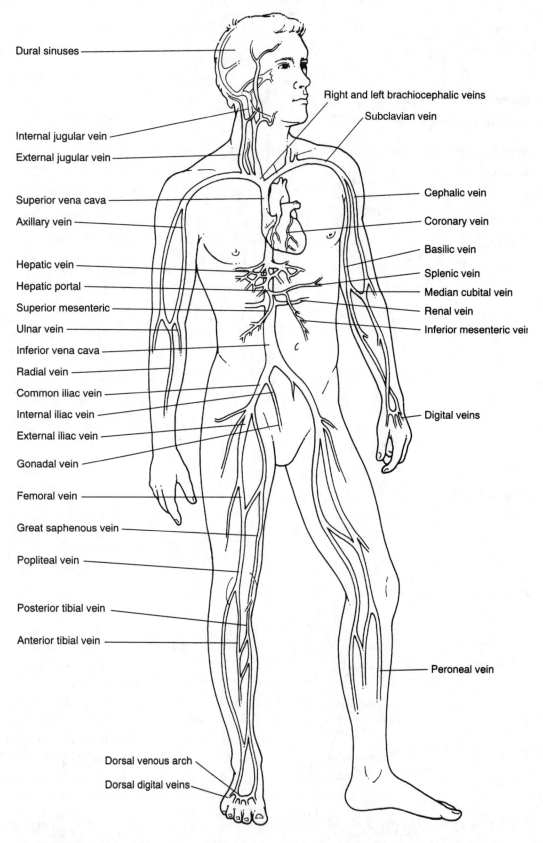

Dural sinuses

Internal jugular vein

External jugular vein

Superior vena cava

Axillary vein

Hepatic vein

Hepatic portal

Superior mesenteric

Ulnar vein

Inferior vena cava

Radial vein

Common iliac vein

Internal iliac vein

External iliac vein

Gonadal vein

Femoral vein

Great saphenous vein

Popliteal vein

Posterior tibial vein

Anterior tibial vein

Dorsal venous arch

Dorsal digital veins

Right and left brachiocephalic veins

Subclavian vein

Cephalic vein

Coronary vein

Basilic vein

Splenic vein

Median cubital vein

Renal vein

Inferior mesenteric vein

Digital veins

Peroneal vein

**FIGURE 15.10**  *The major veins of the human body, excluding the pulmonary veins.*

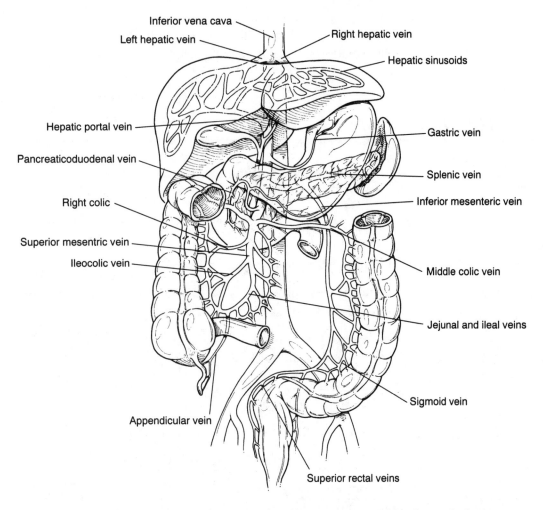

Inferior vena cava

Left hepatic vein

Right hepatic vein

Hepatic sinusoids

Hepatic portal vein

Gastric vein

Pancreaticoduodenal vein

Splenic vein

Right colic

Inferior mesenteric vein

Superior mesentric vein

Ileocolic vein

Middle colic vein

Jejunal and ileal veins

Appendicular vein

Sigmoid vein

Superior rectal veins

**FIGURE 15.11** *The hepatic portal system. The system consists of blood vessels that arise in tissues of the intestine and surrounding digestive organs and carry blood and nutrients to the liver (note the arrows). Nutrients are processed in the liver, then released to the hepatic vein, which carries them to the general circulation and to the tissue cells.*

Other circulations can also be found in the body. For example, the **cerebral circulation** consists of numerous blood vessels arising from arteries extending to the brain. The vessels service the brain, then unite to form veins extending out from the brain. The circulation of vessels in the brain is referred to as the **circle of Willis**.

Another circulatory route is the **hepatic portal system**, which transports blood from the gastrointestinal tract and spleen to the liver (Figure 15.11). The hepatic-portal circulation is a one-way route. Its purpose is to carry nutrients to the liver for processing. The major vessel of the hepatic-portal circulation is the **hepatic portal vein** arising from the gastrointestinal system and flowing toward the liver. After passing through the liver, blood leaves through the hepatic veins, which then unite with the inferior vena cava. Blood in the hepatic-portal system is low in oxygen since it already has been to the gastrointestinal tract. To service its cells the liver receives oxygen-rich blood through the hepatic artery arising from the aorta.

# REVIEW QUESTIONS

**PART A—Completion:** Add the word or words that correctly complete each of the following statements.

1. The heart is enclosed within an area of the thorax known as the _____ .

2. The double sac membrane that covers the heart is the _____ .

3. The heart has three layers of tissue, of which the main constituent is the second layer called the _____ .

4. The layer of the heart tissue lining the heart chambers and making up much of the valve tissue is the _____ .

5. The heart chambers are separated longitudinally by a body of tissue known as the _____ .

6. The receiving chambers of the heart are the _____ .

7. The flat, wrinkled appendage of the atrium is the _____ .

8. The two vena cavae enter the right atrium together with a third vein known as the _____ .

9. The pumping chambers of the heart are the _____ .

10. Blood returns to the heart from the lungs by means of the _____ .

11. The large artery emerging from the left ventricle of the heart is the _____ .

12. Those blood vessels that carry blood away from the heart are _____ .

13. Those blood vessels that carry blood toward the heart are _____ .

14. On the right side of the heart, the valve between the right atrium and right ventricle is the _____ .

15. On the left side of the heart, the bicuspid valve has two flaps and is also known as the _____ .

16. The flaps of the heart valves are anchored to the wall of the ventricles by tissue chords known as _____ .

17. The valves within the pulmonary artery and aorta are referred to as the _____ .

18. Arteries that supply blood to the tissue of the heart are the
_____ .

19. A blockage found in the heart muscle formed by dead cells is a
_____ .

20. Cardiac muscle cells are connected to one another by junctions
called _____ .

21. The sinoatrial node of the heart may be found in the wall of the
_____ .

22. Because the sinoatrial node sets the pace for the nervous activity
of the heart, it is commonly known as the _____ .

23. The nerve fibers, which distribute nerve impulses to tissues of
the heart, are the _____ .

24. The second major node of the heart after the sinoatrial node is
the _____ .

25. Irregular heart rhythms sometimes occur in the tissues and are
known as _____ .

26. The nervous activity of the heart can be exerted by fibers of a
branch of the nervous system known as the _____ .

27. Heart contraction is known by the alternate term of _____ .

28. Relaxation periods of the heart during which no contractions
are occurring are known as _____ .

29. The heart beats each minute approximately _____ .

30. Unusual heart sounds, such as those emitted by poorly functioning valves, are called _____ .

31. The smallest vessels, which carry blood to the cells of the tissues
are the _____ .

32. The innermost layer of the artery is referred to as _____ .

33. When an artery's lumen undergoes a narrowing, the condition
is called _____ .

34. The entry to the capillary is guarded by circular muscles known
as _____ .

35. The union of several capillaries emerging from the cellular environment forms a vessel known as a _____ .

36. The inner layer of the vein often folds inward to form a _____ .

37. Dilated veins cause a condition known as _____ .

38. The pressure of the blood can be measured by an instrument
known as a _____ .

39. A typical blood pressure reading contains two numbers of which the first number is the systolic pressure and the second number is the _____ .

40. A rapid pulse reflects a rapid heart rate and a condition called _____ .

41. A slow pulse reflects a slow heart rate and a condition called _____ .

42. The volume of blood passing through the circulation of an adult is approximately _____ .

43. The regulatory center of the brain that maintains the flow of blood is a cluster of sympathetic neurons known as the _____ .

44. Neurons in arteries of the neck and chest regulate the blood flow and are known as _____ .

45. Two examples of chemicals that affect the blood pressure by inducing vasoconstriction are epinephrine and _____ .

46. Insufficient oxygen and nutrients delivered to the body cells may induce membrane dysfunction and a condition known as _____ .

47. The only artery of the body that carries oxygen-poor blood is the _____ .

48. The only vein of the body that carries oxygen-rich blood is the _____ .

49. The blood vessel circulation within the brain is known as the _____ .

50. The vein that carries nutrients from the gastrointestinal tract to the liver is the _____ .

**PART B—Multiple Choice:** Circle the letter of the item that correctly completes each of the following statements.

1. All the following apply to the heart except
   (A) it is a cone-shaped organ
   (B) it weighs less than a pound
   (C) it is a hollow organ
   (D) it is roughly the size of the person's head

2. The pericardium is the double sac membrane that
   (A) encloses the heart
   (B) line the aorta
   (C) makes up the heart valves
   (D) is found only in the capillaries

3. Most of the cardiac muscle of the heart is found in the
   (A) endocardium
   (B) epicardium
   (C) myocardium
   (D) pericardium

4. The interventricular septum and the intra-atrial septum separate the
   (A) chambers of the heart
   (B) chambers of the lungs
   (C) aorta and pulmonary artery
   (D) bicuspid and tricuspid valves

5. Blood returning to the heart from the body organs enters the
   (A) left atrium through the aorta
   (B) right atrium through the vena cava
   (C) left ventricle by the pulmonary artery
   (D) right ventricle by the pulmonary vein

6. The systemic circuit of the cardiovascular system extends
   (A) from the heart to the lungs
   (B) from heart to the coronary arteries
   (C) from the heart to the body's organs and tissues
   (D) from the gastrointestinal tract to the liver

7. The only vein in the body that transports oxygen-rich blood is the
   (A) coronary vein
   (B) hepatic portal vein
   (C) pulmonary vein
   (D) aortic vein

8. All arteries of the body flow
   (A) to the liver
   (B) to the brain
   (C) away from the lungs
   (D) away from the heart

9. The semilunar valves prevent blood from flowing backwards
   (A) into the atria
   (B) into the ventricles
   (C) into the brain
   (D) into the liver

10. All the following apply to the bicuspid valve except
    (A) it is also called the mitral valve
    (B) it is a semilunar valve
    (C) it is found on the left side of the heart
    (D) it prevents blood from backing into the left atrium

11. The arteries supplying blood to the living tissue of the heart are
    the
    (A) renal arteries
    (B) myocardial arteries
    (C) coronary arteries
    (D) vena cavae

12. A blockage within the heart arteries caused by the death of heart
    muscle cells is known as
    (A) an embolism
    (B) an infarct
    (C) an abscess
    (D) a trochanter

13. Intercalated disks are found
    (A) between the right side and right side of the heart
    (B) between the flaps of the tricuspid valve
    (C) where the aorta joins the pulmonary artery
    (D) between the cardiac muscle cells

14. Which of the following applies to the sinoatrial node?
    (A) it is a mass of muscle cells
    (B) it produces important enzymes
    (C) it generates nerve impulses to contract the heart
    (D) it contains both bicuspid and tricuspid valves

15. The bundle of His
    (A) is found in the aorta
    (B) is a group of Purkinje fibers
    (C) prevents the mitral valve from flapping backward
    (D) is a group of arteries that supply the heart

16. Nervous control of the heart can be exerted by
    (A) nerves from the thoracic region of the spinal column
    (B) the second and third cranial nerves
    (C) by fibers of the sensory somatic system
    (D) by fibers of the autonomic nervous system

17. The condition called arrhythmia is characterized by
    (A) rapid heart contraction
    (B) irregular heart rhythms
    (C) mitral valve prolapse
    (D) semilunar valve dysfunction

18. The terms systole and diastole refer to
    (A) sounds from the heart
    (B) the major artery and vein from and to the heart
    (C) heart contractions and relaxations
    (D) rates of heart pulse

19. Which of the following represents the flow of blood from the heart to the body organs and back to the heart?
    (A) venules to capillaries to veins to arteries
    (B) arteries to capillaries to veins
    (C) capillaries to arterioles to arteries to veins
    (D) veins to arteries to capillaries to arterioles

20. The term vasoconstriction refers to
    (A) increasing the size of the lumen of the blood vessel
    (B) decreasing the size of the lumen of the blood vessel
    (C) delivering oxygen and nutrients to the body tissues
    (D) delivering waste products to the kidney for excretion

21. The blood pressure is measured by an instrument known as a
    (A) electrocardiogram
    (B) electroencephalograph
    (C) sphygmomanometer
    (D) CAT scan machine

22. Blood flowing through a vein tends to
    (A) spurt
    (B) flow smoothly
    (C) carry oxygen to the body cells
    (D) flow at a faster rate than in the artery

23. The pulse rate of an individual averages about
    (A) 10 beats per minute
    (B) 40 beats per minute
    (C) 50 beats per minute
    (D) 70 beats per minute

24. All the following have the ability to regulate blood flow in the body except
    (A) antidiuretic hormone
    (B) epinephrine and norepinephrine
    (C) chemoreceptors
    (D) enzymes from the salivary glands

25. The hepatic portal vein transports blood
    (A) from the heart to the liver
    (B) from the liver to the spleen
    (C) from the gastrointestinal tract to the liver
    (D) from the liver to the gastrointestinal tract

**PART C—True/False: For each of the following statements, mark the letter "T" next to the statement if it is true. If the statement is false, change the underlined word to make the statement true.**

1. The heart is located approximately between the second and fifth ribs and posterior to the vertebral column.

2. The pericardium is a double sac membrane in which the outer membrane is the visceral pericardium.

3. The major constituent of the heart is a layer of cardiac muscle known as the endocardium.

4. The two inferior chambers of the heart are known as the atria.

5. Blood returns from the body through the superior and inferior vena cavae, which empty into the left atrium.

6. Blood returning from the heart muscle enters the ventricular sinus.

7. Blood moves toward the lungs after it leaves the right ventricle.

8. The aorta, the largest artery of the body, receives blood from the right ventricle.

9. The tricuspid valve lies between the left atrium and the left ventricle.

10. The valves found within the pulmonary artery and the aorta are known as semilunar.

11. The atrioventricular valves prevent blood from flowing backward into the ventricles.

12. Dying cells in the heart muscle may form a blockage known as a coronary thrombosis.

13. Impulses for the contraction of heart muscle are generated initially at the atrioventricular node.

14. Fibers known as Purkinje fibers spread out from the SA node and carry impulses to the second major heart node.

15. Some nerve control over the heart can be exerted by fibers of the autonomic nervous system.

16. The condition in which the heart contracts rapidly and irregularly is known as arrhythmia.

17. The relaxation period between heart contractions is correctly known as systole.

18. The heart beats approximately 70–75 times each second.

19. A heart murmur is generally due to unusual heart sounds arising from improper activity of the heart muscle.

20. The smallest heart vessels in the body are known as <u>venules</u>.

21. The narrowing of the lumen of the artery is known as <u>vasodilation</u>.

22. The condition of varicose veins is generally due to improper activity of the vein <u>sphincters</u>.

23. A pulse rate that is more rapid than normal reflects a condition called <u>tachycardia</u>.

24. The carotid bodies and aortic bodies contain neurons called <u>baroreceptors</u> that help regulate the blood flow.

25. The only artery that carries carbon dioxide-rich blood is the <u>pulmonary</u> artery.

**Answers**

## PART A—Completion

1. mediastinum
2. pericardium
3. myocardium
4. endocardium
5. cardiac septum
6. atria
7. auricle
8. coronary sinus
9. ventricles
10. pulmonary veins
11. aorta
12. arteries
13. veins
14. tricuspid valve
15. mitral valve
16. chordae tendinae
17. semilunar valves
18. coronary arteries
19. myocardial infarction
20. intercalated disks
21. right atrium
22. pacemaker
23. Purkinje fibers
24. atrioventricular node
25. arrhythmias
26. autonomic nervous system
27. systole
28. diastole
29. 70–75 times
30. murmurs
31. capillaries
32. endothelium
33. vasoconstriction
34. sphincters
35. venule
36. valve
37. varicose veins
38. sphygmomanometer
39. diastolic pressure
40. tachycardia
41. bradycardia
42. five liters
43. vasomotor center
44. baroreceptors
45. norepinephrine
46. shock
47. pulmonary artery
48. pulmonary vein
49. circle of Willis
50. hepatic portal vein

## PART B—Multiple Choice

| | | | | |
|---|---|---|---|---|
| 1. D | 6. C | 11. C | 16. D | 21. C |
| 2. A | 7. C | 12. B | 17. B | 22. B |
| 3. C | 8. D | 13. D | 18. C | 23. D |
| 4. A | 9. B | 14. C | 19. B | 24. D |
| 5. B | 10. B | 15. B | 20. B | 25. C |

## PART C—True/False

1. anterior
2. parietal
3. myocardium
4. ventricles
5. right
6. coronary
7. true
8. left
9. bicuspid
10. true
11. atria
12. myocardial infarction
13. sinoatrial
14. true
15. true
16. fibrillation
17. diastole
18. minute
19. valves
20. capillaries
21. vasoconstriction
22. valves
23. true
24. chemoreceptors
25. true

# CHAPTER 16

# THE LYMPHATIC AND IMMUNE SYSTEMS

The lymphatic system functions together with the cardiovascular system for the delivery of nutrients and oxygen to the tissue cells and the removal of metabolic waste products. Unlike the cardiovascular system, the lymphatic system is a one-way system, arising in the tissues and extending toward the heart. Its vessels and organs return blood fluid (rather than blood) from the tissues to the circulation for reuse.

Closely associated with the lymphatic system is the immune system. This system is responsible for the body's specific response to foreign microorganisms and chemical molecules. The immune system operates through cells of the lymphatic system, and products of the immune system are usually carried in the lymphatic vessels.

## THE LYMPHATIC SYSTEM

The lymphatic system consists of lymph, the lymphatic vessels, and the lymphoid tissues. The functions of the system are to return fluid (lymph) to the circulatory system from the intercellular tissue spaces (Figure 16.1).

### Lymphatic Vessels

Lymphatic vessels are thin-walled tubes distributed throughout the body. They arise as tubular networks within the tissues and are particularly numerous within the skin tissues, especially the dermis. The microscopic tubes making up the network are called **lymphatic capillaries**. They have a lining of endothelium and, in this regard, they resemble capillaries. However, they are more permeable than blood capillaries.

Lymphatic capillaries lead from the tissues into larger **lymphatic vessels**. Lymphatic vessels, in turn, unite to form still larger vessels, which eventually form the **thoracic duct**, the largest lymphatic vessel of the body. The thoracic duct arises in the abdominal cavity and proceeds up to the thorax anterior to the vertebrae and dorsal to the esophagus. As it nears the neck, the thoracic duct turns to the left and empties its contents into the base of the left subclavian vein. The thoracic duct drains the entire body below the diaphragm and the left half of the body above the diaphragm.

The right half of the body above the diaphragm is drained by another large vessel called the **right lymphatic duct** (Figure 16.2). This duct unites with the right subclavian vein at its base and thereby empties its contents into the cardiovascular system. Since both the thoracic duct and the right lymphatic duct return fluid from the tissues to the cardiovascular system, the fluid of the lymphatic system flows in only one direction.

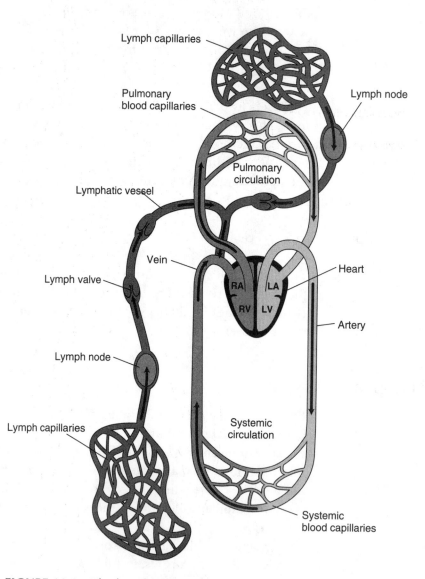

**FIGURE 16.1**   *The lymphatic system as it relates to the circulatory system. Fluid seeps out of the circulatory system in the tissues and, while much of the fluid returns to the system, some fluid enters the lymph capillaries and returns to the circulation by this method. The lymphatic system is therefore a one-way system.*

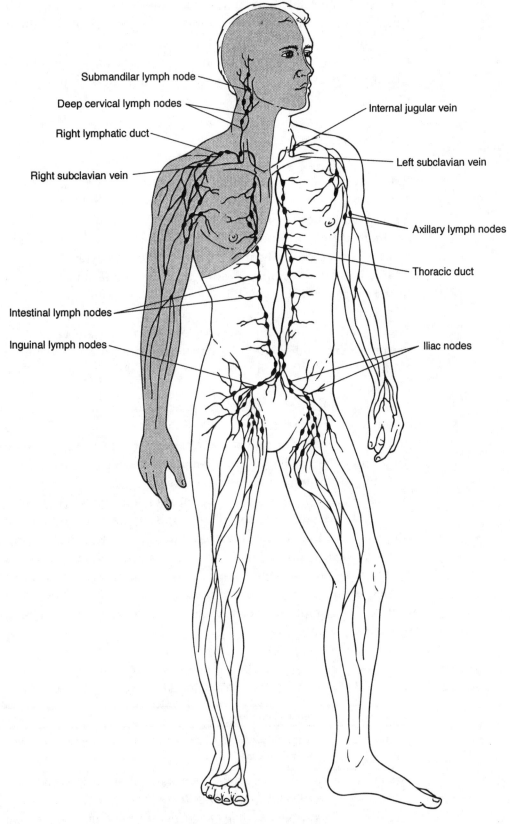

Submandilar lymph node

Deep cervical lymph nodes

Right lymphatic duct

Right subclavian vein

Internal jugular vein

Left subclavian vein

Axillary lymph nodes

Thoracic duct

Intestinal lymph nodes

Inguinal lymph nodes

Iliac nodes

**FIGURE 16.2**    *The human lymphatic system consists of lymph vessels and lymph nodes distributed throughout the body, as illustrated. The shaded portion represents the area drained by the right lymphatic duct, while the unshaded portion is that drained by the thoracic duct.*

Lymphatic vessels have numerous valves. The valves encourage lymph to move in a single direction and operate like the valves of veins (Chapter 15). In the lymphatic vessels the movement of fluid is assisted by pressure exerted on the walls by contracting muscles. The lymphatic vessels are well-adapted for the removal of large molecules, particularly proteins.

## Lymph Nodes

Before lymphatic fluid enters the bloodstream it passes through **lymph nodes**, which are masses of tissue enclosed in capsules (Table 16.1). Lymph nodes provide a filtration mechanism for the lymph before it rejoins the bloodstream. The lymph nodes also contain two types of cells called **T-lymphocytes** and **B-lymphocytes**. These cells are the basic underpinnings of the body's immune system. The two distinct regions of the lymph node are the cortex and medulla. The **cortex**, which is the outer region, contains collections of lymphocytes called follicles. At the center of the follicles are areas called germinal centers. The predominant cells of the germinal cells are B-lymphocytes, and the remainder of the cells of the cortex are T-lymphocytes.

Lymph nodes occur along the pathway of the larger lymphatic vessels and serve as filters for the lymph. Vessels entering the lymph nodes are called **afferent lymphatic vessels**, while vessels leaving the lymph nodes are called **efferent lymphatic vessels** (Figure 16.3).

At the lymph node, extensions of the capsule pass into the interior and separate the lymph node into smaller **lobules**. Within the lobules, a series of **reticular fibers** support the main cells of the lymph nodes, the B- and T-lymphocytes. Spaces within the lymph lobules, called **lymph sinuses**, are regions with relatively few cells. Lymph circulates through these regions. Lymphocytes are arranged densely within the outer part of the node in the cortex. The central region of the lymph node containing fewer lymphocytes is called the **medulla**.

**TABLE 16.1**  *Principal Organs of the Lymphatic System*

| Organ | Primary Functions |
| --- | --- |
| Lymphatic vessels | Carry lymph from peripheral tissues to the veins of the cardiovascular system |
| Lymph nodes | Monitor the composition of lymph; site of cells that engulf pathogens; immune response |
| Spleen | Monitors circulating blood; site of cells that engulf pathogens; site of cells that regulate the immune response |
| Thymus | Controls development and maintenance of T-lymphocytes |

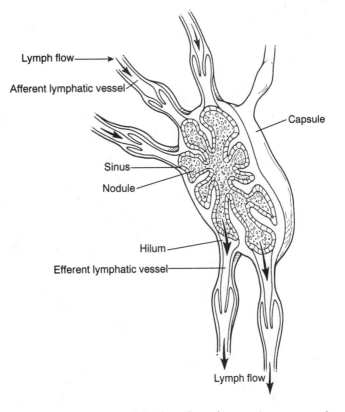

**FIGURE 16.3**   *Structure of the lymph node seen in cross section. The major structures are displayed.*

Lymph nodes are located throughout the body. They are prevalent in the tissues of the neck (the cervical lymph nodes); in the groin (the inguinal lymph nodes); in the bend of the elbow (at the cubital fossa); and behind the knee (in the popliteal fossa). Lymph nodes are also located in the armpits (the axillary lymph nodes) and in the mediastinum (the thoracic region between the lungs). In addition, they follow the course of large blood vessels in the abdominal cavity and the body appendages.

**Tonsils** are aggregates of lymph node tissue located under the epithelial lining of the oral and pharyngeal cavities. The term "tonsil" usually refers to the **palatine tonsils**, which are located at the surface of the palatine bone. Other tonsils include the mass of tissue known as the **pharyngeal tonsil** (also called **adenoids**) at the roof of the pharynx (Chapter 17), and the **lingual tonsils**, which are found in the tissue of the tongue (Figure 16.4). Nodules of lymphoid tissue are found in the wall of the intestinal tract, particularly in the ileum. These nodules of tissue are called **Peyer's patches**.

## The Thymus

Because of its tissue composition, the **thymus** is considered an organ of the lymphatic system. The thymus is located in the upper thorax

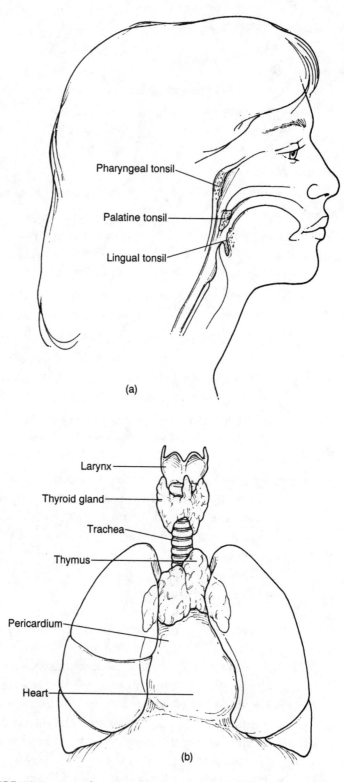

**FIGURE 16.4**   *Lymph system structures. (a) Location of the tonsils in the pharynx. (b) Position of the thymus in the upper thorax.*

in the mediastinum, between the lungs and dorsal to the sternum. During fetal development, the thymus is a relatively large bilobed organ. After birth, the thymus begins to degenerate, and the organ is very small by the teenage years.

The thymus is divided into a number of lobules containing supportive cells and T-lymphocytes, so called because primitive cells are modified in this organ to form the T-lymphocytes. Most T-lymphocytes migrate to the lymph nodes, but some remain in the thymus. The structure of the thymus is similar to the structure of the spleen and lymph nodes, with numerous lobules and elements of the cortex and medulla.

The thymus is also believed to be a gland because it produces and secretes hormones called thymosins (Chapter 13). These hormones are postulated to function in the maturation of T-lymphocytes.

## Spleen

The **spleen** is a lymphoid organ because its functions are consistent with those of the lymphatic system, and its cells are lymphatic cells. The spleen is located in the upper portion of the abdominal cavity, inferior to the diaphragm and on the left side.

The shape of the spleen conforms to the structures it contacts. The spleen is convex where it touches the diaphragm, and concave in three places where it contacts the left kidney, stomach, and large intestine. The area where large blood vessels enter and leave the spleen is called the **hillus**.

Like the lymph nodes, the spleen is surrounded by a capsule of connective tissue, which extends inward to divide the organ into numerous smaller regions (lobules) consisting of cells and small blood vessels. Lymphocytes are packed densely in the cortex, but less so in the medulla. Blood enters the spleen by way of the **splenic artery** (Figure 16.5).

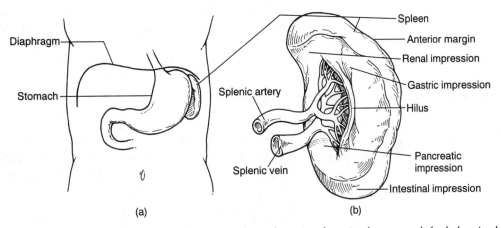

**FIGURE 16.5**  *Details of the spleen. (a) The spleen in place in the upper left abdominal cavity. (b) Some anatomical points of the spleen.*

The spleen has several important functions: It is a reservoir of lymphocytes for the body; it filters blood; it is important in red blood cell and iron metabolism because macrophages in the spleen phagocytize old and broken red blood cells and recycle the iron by sending it to the liver; it serves as a storage depot for blood; and it contains B- and T-lymphocytes for the immune response.

## Lymph

The fluid pumped through the lymphatic vessels is called **lymph**. Lymph is derived from blood. It consists of fluid forced through the semipermeable membrane of the capillary wall by pressure exerted by the heart. Fluid accumulating in the tissue spaces is filled with organic materials released by the cells. The proteins of the fluid are generally unable to pass into the capillaries, so they remain in high concentration within the fluid. In addition, any microorganisms present do not easily pass into capillaries and remain in the fluid.

The tissue fluid entering a nearby lymphatic vessel is lymph. Lymph passes through the lymph nodes, and lymphocytes and monocytes enter the lymph from these organs. This mixture of cleansed fluid and cells enters the circulation.

At the gastrointestinal tract, lymph has a milky consistency. When fats are digested in the digestive system (Chapter 18), the products are fatty acids, glycerol, and other components. While other organic molecules pass into capillaries, fats are reconstituted, then they pass into lymph vessels in the intestinal wall. These lymph vessels are called **lacteals**. Since fats have a milky consistency, the lymph appears milky when it is rich in fats.

An accumulation of tissue fluid in the spaces between the cells is called **edema**. Edema can occur if the lymphatic vessels are blocked, such as in an infection. Edema also occurs if there is a delay in blood movement within the veins or as blood accumulates in the veins. The escape of protein into the intercellular spaces, such as during inflammation, is another possible reason for edema. Protein draws water out of the vessels by osmosis, and the water contributes to the swelling.

## THE IMMUNE SYSTEM

The immune system provides specific resistance to the body during time of disease. The system is comprised of a complex series of cells, chemical factors, and processes in which lymphocytes respond to and eliminate foreign agents or substances known as antigens. Elimination can be by lymphocyte destruction of the antigens or by specialized protein molecules called antibodies.

## Development of the Immune System

The immune system develops during the third month after conception. At this period in fetal development, a set of primitive cells called **stem cells** arise in the bone marrow (Chapter 14). Certain of the stem cells form forerunners of the immune system called **lymphopoietic cells**.

Lymphopoietic cells follow either of two courses of development. Certain lymphopoietic cells pass through the thymus, where they are modified to form T-lymphocytes (or T-cells; "T" for thymus). In this modification, receptor molecules are placed on the cell surfaces. The T-lymphocytes migrate through the circulation and come to rest at the lymphoid tissues such as the spleen, tonsils, and lymph nodes.

The remainder of the immune system also develops from the lymphopoietic cells. Certain cells pass through an organ not yet identified in humans and become B-lymphocytes. In the embryonic chick, the modifying organ is the bursa of Fabricius (hence the name B-lymphocyte; "B" for bursa). The human organ corresponding to the bursa of Fabricius may be the bone marrow, liver, or lymph node tissues of the gastrointestinal tract (Figure 16.6). During formation, B-lymphocytes synthesize antibody molecules and position the molecules on their cell membranes. These antibody molecules will later act as receptor sites and react with foreign substances during the immune response.

When the immune system becomes functional after birth, B-lymphocytes and T-lymphocytes occupy a central position in the system. From their vantage point in lymph nodes and other lymphoid organs, the B- lymphocytes and T-lymphocytes encounter microorganisms that enter the systems of the body.

## Antigens

B-lymphocytes and T-lymphocytes can be stimulated by specific molecules called **antigens**, which are often associated with the surface components of microorganisms. An antigen is a substance, usually a large protein or polysaccharide, that stimulates the immune system. It may be part of a virus or a bacterial flagellum or a mold spore, or it may be a chemical substance such as a cytoplasmic macromolecule. The list of antigens is enormously diverse and includes over a million possible antigens.

Normally, a person's own proteins and polysaccharides do not stimulate an immune response because they are interpreted as **self**. Before birth the proteins and polysaccharides of body cells apparently contact and paralyze immune system cells that might later respond to them. Thus the individual becomes tolerant of "self" and will remain able to respond only to antigens interpreted as **nonself** or "foreign." The paralysis of responsive cells must continue throughout life for tolerance to oneself to remain.

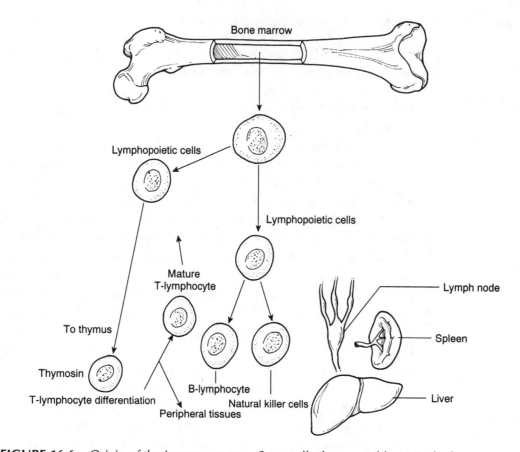

**FIGURE 16.6** *Origin of the immune system. Stem cells (hemocytoblasts) in the bone marrow give rise to lymphopoietic cells, which take either of two courses. Some pass through the thymus and are modified to form T-lymphocytes. Certain mature T-lymphocytes return to the bone marrow to participate in the immune reaction, but most proceed to the peripheral tissues where they congregate. Other lymphopoietic cells are modified in an unknown organ to form B-lymphocytes and natural killer cells. The B-lymphocytes also migrate to the peripheral tissues to join the T-lymphocytes. The lymph nodes and spleen are major depositories of T-lymphocytes and B-lymphocytes.*

Antigens enter the body through a variety of portals, including tiny openings in the mucous membranes of the respiratory tract and openings in the skin when it is penetrated by a wound. Such a wound may be caused by the bite of an arthropod (mosquito, tick, etc.).

## The Immune Process

The immune system reaches maturity several weeks after a person's birth and continues to function until a person's death. To initiate the immune process, foreign organisms or molecules in the body are approached by phagocytes, the white blood cells that specialize in engulfing and destroying foreign materials. Chief among the phagocytes are large, amoeboid cells called **macrophages** (Table 16.2).

**TABLE 16.2** *Macrophages in Various Tissues*

| Name of Macrophage | Tissue |
| --- | --- |
| Alveolar macrophage | Lung |
| Histiocyte | Connective tissue |
| Kupffer's cell | Liver |
| Microglial cell | Neural tissue |
| Osteoclast | Bone |
| Sinusoidal lining cell | Spleen |

Macrophages set the immune process into motion by engulfing and digesting microorganisms. The organism's antigens are preserved and displayed on the surface of the macrophages. Macrophages also have on their surface a set of proteins called **MHC ("major histocompatibility") molecules**. These molecules are present on all body cells and are unique for a particular person. The MHC molecules identify the macrophages as normal body cells. Now the macrophages move off toward the lymph vessels and the lymphoid tissue.

When the macrophage enters the lymph node, it encounters a T-lymphocyte called a **helper T-lymphocyte** (also called a T4 cell). As the two cells meet, the foreign antigens and the MHC antigens on the macrophage react with the receptor sites on the surface of the helper T-lymphocyte. This reaction activates the helper cell to produce and release a series of highly reactive proteins called **lymphokines**.

Lymphokines stimulate either the B-lymphocytes or the T-lymphocyte, depending on the nature of the antigen that initiated the process. The immune system therefore diverges at this point, and two major functional branches are present. One branch is dominated by the T-lymphocyte; the immunity that results is called **cell-mediated immunity**. The second branch is dominated by the B-lymphocyte; the resulting immunity is called **antibody-mediated immunity**.

## Cell-Mediated Immunity

**Cell-mediated immunity (CMI)** is so named because the defense imparted by this branch of the immune system involves a direct interaction between body cells and microorganisms or foreign molecules. The lymphokines released by the helper T-lymphocytes begin the process of CMI by stimulating the rapid multiplication of other T-lymphocytes called **cytotoxic T-lymphocytes**. Cytotoxic T-lymphocytes enter the circulatory system and search for cells displaying the foreign antigens (Figure 16.7).

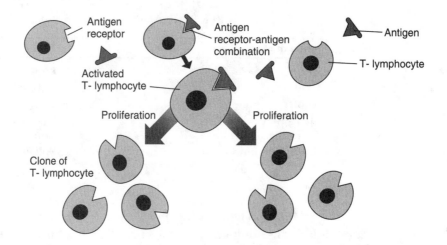

**FIGURE 16.7** *T-lymphocyte activation. In cell-mediated immunity, antigens react only with those T-lymphocytes having the complementary antigen receptor. The antigens avoid reaction with other T-lymphocytes. Once activated, a complex process results in proliferation of a clone of cytotoxic T-lymphocytes. These cells proceed to the infection site and react with microorganisms.*

Most fungus-infected, protozoa-infected, and certain virus- and bacteria-infected cells can be targets of cytotoxic T-lymphocytes. The cytotoxic T-lymphocyte interacts with the infected cell by recognizing the cellular MHC antigens and foreign antigens. It then attacks and destroys the cell, exerting a "lethal hit" on the cell.

Cytotoxic T-lymphocytes are also the source of additional lymphokines. Secreted at the antigen site, the lymphokines draw a supply of fresh macrophages to the infection site and stimulate them to destroy the microorganisms.

Another T-lymphocyte that bears mention is the **natural killer cell**. This cell also attacks microorganisms and infected cells, but it is a less-specialized cell than the cytotoxic T-lymphocyte. Natural killer cells appear to be a primary mechanism of defense of the body against tumor cells.

To prevent the immune process from becoming too exaggerated and destroying normal body cells, another T-lymphocyte comes into play. This is the **suppresser T-lymphocyte** (also known as the T8 cell). The suppresser T-lymphocyte dampens the activity of cytotoxic T-lymphocytes and natural killer cells and slows the immune process as the antigen stimulus lessens.

## Antibody-Mediated Immunity

The second branch of the immune system, **antibody-mediated immunity (AMI),** depends on the activity of B-lymphocytes. As the macrophage moves among the different types of B-lymphocytes in

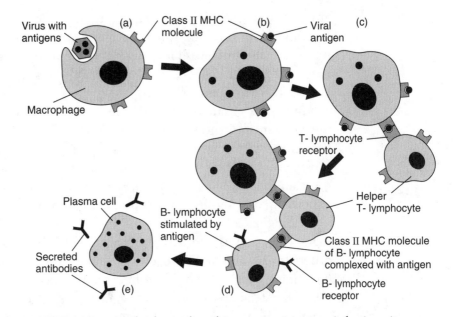

**FIGURE 16.8**   *Antibody-mediated immunity. (a) At an infection site, a virus is engulfed by a macrophage. (b) The macrophage displays the antigens at its surface within MHC molecules and proceeds to the immune system. (c) A helper T-lymphocyte interacts with the macrophage and (d) with a B-lymphocyte. (e) The reaction activates the B-lymphocyte, and it reverts to a plasma cell. The plasma cell produces and secretes antibodies for the immune reaction.*

the lymphoid tissue, it eventually encounters one type that contains the surface antibody molecules corresponding to its antigens. The binding of the antigen and antibody molecules, together with the intervention of a helper T-lymphocyte, activates or "commits" the B-lymphocyte (Figure 16.8).

Once committed, B-lymphocytes undergo cell division to give rise to a colony of cells programmed to produce **antibodies**. Antibody molecules specific for the antigen pour forth from the B-lymphocytes at a rate of more than 2000 per second. Within hours, other biochemical signals convert many of the B-lymphocytes into **plasma cells**, a group of high-activity, antibody-producing cells. The antigens initiating these activities are derived primarily from viruses and bacteria. Other substances such as milk protein, bee venom, food molecules, and ragweed proteins can also stimulate the process in a specialized type of immune reaction called an allergy.

Antibody molecules are proteins. They occur in five types known as IgG, IgM, IgA, IgD, and IgE. The most common of these is **IgG**. It is composed of two long chains and two small chains of amino acids. The antibody molecule has a hinge point where the chains diverge, and the molecule is therefore depicted as a Y (Figure 16.9). One end of the antibody molecule, the variable end, is highly specific for the antigen that elicited its production. For this reason the antibody will interact with that antigen only. The immune system has the capacity to produce a million different kinds of antibodies, one for each different possible antigen. Thus, a measles antibody will react only with a measles virus, a chickenpox antibody only with a chickenpox virus, and so forth.

Antibodies circulate through the body and soon encounter the microorganisms whose antigens stimulated their production. Then they chemically combine with the antigen molecules and neutralize the microorganisms by any of several mechanisms. For example, some antibodies bind to viral surfaces and prevent viral penetration of cells by covering the key receptor sites. Other antibodies combine with antigens of the surface of bacteria and bind the bacteria together in a meshlike pattern that can easily be phagocytized. Still others form a bridge between microorganisms and macrophages to encourage phagocytosis, and others set off a cascading series of reactions that tear apart microbial membranes. The antigen-antibody reaction usually results in destruction of the microorganism and recovery from disease.

As the most common antibody, IgG is closely associated with specific resistance to disease. **IgM** is also involved in resistance (Table 16.3). This molecule is composed of 20 amino acid chains. It is the first antibody to appear in the circulation after infection takes place, and it is responsible for the first interactions with antigens. **IgA**

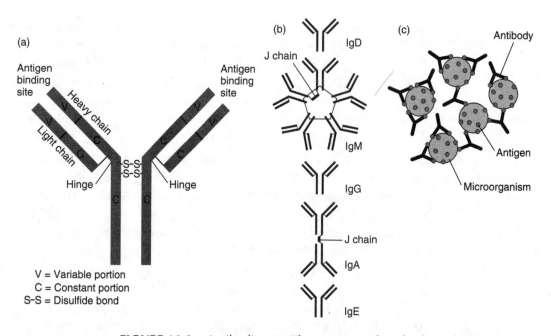

**FIGURE 16.9**    *Antibodies. (a) The structure of antibody molecule showing the four chains that make up the molecule (b) The structures of five different types of antibodies (c) The reaction between antibody molecules and antigens on the surface of a microorganism. Antibody molecules bind the microorganisms together and assist phagocytosis.*

consists of eight amino acid chains. It interacts with microorganisms at the body surface and along the respiratory, gastrointestinal, and other tracts open to the environment. **IgD** functions as the receptor site on B-lymphocytes, and **IgE** is produced during allergic reactions. Both antibodies contain four amino acid chains.

In allergic reactions, IgE fixes itself to basophils and mast cells and encourages the release of histamine, serotonin, and other physiologically active substances. These substances induce smooth muscle contractions and cause the labored breathing, abdominal cramps, hives, and other characteristic signs of allergies.

**TABLE 16.3**  *Characteristics of Five Types of Antibodies*

| Antibody Designation | Percentage of Antibody in Serum | Location in Body | Molecular Weight (daltons) | Number of Four Chain Units | Crosses Placenta | Functions |
|---|---|---|---|---|---|---|
| IgM | 5–10 | Blood, lymph | 900,000 | 5 | No | Principal component of primary response |
| IgG | 80 | Blood, lymph | 150,000 | 1 | Yes | Principal component of secondary response |
| IgA | 10 | Body secretion, body cavities | 400,000 | 2 | No | Protection in body cavities |
| IgE | <1 | Blood, lymph | 200,000 | 1 | No | Role in allergic reactions |
| IgD | 0.05 | Blood, lymph | 180,000 | 1 | No | Receptor site on B-lymphocyte |

# REVIEW QUESTIONS

*PART A—Completion:* **Add the word or words that correctly complete each of the following statements.**

1. The lymphatic system arises in the tissues and extends toward the _____ .

2. The lining of the lymphatic capillaries, like that of the blood capillaries, consists of _____ .

3. The largest lymphatic vessel of the body is the _____ .

4. The thoracic duct empties its contents into a vein known as the _____ .

5. The thoracic duct drains the entire body below the _____ .

6. The valves of the lymphatic vessels operate in a manner similar to the valves of the _____ .

7. Lymphatic fluid is filtered in masses of tissue known as _____ .

8. The two major types of cells in the lymph nodes are T-lymphocytes and _____ .

9. Lymph vessels entering the lymph nodes are referred to as _____ .

10. Extensions of the lymph node capsule pass into the lymph node and separate it into smaller _____ .

11. The main cells of the lymph nodes are supported by a series of _____ .

12. The cells of the lymph nodes are arranged densely within the outer portion of the lymph node called the _____ .

13. At the center of the lymph node, the region is known as the _____ .

14. In the process of immunity, the lymph nodes serve as the sites for the production of _____ .

15. In the neck tissues, the lymph nodes are known as the _____ .

16. Those lymph nodes located in the armpits are called _____ .

17. Lymph nodes may be found in the popliteal fossa, which is located behind the _____ .

18. Aggregates of lymph node tissue located behind the epithelial lining of the oral cavity are called _____ .

19. One of the important tonsils is located at the surface of a bone known as the _____ .

20. Nodules of lymphoid tissue found in the wall of the intestinal tract are called _____ .

21. Primitive cells are modified to form T-lymphocytes within an organ called the _____ .

22. The thymus is located in the body in a cavity called the _____ .

23. The thymus is relatively large during the development of the _____ .

24. The spleen is located in the upper portion of a cavity called the _____ .

25. The area where large blood vessels enter and leave the spleen is known as the _____ .

26. Blood entering the spleen does so by way of the _____ .

27. The spleen is the organ in the body where destruction occurs in the _____ .

28. Lymph consists of fluid derived from the _____ .

29. The lymph contains substances unable to pass into the capillaries such as _____ .

30. Lymph sometimes has a milky consistency due to the presence of _____ .

31. Lymph vessels lining the wall of the ileum and jejunum are known as _____ .

32. An accumulation of tissue fluid in the spaces between the cells is a condition called _____ .

33. Development of the immune system begins about the third month after _____ .

34. B-lymphocytes are so-named because they are formed in the embryonic chick in an organ called the bursa of _____ .

35. During formation, B-lymphocytes position on their cell membranes a number of receptor sites consisting of _____ .

36. Those substances capable of stimulating the immune system are known as _____ .

37. Normally a person's own proteins and polysaccharides do not stimulate the immune system because they are interpreted as _____ .

38. Antigens may enter the bloodstream when they penetrate the skin from a bite by an _____ .

39. The immune system reaches maturity several weeks after a person's _____ .

40. To initiate the immune process, foreign organisms are engulfed by macrophages in the process of _____ .

41. The T-lymphocyte that participates in both major immune processes is the _____ .

42. The immune process in which a direct interaction between body cells and microorganisms takes place is called _____ .

43. The T-lymphocyte that exerts a direct interaction with infected body cells is the _____ .

44. Substances secreted by T-lymphocytes that attract macrophages to an infection site are called _____ .

45. T-lymphocytes that prevent the immune process from becoming too exaggerated are called _____ .

46. Antibodies are produced by cells derived from B-lymphocytes and known as _____ .

47. The organic substance that makes up the sole component of antibody molecules is _____ .

48. Antibodies can occur in various types numbering _____ .

49. The reaction between the antibody molecule and the antigen molecule is said to be highly _____ .

50. Because the amino acid chains of the antibody molecule diverge, the molecule is often depicted in the shape of _____ .

*PART B—Multiple Choice:* **Circle the letter of the item that correctly completes each of the following statements.**

1. The lymphatic system consists of all the following except
   (A) blood
   (B) lymph nodes
   (C) lymphatic vessels
   (D) lymph

2. Which of the following applies to the thoracic duct?
   (A) it drains the entire body above the diaphragm
   (B) it empties its contents into the subclavian vein
   (C) it carries blood into the lymphatic system
   (D) it arises in the vessels of the brain

3. Lymphatic capillaries resemble blood capillaries because lymphatic capillaries
   (A) have the same permeability as blood capillaries
   (B) lead to the vena cava
   (C) have a lining of endothelium
   (D) are thick-walled tubes

4. The fluid that passes through the lymphatic vessels
   (A) flows toward the lungs
   (B) passes from the lymphatic vessels into the veins
   (C) enters the left ventricle of the heart through the right thoracic duct
   (D) moves in a single direction toward the heart

5. The T-lymphocytes and B-lymphocytes are the major cells of the
   (A) lymph nodes
   (B) lymphatic vessels
   (C) adrenal gland
   (D) thymus

6. All the following are important functions of the lymph nodes except
   (A) they serve as sites for production of antibodies
   (B) they remove foreign material phagocytized by macrophages
   (C) they are the sites where antigens stimulate the immune system
   (D) they function in the production of neutrophils, eosinophils, and basophils.

7. Lymph nodes may be located in the human body in the tissues of the
   (A) stomach and brain
   (B) groin and neck
   (C) ventricle and atrium
   (D) thyroid gland and adrenal gland

8. The nodules of lymphoid tissue found in the wall of the intestinal tract are known as
   (A) Hashimoto's nodes
   (B) Grave's region
   (C) DiGeorge's nodes
   (D) Peyer's patches

9. In the human body, the thymus is located
   (A) along the femoral artery
   (B) in the medulla of the brain
   (C) in the mediastinum of the upper thorax
   (D) between the 19th and 20th vertebrae

10. The movement of fluid through the lymphatic vessels is assisted by
    (A) pressure from the right ventricle
    (B) pressure of contracting muscles
    (C) movement of phagocytes such as macrophages
    (D) movement of red blood cells

11. The thymus is largest and most visible
    (A) in the teenage years
    (B) in senior citizen
    (C) between the ages of six and 12 years
    (D) in the fetal stage

12. All the following functions are associated with the spleen except
    (A) it provides a filtration system for blood
    (B) it is the site of red blood cell breakdown
    (C) it is a storage depot for blood
    (D) it is the major site of white blood cell formation

13. Which of the following describes the location of the spleen in the human body?
    (A) anterior to the diaphragm in the lower thoracic cavity
    (B) lateral to the vertebral column along the posterior thoracic wall
    (C) within the pelvic cavity
    (D) posterior to the diaphragm in the upper portion of the abdominal cavity

14. Which of the following is not likely to be found in the lymph?
    (A) red blood cells
    (B) protein molecules
    (C) microorganisms
    (D) macrophages

15. Those lymphatic vessels that are rich in fat
    (A) are found only in the brain
    (B) are known as lacteals
    (C) enter the left atrium of the heart
    (D) are found only within the spleen

16. All the cells of the immune system arise from
    (A) cells in the sinoatrial node
    (B) primitive cells in the bone marrow
    (C) primitive cells in the thymus
    (D) cells located primarily in the pons of the brain

17. T-lymphocytes are so-named because they are produced primarily in the
    (A) thyroid gland
    (B) tissues of the thorax
    (C) tissues stimulated by the trigeminal nerve
    (D) thymus

18. The immune system in humans becomes fully functional
    (A) about three months before birth
    (B) at birth
    (C) six months after birth
    (D) about one year after birth

19. All the following characteristics apply to antigens except
    (A) they may be cellular substances such as macromolecules
    (B) they may consist of proteins or polysaccharides
    (C) the list of possible antigens is very limited
    (D) they stimulate the immune system

20. The cells that set the immune process into motion are phagocytic cells known as
    (A) eosinophils
    (B) macrophages
    (C) plasma cells
    (D) antigens

21. The activity of T-lymphocytes brings about
    (A) the production of antibodies by plasma cells
    (B) the release of antigens by bacteria
    (C) cell-mediated immunity
    (D) the reaction between antigens and antibody molecules.

22. The lymphokines secreted by cytotoxic T-lymphocytes increase the activity of
    (A) red blood cells
    (B) brain cells
    (C) macrophages
    (D) B-lymphocytes

23. Once they have reacted with antigen molecules, the B-lymphocytes of the immune system
    (A) leave the lymphoid tissue and proceed to the antigen site
    (B) change into red blood cells that will secrete toxins
    (C) change into macrophages that will perform phagocytosis
    (D) convert into plasma cells that will secrete antibodies

24. Which of the following is true of antibody molecules?
    (A) there are five different types of antibody molecules
    (B) all antibody molecules are composed of polysaccharide
    (C) an antibody molecule is often depicted as a Y
    (D) one end of the antibody molecule is highly specific for the antigen molecule

25. Antibody molecules neutralize invading microorganisms by all of the following methods except
    (A) they bind bacteria together in a meshlike pattern
    (B) they encourage phagocytosis of microorganisms
    (C) they set off a series of reactions that tear apart microbial membranes
    (D) they rob microorganisms of the oxygen they need for metabolism

**PART C—True/False:** For each of the following statements, mark the letter "T" next to the statement if it is true. If the statement is false, change the underlined word to make the statement true.

1. The tissue lining of the lymphatic capillaries consists of <u>epithelium</u>.

2. The largest lymphatic vessel of the body is the <u>vena cava</u>.

3. Before entering its contents into the cardiovascular system, the right lymphatic duct unites with the <u>right subclavian vein</u>.

4. The lymphatic system is a <u>one-way</u> system for fluid.

5. The two major cells of the lymph nodes are the T-lymphocytes and <u>P-lymphocytes</u>.

6. Those lymph vessels entering the lymph nodes are known as <u>efferent</u> lymph vessels.

7. Within the lymph nodes, lymphocytes are arranged densely within the outer portion of the node called the <u>medulla</u>.

8. The lymph nodes serve the body as the site for the production of <u>antigens</u>.

9. The <u>inguinal</u> lymph nodes are the lymph nodes found in the neck.

10. Lymphoid tissue found in the wall of the intestinal tract is called <u>palatine tonsils</u>.

11. The thymus is the body organ in which <u>B-lymphocytes</u> are produced.

12. The spleen is located in the upper portion of the abdominal cavity, just <u>anterior</u> to the diaphragm.

13. The fluid derived from blood that is pumped through the lymphatic vessels is known as <u>lymph</u>.

14. Lymph sometimes has a milky consistency because of the presence of <u>proteins</u>.

15. An accumulation of tissue fluid in the spaces between the cells causes a condition known as <u>edema</u>.

16. The immune system begins to develop during the third month after <u>birth</u>.

17. The receptor sites positioned on the cell membrane of B-lymphocytes are <u>antigen</u> molecules.

18. An antigen molecule is usually a large protein or <u>nucleic acid</u> molecule.

19. Antigens are chemical molecules interpreted by the body as being <u>self</u>.

20. The immune process begins with phagocytosis of foreign organisms by body cells such as <u>lymphocytes</u>.

21. The immunity imparted to the body by T-lymphocytes is known as <u>cell-mediated</u> immunity.

22. After stimulation by antigens, B-lymphocytes revert to <u>phago-cytic</u> cells, which are responsible for antibody production.

23. To prevent the immune process from becoming too exagger-ated, the body uses its <u>helper</u> T-lymphocytes to slow down the process.

24. All antibody molecules are composed of <u>carbohydrate</u>.

25. The immune system has the capacity to produce a <u>dozen</u> differ-ent kinds of antibody.

**Answers**

## PART A—Completion

1. heart
2. endothelium
3. thoracic duct
4. subclavian vein
5. diaphragm
6. veins
7. lymph nodes
8. B-lymphocytes
9. afferent lymphatic vessels
10. lobules
11. reticular fibers
12. cortex
13. medulla
14. antibodies
15. cervical lymph nodes
16. axillary lymph nodes
17. knee
18. tonsils
19. palatine bone
20. Peyer's patches
21. thymus
22. thorax
23. fetus
24. abdominal cavity
25. hilus
26. splenic artery
27. red blood cells
28. blood
29. proteins
30. fats
31. proteins
32. edema
33. conception
34. Fabricius
35. antibodies
36. antigens
37. self
38. arthropod
39. birth
40. phagocytosis
41. helper T-lymphocytes
42. cell-mediated immunity
43. cytotoxic T-lymphocytes
44. lymphokines
45. suppresser T-lymphocytes
46. plasma cells
47. proteins
48. five
49. specific
50. Y

## PART B—Multiple Choice

1. A
2. B
3. C
4. D
5. A
6. D
7. B
8. D
9. C
10. B
11. D
12. D
13. D
14. A
15. B
16. B
17. D
18. C
19. C
20. B
21. C
22. C
23. D
24. B
25. D

## PART C—True/False

1. endothelium
2. thoracic duct
3. true
4. true
5. B-lymphocytes
6. afferent
7. cortex
8. antibodies
9. cervical
10. Peyer's patches
11. T-lymphocytes
12. posterior
13. true
14. fats
15. true
16. conception
17. antibody
18. polysaccharide
19. nonself
20. macrophages
21. true
22. plasma
23. suppresser
24. protein
25. million

# CHAPTER 17

# THE RESPIRATORY SYSTEM

The respiratory system functions in the exchange of oxygen and carbon dioxide between the cells of the body and the external environment. These processes are collectively called **respiration**. The circulatory system transports oxygen and carbon dioxide between the cells of the body and the respiratory system, where gas exchange takes place.

The respiratory system contains highly branched, hollow tubes forming the air passageways, which constitute the conducting portion of the respiratory system. The smallest branches of these tubes terminate in clusters of microscopic air sacs called **alveoli**, which constitute the respiratory portion of the respiratory system. The passageways conducting air into and out of the alveoli include, in descending order, the nasal cavities, the pharynx, the larynx, the trachea, the bronchi, and the bronchioles (Figure 17.1). The alveoli, bronchi, and bronchioles are organized into two paired lungs, which are surrounded by a two-layered membrane.

Gas exchange takes place at the alveoli, which provide large surface area. These sacs are formed of delicate elastic membranes covered by the extensive capillary network of the pulmonary circulation. Blood rich in carbon dioxide and poor in oxygen enters the lungs through the pulmonary arteries. Blood leaving the lungs through the pulmonary veins is low in carbon dioxide and rich in oxygen. The exchange of gases occurs by diffusion, with the net result that carbon dioxide in the blood is exchanged for oxygen in the alveoli.

## ANATOMY OF THE RESPIRATORY SYSTEM

The respiratory system is composed of numerous organs used to carry air into and out of the lungs.

### The Nose and Nasal Cavities

The **nose** is considered the normal route by which air enters the respiratory system. Air may also enter the system through the mouth because the nasal cavities and mouth meet at the region at the back of the mouth called the **pharynx**. The pharynx is a common passageway for both the respiratory and digestive systems.

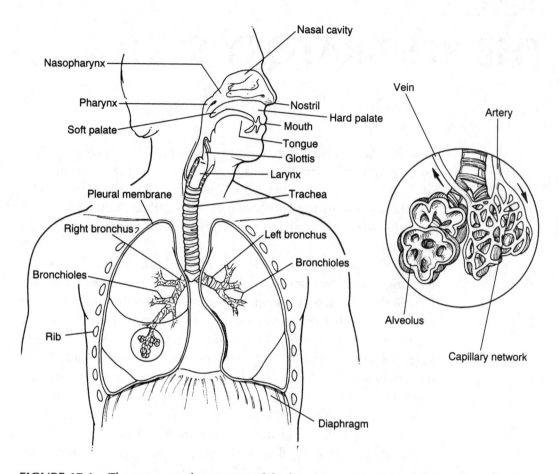

**FIGURE 17.1**   *The organs and structures of the human respiratory system in position in the thoracic cavity. A section of the lung is enlarged to show the details of several alveoli, the microscopic air sacs used for gas exchange.*

The external portion of the nose is composed of cartilage and skin. The internal portion, called the **nasal cavities**, is lined with mucous membrane. The openings of the nasal cavities to the external environment are called the **external nares**, or **nostrils**.

The nasal cavity is divided medially by the **nasal septum**. The nasal cavity is further subdivided into passageways by bony extensions known as the superior, middle, and inferior **nasal conchae** (Figure 17.2). Openings from the nasal cavities called **sinuses** extend into the frontal, sphenoid, ethmoid, maxillary, and other bones of the skull. The sinuses are sites where air is warmed and its velocity is slowed to permit particles to precipitate and olfactory sensations to occur. The linings of sinuses are continuous with linings of the nasal cavity.

The nasal cavity is also associated with the sense of smell (olfaction). Part of the nasal mucosa at the roof of the nasal cavities forms the **olfactory region**. Cells in this region detect various types of molecules and send impulses to the brain by the olfactory nerve of the sensory somatic nervous system. The brain interprets these impulses as smells.

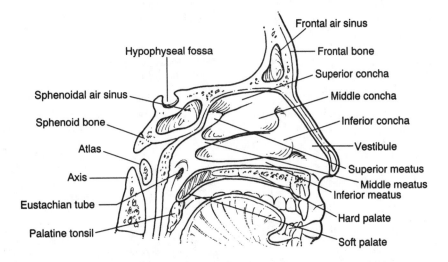

**FIGURE 17.2** *Structures of the human nose with the nasal septum removed. The conchae and the sinuses of frontal and sphenoid bones can be seen. The opening of the Eustachian tube is visible.*

The nose is adapted for warming, moistening, and filtering the air. Blood vessels in the nasal mucosa act as radiators to cool hot air; they also warm cold air. The mucus secreted by the nasal mucosa adds moisture to dry air while trapping fine dust particles and microorganisms. Ciliated cells of the mucosa then move contaminated mucus into the throat where it is swallowed.

Inflammation of the mucosal membranes is called **rhinitis**. Allergies occurring in the nasal chambers are referred to as **allergic rhinitis**. Such things as pollen, feathers, mites, and animal dander can cause this condition. A form of allergic rhinitis caused by pollen grains is commonly referred to as hay fever.

## Pharynx

The **pharynx** is also known as the **throat**. It is a region extending from the nasal cavities to the larynx. The portion of the pharynx immediately behind the nasal cavities and above the soft palate is called the **nasopharynx**. Next dorsally is the area called the **oropharynx**, where the digestive and respiratory passageways meet one another. Next is the **laryngopharynx**, which lies immediately before the larynx.

Two auditory tubes called the **Eustachian tubes** open from the middle ear into the lateral walls of the nasopharynx. The Eustachian tubes are used to equalize the air pressure between the nasopharynx and middle ear. Middle ear infections often result from microorganisms traveling up the Eustachian tubes from the nasopharynx.

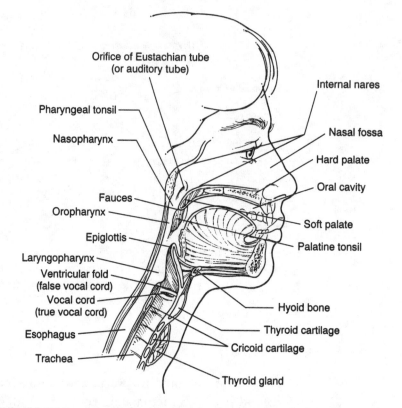

**FIGURE 17.3**   *A sagittal section of the head and neck showing many organs of the respiratory tract relative to other organs of this region. Note the structures of the nose, the hard and soft palates, and the three regions of the pharynx. The position of the esophagus relative to the trachea is shown, and the arrow displays the closure of the epiglottis over the glottis. The tonsils are illustrated in position.*

On the posterior wall of the nasopharynx in the medial region is a mass of lymphatic tissue called the **pharyngeal tonsil** (Figure 17.3). The tonsil protects the respiratory system by trapping airborne infectious agents. When swollen, the tonsil is often referred to as **adenoids**. Adenoids may obstruct the passage of air. Oval masses of lymphatic tissue on the lateral aspects of the pharynx behind the mouth are called the **palatine tonsils**. Their function is similar to that of the pharyngeal tonsil. **Tonsillitis** is inflammation of the palatine tonsils.

The pharynx serves as a passageway for both the digestive and respiratory systems. At its distal end, the pharynx branches into two tubes: the esophagus, which leads to the stomach; and the larynx, which leads to the lungs.

## The Larynx

The **larynx** is a cartilaginous structure connecting the pharynx and trachea at the level of the cervical vertebrae. It is composed of connective tissue containing nine pieces of cartilage arranged in box-like formation. The largest cartilage is the **thyroid cartilage**, also known as the "Adam's apple." The thyroid cartilage is visible in the ventral aspect of the throat and is more pronounced in adult males than adult females.

Another important cartilage is the **cricoid cartilage**, which resembles a signet ring and connects the larynx and trachea. A third cartilage is the **epiglottic cartilage**, or **epiglottis**, a leaf-shaped "lid" at the entry to the larynx. The function of the epiglottis is to seal off the respiratory tract when food passes into the esophagus. The opening to the larynx is called the **glottis**.

The larynx functions as a passageway for air and in the production of sound (Table 17.1). Two sets of heavy membranous folds of tissue project from the lateral walls of the larynx. These folds of tissue are called **vocal cords**. When air is exhaled from the lungs, the vocal cords vibrate and produce sounds that can be modified into words by muscles of the neck, lips, tongue, and cheeks. The length of the vocal cords determines **pitch**, and since females and children have shorter vocal cords, they have voices of a higher pitch.

**TABLE 17.1**   *Organs of the Human Respiratory System*

| Structure | Description | Function |
|---|---|---|
| Nasal cavity | Hollow space within nose | Conducts air to pharynx; mucous lining filters, warms, and moistens air |
| Sinuses | Hollow spaces in bones of the skull | Reduce weight of the skull; serve as resonant chambers; spaces for conditioning of air |
| Larynx | Enlargement at the top of the trachea | Passageway for air; houses vocal cords |
| Trachea | Rigid tube that connects larynx to bronchial tree | Passageway for air; mucous lining filters air |
| Bronchial tree | Branched tubes that lead from the trachea to the alveoli | Conducts air from the trachea to the alveoli; mucous lining filters air |
| Lungs | Soft, cone-shaped organs that occupy most of the thoracic cavity | Contain the air passages, alveoli, blood vessels, and other tissues of the lower respiratory tract |

## The Trachea, Bronchi, and Bronchioles

The larynx opens into a rigid tube called the **trachea** (often called the windpipe). The trachea is approximately four to five inches long in the midline of the neck. It is supported and held open by a series of C-shaped rings of cartilage stacked one upon the other and open at the dorsal aspect. The area between adjacent cartilages and between the tips of cartilage contains connective tissue and smooth muscle. The trachea furnishes an open passageway for incoming and outgoing air. Its ciliated cells also filter air before it enters the bronchi.

The trachea branches into two primary **bronchi** (singular bronchus), which have the same structure as the trachea. The right bronchus is slightly larger and more vertical than the left bronchus (Figure 17.4).

The bronchi become smaller and smaller as they extend into the lungs, and eventually their diameter is reduced to about one millimeter. At this point there is no cartilage in the tubes. The bronchi are now known as **bronchioles**. Bronchioles are composed entirely of smooth muscle supported by connective tissue. They continue to subdivide until they form the smallest air passageways, called the **terminal bronchioles**. The branching and rebranching pattern of the bronchi and bronchioles constitute a conducting network within the lungs referred to as the **bronchial tree**. Terminal bronchioles extend into the alveoli.

Inflammation of the bronchial tree is commonly known as **bronchitis**. Another condition affecting the bronchial tree is **asthma**. Asthma is accompanied by periodic attacks of wheezing and difficult breathing. It is caused by spasms of the smooth muscles, usually triggered by allergens in the environment.

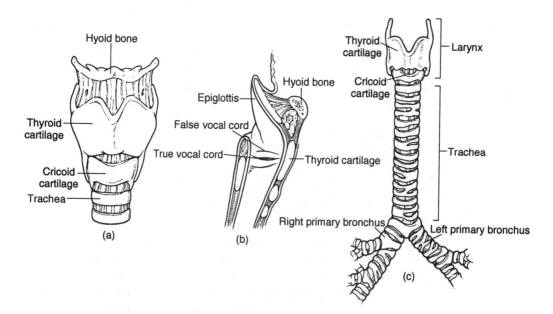

**FIGURE 17.4** *The larynx, vocal cords, and trachea. (a) The larynx is seen from the anterior aspect. The three major cartilages are illustrated. (b) A sagittal section of the larynx as seen from the right side. The vocal cords are folds of tissue within the larynx. Note the lidlike structure of the epiglottis allowing it to close over the larynx during swallowing. (c) The trachea displays the supporting rings of cartilage and branches to form the bronchi. Note that the right bronchus is slightly larger and more vertical than the left bronchus.*

# The Lungs

The lungs are paired organs occupying most of the space of the thoracic cavity. They consist of millions of small, cup-shaped outpockets (sacs) called the **alveoli**. The respiratory membranes of alveoli constitute an extremely thin barrier through which gases can pass by diffusion. There are approximately 300 million alveoli in the lungs of an average adult.

The lungs are separated from one another by a median dividing wall and an area containing the heart and other thoracic organs (for example, the thymus, part of the esophagus, and several large blood vessels) embedded in connective tissue. This area is called the **mediastinum**.

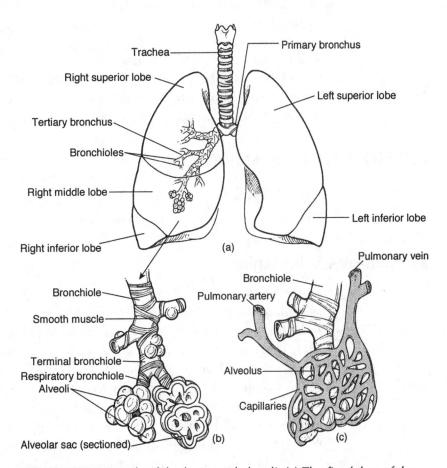

**FIGURE 17.5**  *Details of the lungs and alveoli. (a) The five lobes of the lung are illustrated, and the bronchial tree can be seen extending to all lung areas. (b) An enlarged portion of the terminal end of the bronchial tree showing a bronchiole extending to a group of alveoli. Note the straps of smooth muscle that encircle the bronchiole. (c) A group of alveoli shown as the basic functional unit of the lung. A branch of the pulmonary artery carries blood to the alveoli and a capillary network surrounds the alveoli. Gas exchange occurs here. A branch of the pulmonary vein then carries blood away from the alveoli and back to the left side of the heart.*

The lungs have a somewhat conical shape and an elastic, spongy texture derived from the nature of the alveoli. The right lung is subdivided into three lobes, while the left lung is subdivided into two (Figure 17.5). Each lobe is further divided into smaller **lobules**, and each lobule is serviced by a large bronchiole.

Each lung is surrounded by a two-layered membrane called the **pleura**. The inner layer of the pleura is called the **visceral pleura**. This layer covers the surface of each lung and reaches into the fissures between the lobes of the lung and encloses the mediastinum. The outer layer of the pleura, called the **parietal pleura**, lines the inner surface of the thoracic cavity.

The visceral and parietal pleura are continuous with one another at a point where the primary bronchus, blood vessels, and nerves enter each lung. Therefore, the two layers of the pleura form a collapsed sac. The area within the sac (between the visceral and parietal pleura) is called the **pleural cavity**. Fluid in the cavity keeps the two pleural membranes in close contact with each other and allows them to glide smoothly over each other. The fluid also adheres the two layers of the pleura to one another.

## PHYSIOLOGY OF RESPIRATION

The respiration process is an important physiological mechanism that involves breathing and gas exchange.

## Mechanism for Breathing

In the process of breathing, air moves into and out of the alveoli. Breathing takes advantage of the principle that air flows from a region of high pressure (high density) to a region of low pressure (low density). Therefore, air will enter the lungs if air in the alveoli has a lower pressure than the air in the atmosphere. Air will leave the lungs if air in the alveoli has a higher pressure than the air in the atmosphere.

The pressure changes occurring in the lungs are generated by the activity of skeletal muscles called **respiratory muscles**. The pressure changes depend upon the elasticity of the lungs and the anatomical relationship of the pleura to the lungs. The pressure changes also depend on the presence of a closed, thoracic compartment in which the lungs lie. In addition, pressure changes are generated by the firm attachment of the visceral pleura (lining the lungs) to the parietal pleura (lining the thoracic cavity).

During **inspiration**, contractions occurs in a set of respiratory muscles between the ribs. These muscles are called the **external intercostal muscles**. The muscle contractions raise the ribs upward and outward (Figure 17.6). At the same time, the dome-shaped **diaphragm** contracts and moves downward. Together, these muscle contractions greatly increase the volume of the thorax. With the

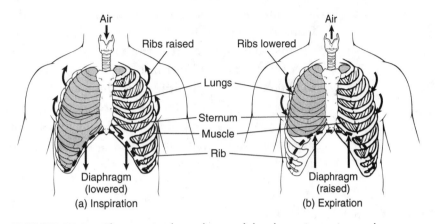

**FIGURE 17.6** *Changes in the volume of the thoracic cavity as the breathing muscles contract. (a) The diaphragm and rib muscles contract and increase the volume of the thoracic cavity. The lungs also expand, thereby increasing their volume and causing air to be inhaled. (b) When the muscles relax, the thoracic cavity returns to a smaller size (note the size differences). Now the lungs relax, thereby decreasing their volume and causing exhalation of air.*

increased volume, the elastic lungs stretch out because they are attached to the walls of the thorax by the pleura. Thus, the lungs expand as the thorax expands.

Expansion of the lungs increases the amount of area (the volume) within the air passageways and the alveoli. Increasing the volume also lowers the air pressure within the alveoli and passageways. And, since air flows from a region of high pressure to one of low pressure, atmospheric air flows freely into alveoli of the lungs.

After air has filled the lungs, gas exchange takes place between the alveoli and blood. The next process to occur is expiration. During **expiration**, the respiratory muscles (the external intercostal muscles and diaphragm) undergo relaxation. With relaxation, the volume of the thorax decreases as it returns to its original shape. This activity compresses the lungs and decreases the volume within the lungs. The volume decrease also raises the pressure of air within the lungs. Since air flows from a region of high pressure to one of low pressure, the air flows freely out of the alveoli, through the passageways, and into the atmosphere.

The process of expiration empties the lungs. It is a passive process, which can be controlled by the body but not as extensively as inspiration.

## Volume of the Lungs

Under resting conditions and during a normal breath, about 500 milliliters (mL) of air enter and leave the lungs. This air volume is called the **resting tidal volume**. After a normal expiration, about

2500 mL of air remain in the lungs. Should the person force air from the lungs, about 1000 mL will still remain. This remaining volume is called the **residual volume**.

A forced inspiration can bring into the lungs an additional 2500 to 3500 mL of air in addition to the 500 mL normally inspired. The largest volume of air that can be brought into the lungs is the **vital capacity** of the lungs. Since maximum inspiration requires exhaustive effort on the part of the muscles, the vital capacity of the lungs is hardly ever reached.

# Control of Breathing

Breathing is controlled by contractions of the respiratory muscles, which are controlled by nerve stimulations. The main area for respiratory muscle control is a portion of the brain called the **respiratory control center**. The respiratory control center is located in the brain stem and includes parts of the medulla oblongata and pons (Figure 17.7). Two groups of neurons within this region control the rhythm of breathing and the respiratory movements of forced breathing. Another area in the control center, called the **pneumotaxic area**, regulates the rate of breathing.

The respiratory centers in the brain stem indirectly monitor the carbon dioxide levels in the bloodstream. Carbon dioxide is a waste product of cellular respiration. As the carbon dioxide concentration increases in arterial blood, the gas diffuses into the cerebrospinal fluid bathing the respiratory center. The carbon dioxide increase in the fluid causes a corresponding increase in the concentration of bicarbonate ions and hydrogen ions (that is, acidity) in the fluid. The high level of hydrogen ions activates the respiratory center, and nerve impulses from the pneumotaxic area are sent to the respiratory muscles. The impulses increase the breathing rate and depth. As carbon dioxide is exhaled from the lungs, the hydrogen ion concentration drops in the blood. This leads to a decrease in hydrogen ion concentration in the cerebrospinal fluid, and the reduced hydrogen ion level removes the activation of the respiratory center.

Other receptors for the respiratory system are located throughout the body. For instance, there are chemical receptors (chemoreceptors) located in the carotid arteries and in the arch of the aorta (Chapter 15). These chemoreceptors monitor the dissolved oxygen content of the blood. When the oxygen level is low, the chemoreceptors are stimulated, and they send impulses to the respiratory control center. The center increases the rate and depth of breathing.

The activity of the respiratory center and the chemoreceptors are involuntary mechanisms for respiratory control. The body can partially override these mechanisms by nerve impulses originating in the cerebral cortex and passing to the respiratory control center. Voluntary control permits breath-holding during swimming or other times.

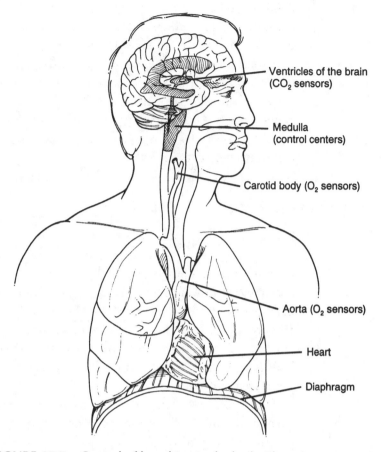

**FIGURE 17.7**  *Control of breathing in the body. Three important centers of control are located in the medulla of the brain, in the carotid arteries, and in the arch of the aorta.*

However, as the levels of carbon dioxide and hydrogen ions increase in body fluids, impulses from the respiratory center overcome the voluntary inhibition of breathing. One is then forced to take a breath. Deep and rapid breathing is called **hyperventilation**.

## Gas Exchange

Oxygen and carbon dioxide are transported to and from the lungs by slightly different mechanisms. For oxygen, about 2 percent of the gas is dissolved in the plasma or in the intracellular fluid of the red blood cells. The remaining 98 percent is carried by hemoglobin molecules within the red blood cells. Each hemoglobin molecule is able to bind loosely to four molecules of oxygen. When oxygen is bound to hemoglobin, the complex is referred to as **oxyhemoglobin**.

For carbon dioxide, a very small fraction (about 7 percent) is carried as a dissolved gas in plasma and in the intracellular fluid of the red blood cells (Figure 17.8). Some carbon dioxide (about 25 to 30 percent) is carried as **carboxyhemoglobin** by hemoglobin molecules, which have previously released their oxygen in the tissues.

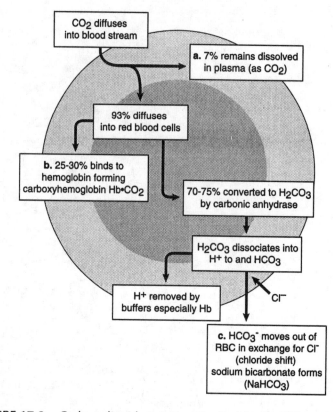

**FIGURE 17.8** *Carbon dioxide transport in the blood. Carbon dioxide enters the bloodstream (upper left) and three different mechanisms account for its transport: (a) as dissolved $CO_2$; (b) as carboxyhemoglobin; and (c) as sodium bicarbonate.*

The remaining carbon dioxide (about 70–75 percent) is carried in the blood in the form of **bicarbonate ions** ($HCO_3^-$), located in the red blood cells. These bicarbonate ions are formed after carbon dioxide enters the blood plasma at the tissues and combines with water to form carbonic acid ($H_2CO_3$). This reaction is catalyzed by **carbonic anhydrase**, an enzyme present in red blood cells. The carbonic acid then dissociates into hydrogen ions and bicarbonate ions. Some bicarbonate ions gather in red blood cells and are the mechanism for carbon dioxide transport. However, many ions diffuse into the surrounding plasma where they associate with sodium ions to form sodium bicarbonate ($NaHCO_3$) for transport purposes. Each time a bicarbonate ion diffuses across the membrane of the red blood cell, a chloride ion diffuses into the RBC. When a large number of $CO_2$ molecules enters the blood, the processes occur swiftly, and there is a rush of chloride ions into the RBCs, a phenomenon called the **chloride shift**.

At the alveoli, oxygen in the air is exchanged for carbon dioxide in the blood. The driving force behind this exchange is a passive process called **diffusion**. Diffusion is the movement of molecules from an area of high concentration to an area of low concentration.

The process does not involve any energy expenditure and is completely passive. Consideration of the oxygen and carbon dioxide pressures are presented in Chapter 21.

At the alveoli, red blood cells pass through a microscopic capillary at the surface of the alveolar sac. Oxygen from the alveolar sac diffuses across the respiratory membrane into the plasma and then into the interior surface of the red blood cells. Diffusion occurs in this direction because the red blood cells are oxygen-poor (deficient in oxygen), while the air in the alveolus is oxygen-rich. Oxygen molecules enter the red blood cells and bind loosely to the hemoglobin molecules for transport to the cells (Figure 17.9).

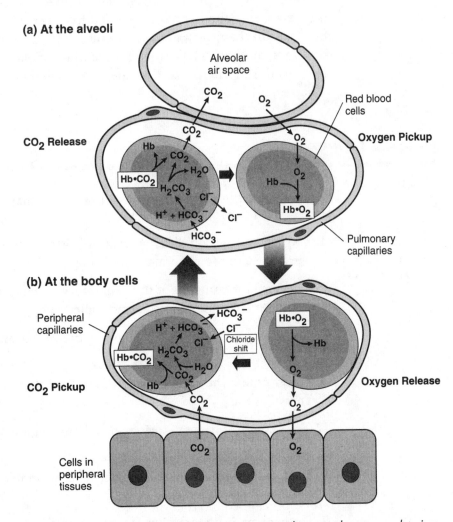

**FIGURE 17.9**   *A summary of gas transport and gas exchange mechanisms in the human body. (a) At the alveoli, $CO_2$ is released and oxygen is picked up. The gases pass across the membranes of the alveolus and the capillary. (b) At the body cells, the opposites occur for the two gases. Note that oxygen is carried bound to the hemoglobin (Hb) of RBCs, while $CO_2$ is carried dissolved in plasma, bound to hemoglobin molecules, and as bicarbonate ions in sodium bicarbonate.*

As oxygen molecules are passing from the alveolar sac into the red blood cells, bicarbonate ions are converting back to carbon dioxide molecules. The carbon dioxide molecules then pass by diffusion from red blood cells to the alveolar sac. Diffusion of carbon dioxide occurs away from the red blood cells because they are carbon-dioxide-rich, while the air in the alveoli is carbon-dioxide-poor.

Once the gas exchange between the blood and air has taken place, the blood cells are free to leave the area of the alveolar sac. The capillaries join together to form pulmonary venules, which join together to form pulmonary veins, which meet to form the main **pulmonary vein** back to the heart.

The pulmonary vein enters the left side of the heart, and the left ventricle then pumps the oxygen-rich blood to the body tissues. At the body tissues the oxygen is released from the red blood cells by the same mechanisms as at the alveoli, and a new load of carbon dioxide is taken up. The oxygen is used in the metabolism of the tissue cells during the process of energy release and ATP formation (Chapter 19).

# REVIEW QUESTIONS

*PART A—Completion:* **Add the word or words that correctly complete each of the following statements.**

1. Oxygen and carbon dioxide are transported between the cells of the body and the respiratory system by the _____ .

2. The smallest branches of the respiratory system tubes terminate in clusters of microscopic air sacs called _____ .

3. Gas exchange takes place between the blood and the atmosphere in structures called _____ .

4. Blood enters the lung from the heart through a major artery known as the _____ .

5. The mouth and nasal cavities meet in a region known as the _____ .

6. The nasal cavities open to the external environment at the external nares, also called the _____ .

7. Passageways in the nasal cavity are separated from one another by bony extensions of the nasal septum known as the _____ .

8. Many bones of the skull contain openings from the nasal cavities referred to as _____ .

9. At the roof of the nasal cavities the sense of smell is associated with an area called the _____ .

10. Three functions of the nose include moistening the air, filtering the air, and _____ .

11. Allergic reactions occurring within the nasal chambers are given the general name _____ .

12. Another name for the pharynx of the body is the _____ .

13. The digestive and respiratory passageways meet one another at a region of the pharynx referred to as the _____ .

14. The lateral walls of the nasopharynx contain openings from tubes from the middle ear known as _____ .

15. The mass of lymphatic tissue on the posterior wall of the nasopharynx in the medial region is the _____ .

16. The largest cartilage of the larynx, also known as the Adam's apple, is the _____ .

17. The cartilage of the larynx that resembles a signet ring and connects the trachea and larynx is the _____ .

18. The opening to the larynx is referred to as the _____ .

19. Vibrations of the vocal chords are due to air exhaled from the _____ .

20. Because men have longer vocal chords, their voices have lower _____ .

21. The trachea is supported by a series of C-shaped rings of _____ .

22. The two primary tubes that branch from the trachea are the _____ .

23. There is no cartilage in the tubes when the bronchi become _____ .

24. Bronchioles are composed entirely of muscle referred to as _____ .

25. The common name for inflammation of the bronchial tree is _____ .

26. The lungs occupy most of the space of the _____ .

27. The number of alveoli in the average adult numbers approximately _____ .

28. The left lung is subdivided into two lobes, while the right lung is subdivided into _____ .

29. The two-layer membrane surrounding each lung is the _____ .

30. The outer layer of pleura lining the inner surface of the thoracic cavity is the _____ .

31. The underlying principle of breathing is that air flows from the region of high pressure to a region of _____ .

32. Pressure changes occurring in the lungs can be traced to the activity of skeletal muscles known as _____ .

33. During inspiration, the ribs are raised upward and outward by a set of respiratory muscles called _____ .

34. During inspiration, contractions cause the downward movement of a dome-shaped muscle known as the _____ .

35. The relaxation of respiratory muscles compresses the thorax and increases the air pressure in the _____ .

36. While inspiration is an active process, the process of expiration is a _____ .

37. The amount of air that enters and leaves the lungs in a normal breath under resting conditions is the _____ .

38. The largest volume of air that can be brought into the lungs during a forced inspiration is the _____ .

39. Breathing is controlled by an area of the brain called the respiratory control center, which includes part of the medulla oblongata and the _____ .

40. The respiratory centers in the brain are regulated indirectly by the bloodstream's level of _____ .

41. The rate of breathing is controlled by an area of the brain known as the _____ .

42. The respiratory center is activated by the fluid level of _____ .

43. While the contraction of skeletal muscles is usually voluntary, the control of breathing is usually _____ .

44. The great majority of oxygen is carried in the body in association with _____ .

45. Approximately 70–75 percent of carbon dioxide is transported in the blood in the form of _____ .

46. Approximately 25–30 percent of the carbon dioxide in the body is carried in association with _____ .

47. The driving force behind the exchange of gases in the alveoli is the process of _____ .

48. Where active transport is an active process for the movement of molecules, diffusion is a _____ .

49. During the exchange of gases, oxygen molecules move in the direction of the _____ .

50. During the diffusion process, carbon dioxide molecules move in the direction of the _____ .

**PART B—Multiple Choice: Circle the letter of the item that correctly completes each of the following statements.**

1. The nasal cavities and mouth meet at a region of the body called the
   (A) nasal conchae
   (B) sinuses
   (C) pharynx
   (D) trachea

2. All the following apply to the sinuses except
   (A) they are bony extensions of the nasal cavity
   (B) they may be found in the frontal, maxillary, vomer, and parietal bones
   (C) their linings are continuous with linings of the nasal cavity
   (D) they are places where air circulates

3. At the roof of the nasal cavities, receptors exist for the sense of
   (A) hearing
   (B) smell
   (C) balance
   (D) touch and feel

4. All the following are functions of the nose except
   (A) it moistens the air
   (B) it serves as a site for warming the air
   (C) it is the place where air is filtered
   (D) it is the site of gas exchange

5. The mucus secreted by the nasal mucosa
   (A) traps microorganisms
   (B) dries the air
   (C) provides nutrients to the nasal cells
   (D) contains digestive enzymes

6. Such things as pollen grains, mites, and feathers may be the cause of
   (A) pneumonia
   (B) tonsillitis
   (C) allergic rhinitis
   (D) emphysema

7. The digestive and respiratory passageways meet one another at the
   (A) larynx
   (B) esophagus
   (C) oropharynx
   (D) nostrils

8. The function of the Eustachian tubes is to
   (A) deliver nutrients to the middle ear
   (B) provide digestive enzymes for carbohydrates
   (C) deliver hormones to the mouth
   (D) equalize air pressure between the pharynx and middle ear

9. The function served by tonsils is to
   (A) produce important hormones for the body
   (B) synthesize red blood cells for the body
   (C) trap airborne infectious agents
   (D) synthesize blood clotting proteins

10. The two tubes at the distal end of the pharynx are the
    (A) Eustachian tube and nostril
    (B) larynx and esophagus
    (C) vena cava and aorta
    (D) small and large intestines

11. The function of the cartilage rings in the trachea is to
    (A) support and hold open the trachea
    (B) provide calcium to the bloodstream
    (C) trap foreign microorganisms in the respiratory tract
    (D) relay nerve impulses to the respiratory tract

12. The epiglottis has the function of
    (A) supporting the trachea
    (B) sealing off the respiratory tract when food passes into the esophagus
    (C) serving as a site for the movement of vocal chords
    (D) increasing the pitch of the voice in females and children

13. The trachea is supported and held open by rings of
    (A) skeletal muscle
    (B) bone
    (C) epithelial tissue
    (D) cartilage

14. The branches that emerge from the trachea and lead to the lungs are the
    (A) alveolar tubes
    (B) bronchi
    (C) terminal bronchioles
    (D) alveolar ducts

15. Spasms of the smooth muscles in the bronchial tree may be triggered by an allergy and may result in
    (A) asthma
    (B) hay fever
    (C) skeletal muscle contraction
    (D) contractions of the cartilage rings

16. Which of the following applies to the right lung?
    (A) it is subdivided into two lobes
    (B) blood reaches it by the pulmonary vein
    (C) it is subdivided to three lobes
    (D) it has its own nerve supply

17. The two-layer membrane known as the pleura
    (A) is found within the trachea
    (B) separates the left and right lungs
    (C) surrounds each lung
    (D) defines the limit of the alveolus

18. The pleural cavity is found
    (A) between the visceral and parietal pleura
    (B) between the thoracic and abdominal cavities
    (C) between the dorsal and ventral cavities
    (D) surrounding the heart

19. Contractions of the respiratory muscles result in a
    (A) decrease in the volume of the thorax
    (B) increase in the volume of the thorax
    (C) increase in the amount of blood flowing through the lungs
    (D) decrease in the amount of blood flowing through the lungs

20. The pressure changes occurring in the lungs during inspiration
    are due to all the following except
    (A) presence of a closed thoracic compartment where the lungs
        lie
    (B) the elasticity of the lungs
    (C) the firm attachment of the pleural membranes
    (D) nerve impulses reaching the lungs from the brain

21. The body has control of
    (A) blood flow to the lungs but not from the lungs
    (B) nerve impulses to the lungs
    (C) inspiration but not expiration
    (D) lung contraction but not lung expansion

22. A normal inspiration of air brings into the lungs about
    (A) 2500 mL of air
    (B) about 10,000 mL of air
    (C) about 500 mL of air
    (D) about 10 mL of air

23. The vital capacity is the
    (A) largest volume of air that can be brought into the lungs
    (B) largest volume of air that can be expelled from the lungs
    (C) amount of air remaining in the lungs after expiration
    (D) amount of air entering the lung during a normal inspiration

24. All the following have an effect on the control of breathing except
   (A) chemical receptors in the carotid arteries
   (B) the respiratory control center in the brain stem
   (C) the level of hydrogen ions in the cerebrospinal fluid
   (D) amount of blood flowing into the lungs

25. Carbon dioxide can be carried in the bloodstream by all the following methods except
   (A) attached to hemoglobin molecules
   (B) as a dissolved gas in plasma
   (C) as bicarbonate ions
   (D) attached to hormone molecules in the blood

*PART C—True/False:* **For each of the following statements, mark the letter "T" next to the statement if it is true. If the statement is false, change the <u>underlined</u> word to make the statement true.**

1. The alveoli are formed of delicate elastic membranes covered by an extensive capillary network of the <u>systemic</u> circulation.

2. The normal route by which air enters the respiratory system is the <u>mouth</u>.

3. Openings from the nasal cavities called <u>sinuses</u> extend into several bones of the skull.

4. The air is warmed, moistened, and filtered as it passes through the <u>mouth</u>.

5. The condition known as allergic rhinitis is also referred to as <u>asthma</u>.

6. The Eustachian tubes are used to equalize the air pressure between the nasopharynx and <u>inner ear</u>.

7. At its distal end, the pharynx branches into two tubes called the <u>esophagus</u> and the larynx.

8. The largest cartilage of the larynx is the <u>cricoid cartilage</u>, also known as the "Adam's apple."

9. The vocal chords, which function in the production of sound, are located within the <u>trachea</u>.

10. The trachea is held open by a series of C-shaped rings of <u>bone</u> stacked upon one another.

11. The bronchioles are composed entirely of <u>skeletal</u> muscle supported by connective tissue.

12. The two-layer membrane that surrounds each lung is known as the <u>peritoneum</u>.

13. The bronchus that is slightly larger and more vertical is the <u>left</u> bronchus.

14. Expansion of the lungs is due in part to contractions taking place in the <u>diaphragm</u> muscle.

15. The expansion of the lungs increases the volume of the lungs and <u>raises</u> the air pressure within the lungs.

16. During the process of expiration, the respiratory muscles undergo <u>contractions</u>, and the thoracic returns to its original shape.

17. The resting tidal volume of the lungs during a normal breath is about <u>500 milliliters</u> of air.

18. The <u>vital capacity</u> is the largest volume of air that can be brought into the lungs.

19. The respiratory control center for the lungs is located in a portion of the <u>heart</u>.

20. The increase of carbon dioxide in the blood fluid causes a corresponding <u>decrease</u> in the concentration of hydrogen ions in the fluid.

21. Nerve impulses originating in the <u>cerebellum</u> can override the activity of the respiratory center in regulating breathing.

22. The <u>least</u> amount of carbon dioxide is transported in the blood as bicarbonate ions.

23. The passive process known as <u>osmosis</u> accounts for the movement of oxygen and carbon dioxide gas across the capillary membrane.

24. For transport to the body's cells, oxygen molecules bind loosely to <u>hemoglobin</u> molecules contained in red blood cells.

25. As they leave the area of the alveoli, capillaries join together to form pulmonary <u>venules</u>.

**Answers**

## PART A—Completion

1. circulatory system
2. alveoli
3. alveoli
4. pulmonary artery
5. pharynx
6. nostrils
7. nasal conchae
8. sinuses
9. olfactory region
10. warming the air
11. allergic rhinitis
12. throat
13. oropharynx
14. Eustachian tubes
15. pharyngeal tonsils
16. thyroid cartilage
17. cricoid cartilage
18. glottis
19. lungs
20. pitch
21. cartilage
22. bronchi
23. bronchioles
24. smooth muscle
25. bronchitis
26. thoracic cavity
27. 300 million
28. three lobes
29. pleura
30. parietal pleura
31. low pressure
32. respiratory muscles
33. external intercostal muscles
34. diaphragm
35. lungs
36. passive process
37. resting tidal volume
38. vital capacity
39. pons
40. carbon dioxide
41. pneumotaxic area
42. hydrogen ions
43. involuntary
44. hemoglobin molecules
45. bicarbonate ions/sodium bicarbonate
46. hemoglobin molecules
47. diffusion
48. passive process
49. red blood cells
50. alveolar sac

## PART B—Multiple Choice

| | | | | |
|---|---|---|---|---|
| 1. C | 6. C | 11. A | 16. C | 21. C |
| 2. B | 7. C | 12. B | 17. C | 22. C |
| 3. B | 8. D | 13. D | 18. A | 23. A |
| 4. D | 9. C | 14. B | 19. B | 24. D |
| 5. A | 10. B | 15. A | 20. D | 25. D |

## PART C—True/False

1. pulmonary
2. nose
3. true
4. nose
5. hay fever
6. middle ear
7. true
8. thyroid cartilage
9. larynx
10. cartilage
11. smooth
12. pleura
13. right
14. true
15. lowers
16. relaxations
17. true
18. true
19. brain
20. increase
21. cerebral cortex
22. most
23. diffusion
24. true
25. true

# CHAPTER 18

# THE DIGESTIVE SYSTEM

The digestive system has two important functions: breaking down large food molecules into small molecules and absorbing small molecules into the body. The system consists of the gastrointestinal tract and a number of accessory organs. Organs of the gastrointestinal tract include the mouth, esophagus, stomach, and small and large intestines. The accessory organs include the salivary glands, liver, and pancreas (Figure 18.1).

The gastrointestinal tract is a muscular tube about 29.5 ft. (9 m) in length. Its walls are composed of four distinct layers. The innermost layer is called the mucous membrane, or **mucosa**. This layer is

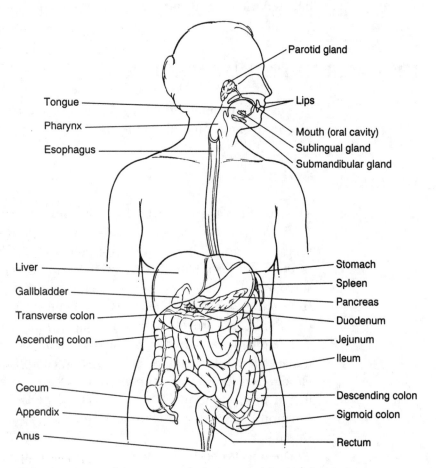

**FIGURE 18.1**   *The organs and structures of the human digestive system relative to other structures of the body as seen from the ventral position.*

composed of epithelium lying on an underlying base of connective tissue, with a small amount of smooth muscle. The mucosa contains the glands of the gastrointestinal system. The glands secrete enzymes to digest food molecules and mucus to protect the tissues of the gastrointestinal tract.

The next deeper layer of the gastrointestinal tract is the **submucosa**. This layer contains blood vessels, lymphatic vessels, and nerves. The next layer consists of two sublayers of **smooth muscle**. Certain smooth muscles encircle the gastrointestinal tract and have the ability to contract, thereby decreasing the diameter of the tract. Other muscles run lengthwise along the gastrointestinal tract. When these muscles contract, they shorten the length of the tube.

The outer (deepest) layer of the gastrointestinal tract is called the serous layer, or **serosa**. This layer is composed of visceral peritoneum. The visceral peritoneum is continuous with the parietal peritoneum lining the abdominal cavity. Cells of the visceral peritoneum (serosa) secrete fluid to keep the outer surface moist and permit the organs to slide freely past one another. The space between the visceral and parietal peritoneum is the peritoneal cavity.

## THE GASTROINTESTINAL TRACT

The gastrointestinal tract is the main tube of the digestive system. It extends from the mouth to the anus.

### The Mouth

The mouth is the organ of the gastrointestinal tract that receives food and prepares it for digestion by mechanically reducing the food mass and mixing it with saliva (Table 18.1). The major portion of the mouth is a chamber called the **oral cavity**. The oral cavity is surrounded by the lips, cheeks, tongue, and hard and soft palates. Its functions are to analyze foods, mechanically process foods, lubricate foods, and provide a place for digestion by salivary enzymes.

The **tongue** is connected to the floor of the oral cavity by a fold of tissue called the **frenulum**. The tongue is composed of skeletal muscle covered by a mucous membrane. On the sides of the tongue within the tongue papillae are the **taste buds** (Chapter 13). The function of the tongue is to work with the saliva to form the food into a bolus.

The mechanical destruction of large food masses is brought about by the teeth (Figure 18.2). **Teeth** are of two types. The **deciduous** teeth number 20 and are referred to as "baby" teeth. They are shed by the age of six and are replaced by **permanent** teeth, which number 32.

**TABLE 18.1** *Digestive Organs and Their Functions*

| Organ | Important Functions |
| --- | --- |
| Mouth | Mixes food with salivary secretions; taste, chewing |
| Salivary glands | Produce buffers and enzymes that begin digestion |
| Pharynx | Passageway shared with respiratory system, leads to esophagus |
| Esophagus | Delivers food to stomach |
| Stomach | Secretes acids and digestive enzymes that break down proteins |
| Small intestine | Secretes enzymes and other factors for nutrient digestion; absorbs nutrients |
| Liver | Secretes biles (required for lipid digestion), synthesizes blood proteins, stores lipid carbohydrate reserves |
| Gallbladder | Stores biles for release into small intestine |
| Pancreas | Secretes digestive enzymes and buffers into small intestine |
| Large intestine | Removes water from nondigested material; stores wastes |
| Anus | Opening to exterior for discharge of feces |

The teeth are classified into four types: **incisors**, which are used for biting off large pieces of food; **cuspids**, or canines, which are cone-shaped teeth used for grasping and tearing food; **bicuspids**, or premolars, which are flat teeth specialized for grinding; and **molars**, which are also flat teeth used for grinding foods.

The basic structure of a tooth includes an enamel covered **crown**, a **root** covered by a material called cementum, and the **neck** of the tooth, which joins the crown of the tooth to the root. A tooth contains two major components known as enamel and dentin. **Enamel** is the hardest substance in the body and is found at the outside surface of the tooth. It consists basically of calcium salts in a chemical substance called hydroxyapatite. The second component, **dentin**, is softer than the enamel and makes up the largest portion of the tooth. It is located beneath the enamel and surrounds the tooth's pulp. The **pulp** contains the blood vessels, nerves, and connective tissues of the tooth.

There are three **salivary glands** secreting saliva into the oral cavity. Technically, these glands are accessory organs of the digestive system. The first salivary gland is the **parotid gland** (Figure 18.3). This paired gland lies in the facial tissue below the ears. It is the largest salivary gland and is drained by **Stensen's duct** into the inside area of the cheek.

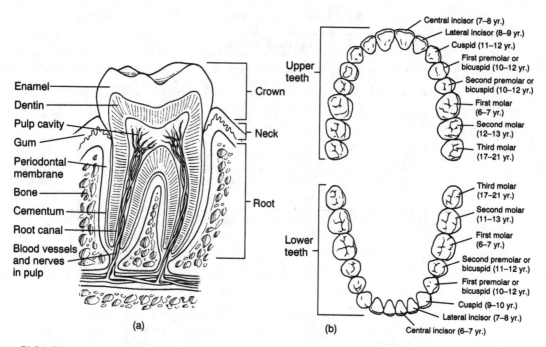

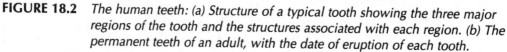

**FIGURE 18.2** *The human teeth: (a) Structure of a typical tooth showing the three major regions of the tooth and the structures associated with each region. (b) The permanent teeth of an adult, with the date of eruption of each tooth.*

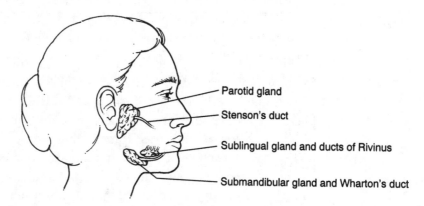

**FIGURE 18.3** *The three salivary glands in position with their associated ducts. The submandibular gland is also called the submaxillary gland.*

The second salivary gland is the **submandibular,** or **submaxillary gland,** located near the inner surface of the mandibles. It is also a paired gland and is drained by **Wharton's duct** into the floor of the mouth. The third salivary gland is the **sublingual gland.** Like the other two salivary glands, it is a paired gland and is found in front of the submandibular glands on the floor of the mouth, under the tongue. The secretions of the sublingual glands go into the floor of the mouth through several tubes called the **ducts of Rivinus** (Table 18.2).

**TABLE 18.2** *Characteristics of the Salivary Glands*

| Name of Gland | Location | Duct Opening |
| --- | --- | --- |
| Parotid | Below and in front of ear | On inside of cheek, opposite upper second molar tooth; Stenson's duct |
| Submaxillary (Submandibular) | Posterior part of floor of mouth | Floor of mouth, at sides of frenulum; Wharton's duct |
| Sublingual | Anterior part of floor of mouth, under tongue | Several ducts open into floor of mouth |

The secretion of the salivary glands, the saliva, helps moisten food particles, bind them together, and begin the breakdown of carbohydrate molecules. The serous cells of the salivary glands produce a digestive enzyme called **amylase**. Amylase digests starch molecules and glycogen molecules into two-monosaccharide units called **disaccharides**. The disaccharide produced by the breakdown of starch and glycogen is **maltose**. Mucous cells of the salivary glands produce mucus. Mucus is a thick, viscous liquid that binds food particles together and lubricates the gastrointestinal tract.

The roof of the oral cavity is formed by the palate. The palate consists of a hard anterior part (**hard palate**) and a soft posterior part (**soft palate**). A cone-shaped projection called the **uvula** extends downward from the soft palate.

Located on the soft palate are a set of lymphatic tissues called **tonsils**. There are two types of tonsils: **palatine tonsils** and the **pharyngeal tonsil** (**adenoids**). Both palatine and pharyngeal tonsils are found on the posterior wall of the pharynx. The tonsils are part of the lymphatic system and are composed of lymphatic tissues (Chapter 16). They function in the process of body defense.

## The Esophagus

The esophagus is a straight, collapsible, muscular tube connecting the pharynx to the stomach. It is approximately 10 inches (25 cm) in length, and it passes through the diaphragm to the stomach. The four tissue layers of the gastrointestinal tract wall are first observed in the esophagus. In the superior third of the esophagus, the muscle is skeletal muscle. The muscle type then changes as the esophagus descends, and at its lower third it is entirely smooth muscle.

For food to pass from the mouth to the stomach, it must be swallowed. The process of swallowing is called **deglutition** (Figure 18.4). It is an involved process requiring the coordinated activities of the tongue, soft palate, esophagus, and pharynx. The first phase occurs in the mouth and is voluntary. The food is chewed and mixed with

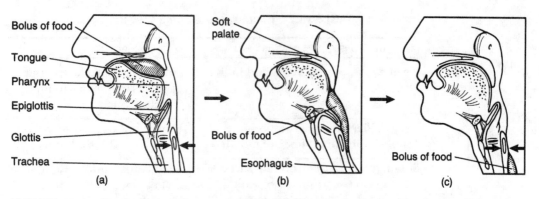

**FIGURE 18.4** *Deglutition and peristalsis. (a) In the swallowing process, the tongue rises and presses the bolus of food against the hard palate, thereby forcing it ahead (b) The bolus of food slides past the epiglottis closed over the trachea and enters the esophagus (c) The muscle of the esophagus contract (arrows) pushing the bolus ahead and toward the stomach*

saliva to form a **bolus** and is then forced to the pharynx by the tongue. Now it is ready for the involuntary phase of swallowing.

During the involuntary phase, the pharyngeal muscles contract and force the bolus into the esophagus. The nerves of the autonomic nervous system then control the involuntary phase by inducing muscular contractions in a process called **peristalsis**. Peristalsis is a series of wavelike contractions of the muscles of the esophagus. The longitudinal muscles contract, then the circular muscles contract. The series of muscular contractions forces the bolus of food along until it reaches a circular muscle at the upper end of the stomach. The muscle is called the **cardiac sphincter**.

## The Stomach

The stomach begins at the cardiac sphincter and ends at a circular muscle at its far end called the **pyloric sphincter** (Figure 18.5). It is a C-shaped organ lying in the upper left quadrant of the abdomen. Its major regions are the **cardia** (closest to the heart), **fundus, body** (or main part), and **pylorus** (the narrow far region). The internal surface of the stomach has folds called **rugae**, which are obvious when the stomach is empty and shrunken. When the stomach is distended, the rugae cannot be observed.

The convex lateral surface of the stomach is referred to as the **greater curvature** of the stomach. The concave, medial surface of the stomach is called the **lesser curvature**. A double layer of peritoneum, the **lesser omentum**, extends from the liver to the lesser curvature. Another extension of the peritoneum, the **greater omentum**, covers the abdominal organs and attaches to the posterior body wall.

The stomach is a storage organ for food and a location for the chemical breakdown of certain molecules. The muscle layers churn food and physically break the bolus of food down into a soupy mixture called **chyme**. Within the mucosa of the stomach are located deep gastric pits, which receive the products of the gastric glands. The gastric glands secrete gastric juice, which contains a variety of substances.

Among the components of the gastric juice are protein-digesting enzymes. These enzymes are the products of **chief cells** within the gastric glands. Another component of the gastric juice is **hydrochloric acid**, which is necessary to activate the protein-digesting enzymes. A third product within the gastric juice is **mucus**, a product of the mucous cells. Mucus protects the stomach wall from self-digestion. A final substance, called **intrinsic factor**, is needed for the absorption of vitamin $B_{12}$ from the stomach.

The major protein-digesting enzyme operating in the stomach is **pepsin** (Table 18.3). However, pepsin is not present in the gastric juice. Rather the gastric juice contains the precursor substance, **pepsinogen**. In the presence of hydrochloric acid, pepsinogen converts to pepsin. Pepsin then breaks large proteins into small proteins referred to as **peptides**. Another protein-digesting enzyme, **rennin**, is available in the infant stomach, but not in the adult stomach.

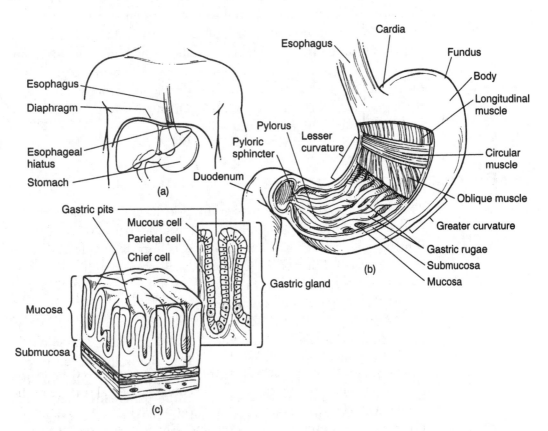

**FIGURE 18.5** *Aspects of the human stomach: (a) The stomach in place in the upper abdominal cavity in the epigastric region. (b) The major anatomical regions and muscle layers of the stomach. Note the inner folds (rugae) that become taut when the stomach is filled. (c) A closeup of the stomach tissue showing the various cells of the gastric wall.*

**TABLE 18.3** *The Characteristics of Several Components of Gastric Juice*

| Component | Source | Function |
|---|---|---|
| Pepsinogen | Chief cells of the gastric gland | Converts to pepsin |
| Pepsin | Formed from pepsinogen in the presence of hydrochloric acid | Protein-digesting enzyme capable of breaking down nearly all types of protein |
| Hydrochloric acid | Parietal cells of the gastric glands | Provides acidic environment; needed for converting pepsinogen into pepsin |
| Mucus | Goblet cells and mucous glands | Provides viscous, alkaline protective layer on the stomach wall |
| Intrinsic factor | Parietal cells of the gastric glands | Encourages the absorption of vitamin $B_{12}$ |

The secretion of gastric juice is controlled by nerves of the parasympathetic nervous system. Also, the hormone **gastrin** is secreted by cells within the stomach lining. Gastrin regulates the gastric glands to produce pepsinogen as well as hydrochloric acid and mucus. The cells that produce gastrin are referred to as enteroendocrine cells.

Some absorption occurs through the lining of the stomach. Substances absorbed include small amounts of water, glucose, ions, and alcohol. The remainder of the food is prepared for further digestion by reduction to chyme. Peristaltic contractions force the chyme through the pyloric sphincter into the small intestine where the major amount of digestion takes place.

## The Duodenum

The small intestine extends from the pyloric sphincter to a circular muscle called the ileocecal sphincter. It is divided into three regions: the first part is the **duodenum**, which is approximately 12 inches long; the second part is the **jejunum**, which is approximately 10 feet long; and the third part is the **ileum**, which is approximately 10 feet long. Most digestion takes place in the duodenum.

The duodenum receives spurts of chyme released from the stomach through the pyloric sphincter. Enzymes enter the lumen of the duodenum through ducts from glands in the walls located in pits called the intestinal crypt or **crypts of Lieberkühn**. In addition, the pancreas contributes enzymes through the **pancreatic duct**, which enters the duodenum. Bile produced by the liver enters the duodenum through the **common bile duct**. The submucosa of the duodenum has nodules of lymphatic tissue called **Peyer's patches**. Also, it contains a number of mucous glands called **Brunner's glands**. The alkaline mucus from these glands helps neutralize the acid of the chyme.

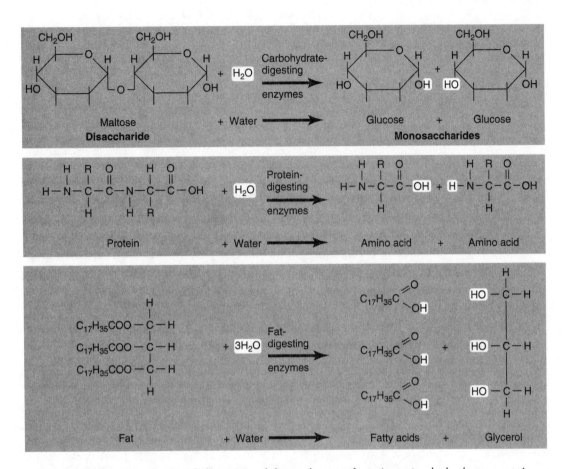

**FIGURE 18.6**   *The chemical digestion of three classes of nutrients (carbohydrate, protein, and lipid) as it occurs in the duodenum. In all three cases, water is used in the digestion, and the products of the digestions contain a part of the water molecule.*

A wide variety of enzymes enters the lumen of the duodenum to break down the various organic substances in food. For example, the pancreatic juice contains **trypsin** and **peptidases**, which digest proteins and peptides into their amino acids (Figure 18.6). Also in the pancreatic juice are a series of carbohydrate-digesting enzymes including **amylase**, **lactase**, **sucrase**, and **maltase**. Fats are broken down by enzymes called **lipases**, and nucleic acids are digested by enzymes called **nucleases**.

In addition, the pancreatic juice contains **bicarbonate ions**, which raise the pH of the intestinal juice and neutralize the acidity introduced in the stomach. This is important because enzymes act in a neutral pH environment.

Another important component of the digestion process is **bile**, which is produced by the liver and stored in the gall bladder. Bile enters the duodenum through the common bile duct at about the same location as the pancreatic duct. Bile has no enzyme activity. Rather, it acts to break down large fat globules into smaller globules, which can be easily digested by lipase enzymes (Table 18.4). This process is called **emulsification**. Bile also increases the absorption of fats as well as fat-soluble vitamins, including vitamins K, D, and A.

In digestion, bile emulsifies fats into tiny droplets known as **micelles**. The micelles serve as carriers for the fatty acids and mono-glycerides of fats. Micelles collide with the membranes of villus cells during absorption and release their products into the cells. The products of fat digestion then pass into the lacteals as triglycerides contained within microscopic droplets called **chylomicrons**.

The release of pancreatic juices is controlled by branches of a cranial nerve as well as by hormones produced by cells of the small intestine. Two important hormones are **secretin** and **cholecystokinin**. Both hormones are essential for the digestive process because they control the release of pancreatic, liver, and gall bladder products. The release of bile into the duodenum, for example, is regulated by the hormone cholecystokinin.

**TABLE 18.4**   *The Sources and Enzymes of Human Digestion*

| Source | Fluids | Enzyme | Substrate | Product | Site of Action |
|---|---|---|---|---|---|
| Salivary glands | Saliva | Salivary amylase | Starches | Maltose | Mouth |
| Stomach glands | Gastric juice | Pepsin | Proteins | Peptides | Stomach |
| | | Renin | Milk protein | Clotted protein | Stomach |
| | | Hydrochloric acid | Many foods | Smaller units | Stomach |
| Pancreas | Pancreatic juice | Pancreatic amylase | Starches | Maltose | Small intestine |
| | | Lipase | Fats | Fatty acids, glycerol | Small intestine |
| | | Trypsin | Proteins | Peptides | Small intestine |
| | | Chymotrypsin | Proteins | Peptides | Small intestine |
| | | Carboxypetidase | Proteins | Peptides | Small intestine |
| | | Nucleases | DNA and RNA | Nucleotides | Small intestine |
| Small intestine | Intestinal juice | Maltase | Maltose | Glucose | Small intestine |
| | | Lactase | Lactose | Glucose, galactose | Small intestine |
| | | Sucrase | Sucrose | Glucose, fructose | Small intestine |
| | | Aminopeptidase | Peptides | Amino acids | Small intestine |
| | | Dipeptidase | Dipeptidase | Amino acids | Small intestine |
| | | Phosphatase | Nucleotides | Smaller units | Small intestine |
| Liver | Bile | Bile salts | Large fat droplets | Emulsified fats | Small intestine |

## The Jejunum and Ileum

A small amount of digestion and most absorption occurs in the jejunum and ileum. The surface area is increased by thousands of villi and microvilli. **Villi** are fingerlike projections of the mucosa, while **microvilli** are microscopic projections of the membranes of the mucosal cells. Within the villus lies a rich series of **capillaries** as well as lymph vessels called **lacteals**. The capillaries receive the products of protein, carbohydrate, and nucleic acid digestion. The lacteals receive the products of fat digestion (Figure 18.7).

The process of absorption through the membranes of the epithelial cells of the intestinal wall, through the interstitial fluid, and into the capillaries occurs largely by the process of active transport using carrier molecules and ATP. Diffusion, facilitated diffusion, and pinocytosis also occur. For lipids, diffusion is the important process of

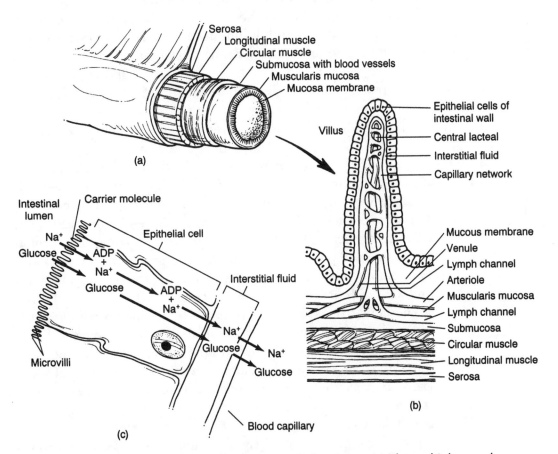

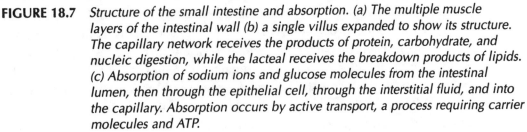

**FIGURE 18.7** *Structure of the small intestine and absorption. (a) The multiple muscle layers of the intestinal wall (b) a single villus expanded to show its structure. The capillary network receives the products of protein, carbohydrate, and nucleic digestion, while the lacteal receives the breakdown products of lipids. (c) Absorption of sodium ions and glucose molecules from the intestinal lumen, then through the epithelial cell, through the interstitial fluid, and into the capillary. Absorption occurs by active transport, a process requiring carrier molecules and ATP.*

molecular movement. The blood capillaries absorb monosaccharides, amino acids, and short-chained fatty acids. Long-chained fatty acids are resynthesized to form triglycerides, and the triglycerides pass by diffusion into the lacteals, as noted earlier. Water, electrolytes such as sodium ions, and vitamins are absorbed into the blood capillaries.

## The Large Intestine

The large intestine consists of the cecum, ascending colon, transverse colon, descending colon, sigmoid colon, rectum, anal canal, and anus. The large intestine is so-named because its diameter is noticeably larger than the diameter of the small intestine. It is subdivided into numerous small pouches called **haustra** (singular haustrum) and measures about 5 to 6 feet in length, with an average diameter of 2.5 inches. The large intestine is also referred to as the **colon**.

The first two or three inches of the large intestine is the cecum, located where the small intestine meets the large intestine in the lower right quadrant of the abdomen. Extending from the cecum is a short wormlike extension of tissue called the **vermiform appendix**. The appendix is a vestigial structure that may become inflamed and require removal.

The **ascending colon** lies in the vertical position on the right side of the abdomen extending upward toward the lower border of the liver (Figure 18.8). Nondigested material enters the ascending colon from the ileum through the **ileocecal valve**. The **transverse colon** extends horizontally across the abdomen near the stomach and spleen. The transverse colon becomes the **descending colon** at the splenic flexure, a 90° bend where the colon turns downward. The descending colon lies in a vertical position on the left side of the abdomen. It joins with the **sigmoid colon**, an S-shaped structure that descends downward and joins with the rectum. The last seven or eight inches of the gastrointestinal tract is called the **rectum**. The rectum terminates at the anal canal, which terminates at the opening called the **anus**.

The functions of the large intestine are to reabsorb water and ions and compact the feces. Approximately 300 to 400 ml of water are reabsorbed daily. The release of large amounts of water is called **diarrhea**. The major ion absorbed in the large intestine is sodium. No chemical digestion occurs in the large intestine.

Another important function of the large intestine is to absorb **vitamins**. Vitamins are substances needed in microscopic amounts for metabolic processes (Chapter 19). These vitamins are commonly produced by bacteria normally residing in the intestine. The large intestine also stores and compacts the materials not digested by the body and forms the **feces**. The release of feces is called **defecation**. Technically speaking, defecation is not an excretory function, but the release of materials not digested by the body.

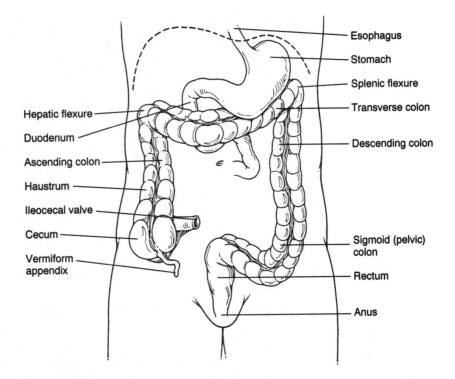

**FIGURE 18.8**  *Ventral view of the structure of the large intestine in position in the abdominal cavity and relative to the stomach. The small intestine from the duodenum to the cecum has been removed. The liver and pancreas have also been removed. The dotted line represents the position of the diaphragm.*

The feces consists of water, inorganic salts, bacteria, epithelial cells shed by the gastrointestinal tract, and undigested food. Defecation is a reflex action aided by voluntary contractions of various muscles.

## THE ACCESSORY ORGANS

The accessory organs support the activities of the digestive system. They include the liver and the pancreas.

### The Liver

The liver is the largest gland in the body and one of the most important accessory organs to the gastrointestinal tract. The liver lies under the diaphragm and is divided into four lobes known as the **right, left, cordate,** and **quadrate lobes** (Figure 18.9). The lobes of the liver are further divided into **lobules,** which contain the liver cells (**hepatocytes**) and cells of the reticuloendothelial system called **Kupffer's cells.**

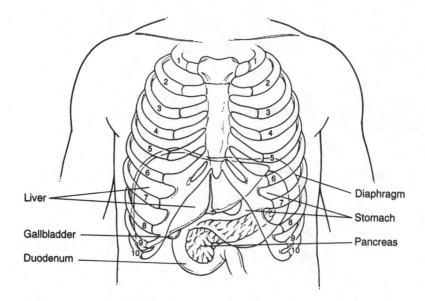

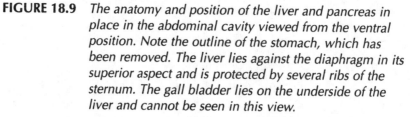

**FIGURE 18.9** *The anatomy and position of the liver and pancreas in place in the abdominal cavity viewed from the ventral position. Note the outline of the stomach, which has been removed. The liver lies against the diaphragm in its superior aspect and is protected by several ribs of the sternum. The gall bladder lies on the underside of the liver and cannot be seen in this view.*

The liver receives material absorbed from the digestive system through a subdivision of the circulatory system called the **hepatic portal system**. This system is formed by venules and veins draining blood from various regions of the digestive system and merging to form a single hepatic portal vein. The **hepatic portal vein** carries blood from the capillary beds of the digestive system to the liver, where it branches to form a second capillary bed. Materials are also delivered from the circulatory system to the liver by the hepatic artery. Materials leave the liver for the circulation by the hepatic vein, which joins the inferior vena cava. The liver is located under the diaphragm and occupies most of the right hypochondrium of the abdominal cavity. Its secretions reach the small intestine by the **hepatic duct**, which leads to the common bile duct, as noted below.

The liver has many vital functions related to the digestive process. One of the important functions is the production of **bile**, a yellow, brownish, or olive-green liquid. Bile has a pH of 7.6 to 8.6 and consists of water and bile salts, cholesterol, a phospholipid called lecithin, bile pigments, and several kinds of ions. The principal bile pigment is **bilirubin**, a substance derived from the heme portion of hemoglobin from broken red blood cells (Chapter 14). Bilirubin is later digested in the intestine by bacteria and one of its breakdown products called urobilinogen gives feces part of its color. Some of the urobilinogen also finds its way to the urine (Chapter 20).

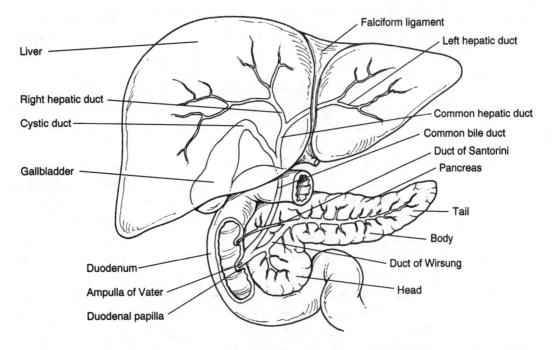

**FIGURE 18.10**  *The structure and duct systems of the liver and pancreas as seen from the ventral position. The various ducts from the gall bladder, liver, and pancreas all come together at the ampulla of Vater in the duodenal papilla.*

The bile produced by the liver is stored in a pear-shaped sac called the **gall bladder**. The gall bladder is located on the visceral surface of the liver and is drained by the **cystic duct** (Figure18.10). The cystic duct delivers bile to the duodenum through the common bile duct. The common bile duct is formed by the union of the cyctic duct from the gall bladder and the hepatic duct from the liver. The gall bladder stores and concentrates bile until it is needed in the digestive process.

Another function of the liver concerns carbohydrate metabolism. When the level of glucose in the blood is high, enzymes in the liver convert glucose to glycogen. This process is called **glycogenesis**. When the supply of blood glucose is low, enzymes in the liver cells convert glycogen into glucose. This process is called **glyogenolysis**. The enzymes of the liver are also able to convert amino acids into carbohydrate molecules for energy use when the level of blood carbohydrate is low, a process known as **gluconeogenesis** (Chapter 19).

With respect to fat metabolism, the liver is able to break down fatty acids to smaller molecules such as **acetyl coenzyme A**. Molecules of acetyl coenzyme A can then be further processed in metabolism (Chapter 19) to release the energy present in the chemical bonds.

In protein metabolism, enzymes of the liver perform a process called **deamination**. Deamination involves the removal of the amino groups from amino acids. The resulting molecules can then be used for energy metabolism or converted to carbohydrates or fats. The amino groups that result from amino acids are used to synthesize a toxic substance called **urea**. Urea is eventually removed from the bloodstream in the kidney and is the major component of urine (Chapter 20).

Other forms of protein metabolism are also performed in the liver. For example, cells of the liver synthesize most **plasma proteins** such as albumin, globulins, and prothrombin and fibrinogen used in blood clotting.

Cells of the liver also remove drugs and hormones from the blood. For example, liver cells can remove drugs such as penicillin and sulfa drugs from the blood and excrete them to the bile. Liver enzymes can also alter the chemical structure of certain steroid hormones such as estrogens and aldosterone.

Vitamin storage is another function of the liver. The liver stores vitamins such as A, $B_{12}$, D, E, and K, as well as minerals such as iron and copper. In liver cells, a protein called **apoferretin** combines with iron ions to form **ferretin**, and iron is stored in the liver in this way. The Kupffer's cells of the liver perform phagocytosis and remove old red and white blood cells from the bloodstream for destruction. Finally, the liver participates in the activation of vitamin D for use in the body.

## The Pancreas

The second accessory organ in the digestive system is the pancreas. The pancreas has an important role in the endocrine system (Chapter 13) as well as in the digestive system. It is an oblong gland about five inches long and one inch thick.

The pancreas lies posterior to the greater curvature of the stomach and is connected by two ducts to the duodenum. The larger duct is called the pancreatic duct, or **duct of Wirsung**; the second duct is called the accessory duct, or **duct of Santorini**. The duct of Wirsung combines with the common duct from the liver and gall bladder and enters the duodenum at a common location called the **ampulla of Vater**. The duct of Santorini enters the duodenum about one inch above the ampulla of Vater.

The cells of the pancreas functioning in digestion are called **acini**. They constitute the exocrine portion of the pancreas and secrete pancreatic juice. **Pancreatic juice** is a clear, colorless liquid containing water, salts, bicarbonate ions, and enzymes. The bicarbonate ions gives pancreatic juice a slightly alkaline pH, which neutralizes the acidity of the digestive material in the stomach.

Many enzymes in the pancreatic juice function in digestion. For instance, pancreatic amylase digests carbohydrates; trypsin, chymotrypsin, and carboxypeptidase digest proteins; lipase digests fats; and ribonuclease and deoxynuclease digest nucleic acids. Secretions of the pancreas are controlled by the hormones secretin and cholecystokinin.

# REVIEW QUESTIONS

**PART A—Completion:** Add the word or words that correctly complete each of the following statements.

1. The two main functions of the digestive system are digestion and
   _____ .

2. The digestive system consists of the gastrointestinal tract, which is a muscular tube whose length is _____ .

3. The innermost layer of the gastrointestinal tract is a mucous membrane known as the _____ .

4. The submucosa of the gastrointestinal tract contains lymphatic vessels, blood vessels, and _____ .

5. The major types of muscle in the gastrointestinal tract is _____ .

6. The outer layer of the gastrointestinal tract is known as the serous layer or _____ .

7. The major portion of the mouth consists of a chamber known as the _____ .

8. Connections of the tongue to the floor of the mouth are made by a fold of tissue known as the _____ .

9. On either side of the tongue within the tongue papillae are a series of _____ .

10. One of the major functions of the tongue is to work with saliva to form food into a mass called the _____ .

11. The "baby" teeth are more correctly called _____ .

12. In a permanent set of teeth, the full number is _____ .

13. Those teeth specialized for grasping and tearing food are _____ .

14. Large pieces of food are bitten off by teeth known as _____ .

15. The three basic regions of a tooth include the crown, the neck, and the _____ .

16. The hardest substance in the body is found at the outside surface of the tooth and is known as the _____ .

17. The largest portion of the tooth is made up of _____ .

18. The blood vessels, nerves, and connective tissues of the tooth are located within the _____ .

19. Within the cheeks of the head, below the ears is the largest salivary gland known as the _____ .

20. The salivary gland drained by Wharton's duct is the submaxillary gland also known as the _____ .

21. Lying under the tongue in the floor of the mouth is the salivary gland called the _____ .

22. A major enzyme found within the saliva assists the breakdown of carbohydrates and is known as _____ .

23. The lymphatic vessels located on the soft palate are called _____ .

24. The digestion of starch in the mouth results in disaccharides known as _____ .

25. Food passes into the stomach from the pharynx through a tube known as the _____ .

26. Swallowing is a process more correctly known as _____ .

27. The series of wavelike contractions that brings food into the stomach is called _____ .

28. The circular muscle at the beginning of the stomach is the cardiac sphincter, while the circular muscle at the end of the stomach is the _____ .

29. The narrow, far region of the stomach is the _____ .

30. The cells within the gastric glands that produce digestive enzymes are known as _____ .

31. The most important acid in digestion taking place in the stomach is _____ .

32. The enzymes of the stomach do not digest the stomach wall because the wall is protected by _____ .

33. The main protein-digesting enzyme in the stomach is known as _____ .

34. The digestion of proteins in the stomach results in a series of _____ .

35. The hormone regulating the activity of gastric glands is _____ .

36. In the stomach, food is converted to a soupy mixture known as _____ .

37. A variety of enzymes enters the duodenum from a large gland known as the _____ .

38. The enzyme trypsin displays its activity on organic substances called _____ .

39. Nucleic acids are digested into their component nucleotides by enzymes known as _____ .

40. The acidity of the small intestine's contents is neutralized by _____ .

41. Before fats can be digested into their component fatty acids, they must be broken into smaller globules by the liver substance called _____ .

42. The liver enzyme responsible for digesting fats is known as _____ .

43. The second part of the small intestine and the place where most absorption occurs is the _____ .

44. The products of fat digestion are absorbed into lymphatic vessels called _____ .

45. The major method for the transport of substances from the small intestine to the blood vessels is known as _____ .

46. The short wormlike extension of tissue where the small and large intestines meet is called the _____ .

47. The last few inches of the gastrointestinal tract that terminates at the anus is the _____ .

48. A major function of the large intestine is to reabsorb ions and _____ .

49. The largest gland in the body and one of the most important accessory structures to the digestive system is _____ .

50. For use in the digestive process, bile is stored in a sac called the _____ .

**PART B—Multiple Choice:** Circle the letter of the item that correctly completes each of the following statements.

1. All the following are functions of the oral cavity except
   (A) to mechanically process foods
   (B) to digest proteins
   (C) to lubricate foods
   (D) to digest certain carbohydrates

2. A fold of tissue called the frenulum connects the
   (A) pancreas to the stomach
   (B) small and large intestines
   (C) gall bladder to the liver
   (D) tongue to the floor of the oral cavity

3. The two types of teeth are
   (A) deciduous teeth and permanent teeth
   (B) gastroid and pyloric teeth
   (C) parotid and colonic teeth
   (D) front and back teeth

4. Enzymes secreted by the salivary glands
   (A) break fats into smaller globules
   (B) provide enzymes for carbohydrate digestion
   (C) are stimulated by gastric hormones
   (D) empty their contents into the roof of the mouth

5. Passage of a bolus of food down the esophagus is assisted by
   (A) enzymes and acids
   (B) striated muscles and tissues
   (C) peristalsis and gravity
   (D) the pyloric sphincter

6. All the following are regions of the stomach except
   (A) the fundus
   (B) the cardia
   (C) the pylorus
   (D) the hilus

7. The lesser and greater omentums are the
   (A) entrances and exits to the stomach
   (B) curvatures of the stomach
   (C) extensions of the peritoneum near the stomach
   (D) glands that empty into the stomach

8. The precursor substance pepsinogen is converted to pepsin
   (A) in the duodenum
   (B) in the presence of hydrochloric acid
   (C) in the pancreas
   (D) only when high concentrations of salt are present

9. The hormone that regulates the activity of the gastric glands is known as
   (A) progesterone
   (B) androgen
   (C) gastrin
   (D) TSH

10. Bicarbonate ions to neutralize stomach acidity are provided to the duodenum by
    (A) the pancreas
    (B) hormones that are liberated in the lining of the gastrointestinal tract
    (C) bile from the liver
    (D) the salivary glands

11. All the following are regions of the small intestine except
    (A) the jejunum
    (B) the ilium
    (C) the duodenum
    (D) the pylorus

12. Bile, which is formed in the liver, assists the
    (A) breakdown of proteins
    (B) absorption of water
    (C) emulsification of fats
    (D) formation of feces

13. Enzymes called nucleases break down nucleic acids
    (A) only in the presence of acid
    (B) into nucleotides
    (C) within the stomach
    (D) in the large intestine

14. Most absorption occurs
    (A) in the esophagus
    (B) in the jejunum
    (C) in the lining of the stomach
    (D) all along the gastrointestinal tract

15. The absorption of the products of fat digestion take place into vessels of the
    (A) circulatory system
    (B) lymphatic system
    (C) hepatic portal system
    (D) venous system

16. Two major methods for absorption of the products of digestion are
    (A) phagocytosis and pinocytosis
    (B) osmosis and phagocytosis
    (C) osmosis and pinocytosis
    (D) active transport and diffusion

17. The large intestine is so-named because it exceeds the small intestine in
    (A) diameter
    (B) length
    (C) number of enzymes produced
    (D) amount of muscle present

18. Where the small intestine meets the large intestine, there is a section known as the
    (A) rectum
    (B) sigmoid colon
    (C) cecum
    (D) fundus

19. An important function of the large intestine is to
    (A) break down proteins
    (B) break down carbohydrates
    (C) absorb vitamins
    (D) absorb nucleotides

20. All the following are commonly found in the feces except
    (A) ATP molecules
    (B) bacteria
    (C) inorganic salts
    (D) epithelial cells

21. The liver receives materials absorbed from the gastrointestinal tract through a subdivision of the circulatory system called the
    (A) venous system
    (B) lymphatic system
    (C) renal system
    (D) hepatic portal system

22. All the following are normally found in the bile except
    (A) cholesterol
    (B) pigments
    (C) various ions
    (D) proteins

23. The process of glycogenesis involves the
    (A) breakdown of glucose
    (B) conversion of amino acids to carbohydrates
    (C) synthesis of glycogen
    (D) breakdown of glycogen

24. The process of deamination results in the
    (A) breakdown of amino acids
    (B) synthesis of glucose
    (C) formation of fat molecules
    (D) formation of glycogen molecules

25. The pancreas is located close to the
    (A) sigmoid colon
    (B) stomach
    (C) cecum
    (D) appendix

*PART C—True/False:* **For each of the following statements, mark the letter "T" next to the statement if it is true. If the statement is false, change the <u>underlined</u> word to make the statement true.**

1. The outer layer of the gastrointestinal tract is composed of the visceral <u>peritoneum</u>.

2. The tongue is composed of <u>smooth</u> muscle covered by a mucous membrane.

3. Those teeth which are flat and specialized for grinding are called <u>incisors</u>.

4. The hardest substance in the body and the substance found at the outer surface of the tooth is the <u>dentin</u>.

5. The parotid gland is the largest salivary gland and is drained by Stenson's duct into the oral duct.

6. The salivary gland located under the tongue is the submandibular gland.

7. The palatine and pharyngeal tonsils are located on the posterior wall of the pharynx.

8. The approximate length of the esophagus is 30 inches.

9. Peristalsis brings the bolus of food through the esophagus to the pyloric sphincter.

10. The internal surface of the stomach has many folds called rugae.

11. The only organic material digested in the stomach is fat.

12. Enzymes that function in the stomach are produced primarily by the chief cells.

13. For the absorption of B vitamins from the stomach, a substance called intrinsic factor is required.

14. In the stomach a bolus of food is converted to a soupy liquid known as kinin.

15. Most digestion in the body goes on in an organ called the large intestine.

16. The enzymes trypsin and peptidase are responsible for the breakdown of carbohydrates.

17. The common bile duct enters the duodenum at about the same location as renal duct.

18. The products of protein, carbohydrate, and nucleic acid digestion enter the lacteals.

19. Much of the process of absorption takes place by the process of active transport.

20. Fingerlike projections of the mucosa of the jejunum are called villi.

21. The small and large intestine meet in the lower right quadrant of the thorax.

22. There are ascending, transverse, and descending portions of the small intestine.

23. Two important cells of the liver are the hepatocytes and the Kupffer's cells.

24. The duct that drains the gall bladder is known as the hepatic duct.

25. The proteins fibrinogen and prothrombin that are used in blood clotting are synthesized in the pancreas.

**Answers**

## PART A—Completion

1. absorption
2. nine meters
3. mucosa
4. nerves
5. smooth muscle
6. serosa
7. oral cavity
8. frenulum
9. taste buds
10. bolus
11. deciduous teeth
12. thirty-two
13. cuspids
14. incisors
15. roots
16. enamel
17. dentin
18. pulp
19. parotid
20. submandibular gland
21. sublingual gland
22. amylase
23. tonsils
24. maltose
25. esophagus
26. deglutition
27. peristalsis
28. pyloric sphincter
29. pylorus
30. chief cells
31. hydrochloric acid
32. mucus
33. pepsin
34. peptides
35. gastrin
36. chyme
37. pancreas
38. proteins
39. nucleases
40. bicarbonate ions
41. bile
42. lipase
43. jejunum
44. lacteals
45. active transport
46. appendix
47. rectum
48. water
49. liver
50. gall bladder

## PART B—Multiple Choice

| | | | | |
|---|---|---|---|---|
| 1. B | 6. D | 11. D | 16. D | 21. D |
| 2. D | 7. C | 12. C | 17. A | 22. D |
| 3. A | 8. B | 13. B | 18. C | 23. C |
| 4. B | 9. C | 14. B | 19. C | 24. A |
| 5. C | 10. A | 15. B | 20. A | 25. B |

## PART C—True/False

1. true
2. skeletal
3. bicuspids
4. enamel
5. true
6. sublingual
7. true
8. ten
9. cardiac
10. true
11. protein
12. true
13. true
14. chyme
15. duodenum
16. proteins
17. pancreatic
18. capillaries
19. true
20. true
21. abdomen
22. large
23. true
24. cystic
25. liver

# CHAPTER 19

# METABOLISM AND NUTRITION

Metabolism is concerned with all the physical and chemical reactions taking place in the cell. The two major aspects of metabolism are **anabolism** (the synthesis of complex molecules) and **catabolism** (the breakdown of complex molecules). The reactions of anabolism usually require an input of energy, while those of catabolism result in an energy output (Table 19.1). The energy is trapped in the high-energy molecule adenosine triphosphate, or ATP.

The reactions of metabolism generally proceed along a **metabolic pathway**, which is a sequence of chemical reactions in which substrates are broken down to end products by the activity of enzymes. Many of the reactions are oxidation or reduction reactions. An **oxidation reaction** is one in which the substrate loses electrons and becomes oxidized, while a **reduction reaction** is one in which the substrate gains electrons and is reduced (Figure 19.1). Every oxidation reaction is accompanied by a reduction reaction because electrons do not exist in a free state. An oxidation can also imply the removal of a hydrogen atom, while a reduction can imply the addition of a hydrogen atom.

The biochemistry of metabolism is centered in the synthesis and breakdown of carbohydrates, fats, proteins, and nucleic acids. Protein synthesis from amino acids is discussed in Chapter 3. This chapter will be concerned principally with carbohydrate breakdown as an energy-yielding process and the metabolism of fats and amino acids.

**TABLE 19.1** *A Comparison of Two Key Aspects of Cellular Metabolism*

| Catabolism | Anabolism |
| --- | --- |
| Breakdown of large molecules | Buildup of small molecules |
| Energy is generally released | Energy is generally required |
| Products are small molecules | Products are large molecules |
| Glycolysis, Krebs cycle, electron transport | Photosynthesis, protein synthesis |
| Mediated by enzymes | Mediated by enzymes |
| Reactions converge to major pathways | Reactions diverge from basic pathways |

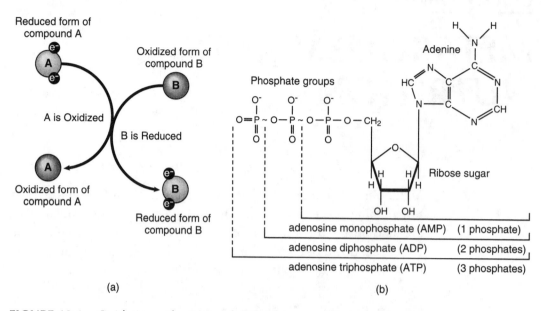

FIGURE 19.1   *Oxidation reduction and ATP. (a) In an oxidation-reduction reaction, two compounds are involved. Compound A loses two electrons to compound B. Compound A becomes oxidized, and compound B is reduced by the reaction. (b) The structure of adenosine triphosphate (ATP), an immediate energy source for the human body on metabolism. Energy is released when the enzyme adenosine triphosphatase cleaves away the terminal phosphate group to yield ADP and a phosphate group.*

## ADENOSINE TRIPHOSPHATE

When a cell needs energy, it commonly utilizes a molecule of adenosine triphosphate (ATP). This molecule consists of an adenosine portion bonded to three phosphate groups. In an ATP molecule, approximately 7.3 kilocalories of energy are present where the terminal phosphate group attaches to the remainder of the molecule. To release the energy, the enzyme adenosine triphosphatase cleaves away the terminal phosphate group as a phosphate ion. In addition to energy, the products of ATP breakdown are **adenosine diphosphate (ADP)** and the **phosphate ion**.

Adenosine diphosphate and phosphate ion can be reconstituted to form ATP, much like a battery can be recharged. To accomplish this synthesis, energy must be available. This energy is made available in human cells through the process of **cellular respiration.**

Cellular respiration takes place by a complex set of processes in the cell. These processes utilize a group of organic molecules called **coenzymes**. Coenzymes are portions of enzyme molecules essential for the activity of the enzyme. Two important coenzymes in human cells are: nicotinamide adenine dinucleotide (**NAD**) and flavin adenine dinucleotide (**FAD**). Both molecules are structurally similar to ATP. NAD has a nitrogen-containing ring called nicotinic acid. In FAD, the chemically active portion is a group called the flavin group.

The vitamin **riboflavin** is used by the body to produce this flavin group.

All **coenzymes** accept electrons and pass them on to other coenzymes or other molecules. The reactions are oxidation-reduction reactions. Other molecules participating in these reactions are iron-containing pigments called **cytochromes**. Together with the coenzymes, cytochromes accept and pass on electrons in the **electron transport system**. The passage of energy-rich electrons among cytochromes and coenzymes drains the energy from the electrons, and the energy is used to unite ADP molecules and phosphate ions to form ATP molecules.

The synthesis of ATP using energy from electron transport involves a complex process referred to as **chemiosmosis**. In chemiosmosis a steep proton gradient is established between the membrane-bound compartments of the cell's mitochondria. This gradient forms when large numbers of protons (hydrogen ions) are pumped into membrane-bound compartments of the mitochondria. The energy used to pump the protons is the energy released from the electrons during the electron transport system.

After protons have accumulated within compartments of the mitochondria, they suddenly reverse their direction and escape back across the membranes out of the compartments. The escaping protons release their energy, and the energy is used by enzymes to unite ADP with phosphate ions to form ATP. The movement of protons is called chemiosmosis, because it involves a movement of chemicals (protons) across a semipermeable membrane.

# CARBOHYDRATE METABOLISM

Glucose is the principal carbohydrate available as an energy source to humans. Other carbohydrates consumed by humans include fructose, galactose, sucrose, lactose, maltose, and starch, all of which are converted to glucose or a related compound for use in energy metabolism

During cellular respiration, cells take carbohydrates into the cytoplasm and break them down to release their energy. In the process, carbon dioxide and water are given off as waste products, and oxygen gas is required as an electron acceptor. The process involves four subdivisions: glycolysis, the Krebs cycle, the electron transport system, and chemiosmosis. In **glycolysis**, glucose molecules are converted to pyruvic acid; in the **Krebs cycle**, pyruvic acid molecules are broken down further, and the energy in the molecules is used to form high-energy compounds such as NADH; in the **electron transport system**, electrons are transported among coenzymes and cytochromes and the energy is released; and in **chemiosmosis**, the energy is used to pump protons across a membrane and provide the energy for ATP synthesis (Figure 19.2).

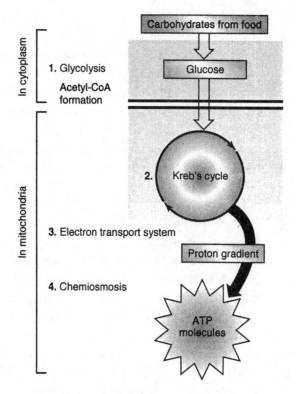

In cytoplasm
1. Glycolysis
   Acetyl-CoA
   formation

In mitochondria
2. Kreb's cycle
3. Electron transport system
4. Chemiosmosis

Carbohydrates from food
Glucose
Proton gradient
ATP molecules

**FIGURE 19.2**   *An overview of carbohydrate metabolism illustrating the relationships of the four major subdivisions. Carbohydrates from foods result in glucose, and glucose is then metabolized through the processes (numbered 1 to 4) to release energy, which is stored in ATP molecules. Some of the biological chemistry take place in the cytoplasm, others in the mitochondrion of the cell.*

# Glycolysis

In glycolysis a glucose molecule is metabolized through a multistep pathway to two molecules of pyruvic acid. The process occurs in the cytoplasm of cells. At least six enzymes operate in the metabolic pathway.

In the first and third steps of the pathway, ATP molecules are employed to energize the chemical reactions. Further along in the process, the 6-carbon glucose molecule is converted into an intermediary compound, which splits into two 3-carbon compounds. The latter undergo additional conversions and eventually form **pyruvic acid** at the conclusion of the process (Figure 19.3).

Toward the latter portion of glycolysis, four ATP molecules are synthesized using the energy released during the chemical reactions (Figure 19.3). Since four ATP molecules are synthesized and two ATP molecules were inserted into the process, a net gain of two ATP molecules results from glycolysis.

Also during the process of glycolysis, another chemical reaction yields high-energy electrons and a hydrogen atom (H). These are

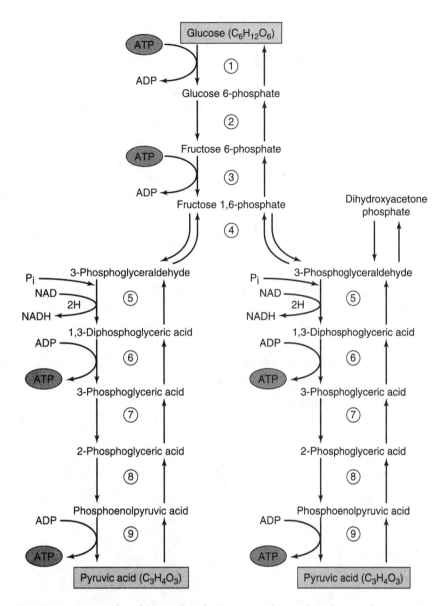

**FIGURE 19.3** *Glycolysis. Glycolysis is a multistep biochemical process in which a molecule of glucose is converted to two molecules of pyruvic acid. Note that in the process, two molecules of ATP are used (reactions 1 and 3), and four molecules of ATP are produced (reactions 6 and 9—each reaction occurring two times), for a net gain of two ATP molecules. In reaction 5, two molecules of NADH are formed for use in electron transport.*

transferred to the coenzyme molecule NAD, thereby changing it to **NADH**. The reduced coenzyme (NADH) will later be used in the electron transport system. During glycolysis two NADH molecules are produced. Glycolysis does not require oxygen, so the process is considered anaerobic.

When anaerobic conditions occur in **muscle cells**, an enzyme converts the pyruvic acid of glycolysis to **lactic acid**. This chemical reaction frees up the NAD for reuse in glycolysis, while providing the cells with two ATP molecules. Eventually, the lactic acid causes intense fatigue and the muscle cell stops contracting (Chapter 8).

## The Krebs Cycle

After glycolysis, cellular respiration involves another multistep process called the **Krebs cycle**, also called the citric acid cycle and the tricarboxylic acid (TCA) cycle. The Krebs cycle utilizes the two molecules of pyruvic acid formed in glycolysis. The cycle yields high-energy molecules of NADH and FADH, as well as some ATP. It occurs in the mitochondrion of a cell. This cellular organelle possesses inner and outer membranes, organized to compartments. The inner membrane is folded over itself many times to form **cristae**. Along the cristae are the important enzymes necessary for the proton pump and ATP production.

Before entering the Krebs cycle, the pyruvic acid molecules are processed. An enzyme acts on the pyruvic acid molecule and releases one carbon atom as a carbon dioxide molecule. The remaining two carbon atoms (the acetyl group) combine with a coenzyme known as coenzyme A to form acetyl-CoA (Figure 19.4). In the process, electrons and a hydrogen ion are transferred to NAD to form high-energy NADH.

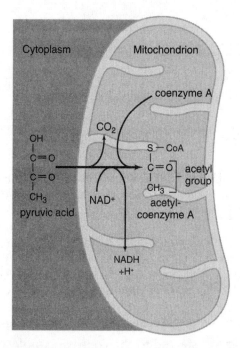

**FIGURE 19.4**   *Formation of acetyl-CoA. Pyruvic acid is formed from glucose molecules during glycolysis taking place in the cytoplasm. The pyruvic acid molecule is transported to the mitochondrion, where a carbon atom is lost as carbon dioxide and the remainder of the molecule (an acetyl group) unites with a coenzyme A molecule. In the process, an NADH molecule is formed for electron transport.*

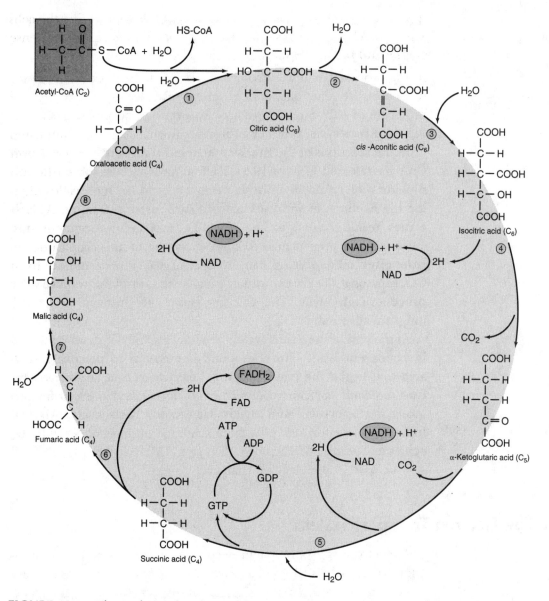

**FIGURE 19.5**   *The Krebs cycle. The Krebs cycle is a complex set of biochemical reactions occurring in the mitochondrion of the cell. Note that a molecule of acetyl-CoA enters the cycle at the upper left to initiate the process. In reactions 4, 5, and 8, a series of NADH molecules are formed for use in electron transport. An FADH₂ molecule is formed in reaction 6, and an ATP molecule is synthesized during this reaction. Molecules of carbon dioxide result in reactions 4 and 5. Another molecule resulted during the formation of acetyl-CoA (Figure 19.4). These three molecules of CO₂ represent the carbons of pyruvic acid from glycolysis and originally from glucose. The CO₂ molecules are expelled from the body during expiration in the lungs.*

Acetyl-CoA is now ready for entry to the Krebs cycle. It unites with a 4-carbon acid called oxaloacetic acid. The combination results in a 6-carbon acid called **citric acid**. Citric acid now undergoes a series of enzyme-catalyzed conversions. The conversions involve up to 10 chemical reactions and are brought about by enzymes. In many of the steps, high-energy electrons are released to NAD molecules. The NAD molecules also acquire hydrogen ions and become NADH molecules. In one of the steps, FAD serves as the electron acceptor. In the process it acquires two hydrogen ions to become $FADH_2$. Also in one of the reactions, enough energy is released to synthesize a molecule of ATP. Since there are two pyruvic acid molecules entering the Krebs cycle, two ATP molecules form.

In the reactions of the Krebs cycle, the two carbon atoms of acetyl-CoA are released (Figure 19.5). Each atom is used to form a carbon dioxide molecule. Since there are two acetyl-CoA molecules entering the Krebs cycle and each has two carbons atoms, four $CO_2$ molecules result. Added to the two $CO_2$ molecules formed in the conversion of pyruvic acid to acetyl-CoA, the total is six $CO_2$ molecules given off as waste gas in the Krebs cycle. The six molecules of $CO_2$ represent the six carbons of glucose that originally entered the process of glycolysis. The $CO_2$ molecules are transported to the lungs for disposal.

At the end of the Krebs cycle, the last chemical compound formed is oxaloacetic acid. This compound is identical to the oxaloacetic acid that began the cycle. It is now ready to accept another acetyl-CoA molecule to begin another turn of the cycle. Note that for two molecules of pyruvic acid metabolized in the Krebs cycle, two ATP molecules have formed, plus there is a large number of NADH molecules and some $FADH_2$ molecules. The NADH and the $FADH_2$ will now be used in the electron transport system.

## The Electron Transport System

The electron transport system occurs along the cristae of the mitochondria, which is where the participating cytochromes and coenzymes are located. In the electron transport system, NADH and $FADH_2$ molecules are used from the Krebs cycle and glycolysis. The molecules give up their electrons to a series of iron-containing pigments (cytochromes) and other coenzymes. The cytochromes and coenzymes transport the electrons among one another, and the energy in the electrons is gradually lost. But it is not lost to the environment. Rather, the energy from electron passages is used to pump protons across the mitochondrial membrane into the outer compartment of the mitochondria. Each NADH molecule contains enough energy to transfer six protons into the outer compartment. Each $FADH_2$ molecule has enough energy to transfer four protons.

The electrons passed among the cytochromes and coenzymes ultimately are taken up by an oxygen atom. Having acquired two electrons, the oxygen atom becomes negatively charged. To balance the charges, the atom takes on two protons from the solution to form a molecule of **water** ($H_2O$). Water is thus an important waste product of the metabolism (Figure 19.6).

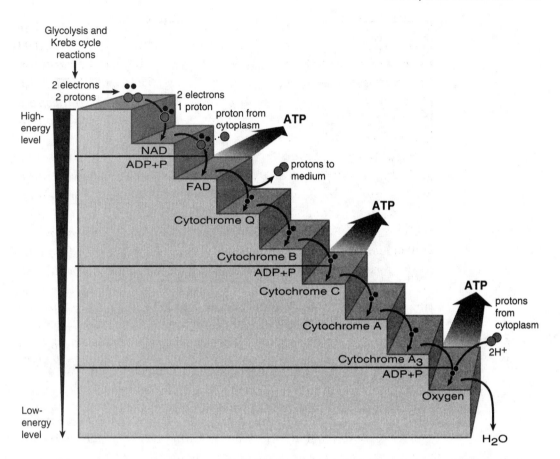

**FIGURE 19.6** *Electron transport. During this process, electrons are liberated from reactions taking place in glycolysis and the Krebs cycle. The electrons pass among NAD, FAD, and other cytochromes and gradually lose their energy. The energy is used to synthesize ATP molecules in the process of chemiosmosis.*

As a final electron receptor, oxygen is responsible for removing electrons from the system. If oxygen were not available at the end of the transport system, electrons could not be released from the coenzymes and cytochromes, and they would be unable to function any further. Then the energy in electrons could not be released; the proton pump could not be established; and ATP could not be produced. In humans, breathing is the essential process that brings oxygen into the body for delivery to the cells for use in cellular respiration.

## Chemiosmosis

The actual production of ATP in cellular respiration takes place through chemiosmosis. As noted previously, chemiosmosis involves proton pumping across the membranes of mitochondria to establish a proton gradient. Once the gradient is established, protons pass down the gradient through particles designated F1. In these particles

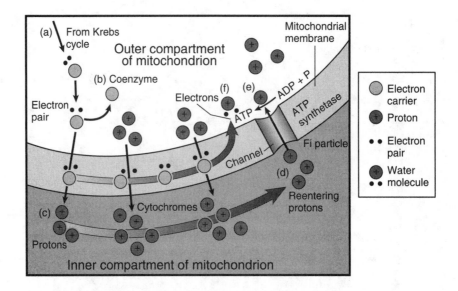

**FIGURE 19.7**   *Chemiosmosis. Electron-carrying coenzyme molecules from the Krebs cycle (a) enter the process. The electrons are lost from the coenzyme (b), and they pass among a series of cytochromes. The energy is used to pump protons across the mitochondrial membrane (c) into the inner compartment, and the protons gather in this compartment. Suddenly, they flow back across the membrane (d) to the outer compartment. In the passage, the energy from the electron flow is used (e) to synthesize ATP molecules.*

enzymes use the energy of the protons to generate ATP, using ADP and phosphate ions as the starting points (Figure 19.7).

The energy production in cellular respiration during chemiosmosis is substantial. It is widely agreed among scientists that a total of 34 molecules of ATP can be produced during cellular respiration as a result of the reactions of the Krebs cycle, the electron transport system, and chemiosmosis. Two ATP molecules are formed during the Krebs cycle, and two molecules of ATP are produced in glycolysis for a total of 38 molecules of ATP. These ATP molecules may then be utilized in the cell for cellular needs (Table 19.2). However, they cannot be stored for long periods of time, and cellular respiration must continue constantly in order to regenerate the ATP molecules as they are used.

## Physiology of Glucose Metabolism

The glucose molecules used in cellular respiration are absorbed from the small intestine into the bloodstream (Chapter 18). They and other monosaccharides, such as fructose and galactose, are transported to the liver by the hepatic portal vein. In the liver, fructose and galactose are converted to glucose, and the glucose molecules can then be used in cellular respiration.

**TABLE 19.2**   *Some Characteristics of Pathways of Cellular Metabolism*

| Form | Pathway | Location | Reactants | Products |
|------|---------|----------|-----------|----------|
| Glycolysis | Embden-Meyerhoff Pathway | Cytoplasm | Glucose | 2 pyruvic acid, 8 ATP (if oxygen present) 2 ATP (if not present) |
| Fermentation | Alcoholic | Cytoplasm | Pyruvic acid | Ethanol, $CO_2$ |
| | Lactic acid | Cytoplasm | Pyruvic acid | Lactic acid |
| Aerobic Respiration and Chemiosmosis | Acetyl-CoA formation | Mitochondria | Pyruvic acid | NADH, $CO_2$, acetyl-CoA |
| | Krebs cycle (2 turns) | Mitochondria | Acetyl-CoA | 2 ATP, 6 NADH, 4 $CO_2$, 2 $FADH_2$ |
| | Electron transport | Mitochondrial membranes | 10 NADH, 2 $FADH_2$ | 34 ATP |

At the tissue cells, the hormone **insulin** facilitates the transfer of glucose molecules across cell membranes by increasing the affinity of membrane carrier molecules for glucose molecules. In the absence of insulin, type I diabetes, or insulin-dependent diabetes, occurs in the patient.

Glucose molecules are also stored in the liver as **glycogen** when the level of blood glucose is high. The process of glycogen formation is called **glycogenesis**. When the level of blood glucose is low, glycogen breaks down and releases glucose molecules into the bloodstream. This process is **glycogenolysis**. The hormones glucagon and epinephrine accelerate glycogenolysis.

Glucose molecules can also be constructed in the liver from non-carbohydrate sources. For example, amino acids can be used to form glucose molecules by an intricate process involving reversal of some of the steps of glycolysis and the Krebs cycle. This process of glucose formation is called **gluconeogenesis** (Figure 19.8). Certain fatty acids as well as glycerol molecules and lactic acid can be changed into glucose molecules through gluconeogenesis.

## FAT AND PROTEIN METABOLISM

In the digestive process, fats and certain lipoproteins are broken down into fatty acids and glycerol molecules. In the intestinal mucosa, fatty acids and glycerol molecules form **chylomicrons**, which are microscopic droplets of fat materials that enter the lacteals, and eventually, the blood circulation. Chylomicrons are composed of fatty acids, cholesterol, and phospholipids.

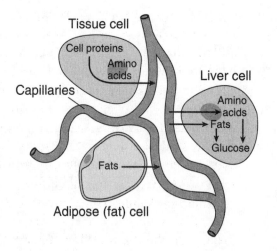

**FIGURE 19.8**   *Gluconeogenesis. The process of gluconeogenesis is one in which noncarbohydrate molecules are converted to glucose molecules for use in energy metabolism. As the illustration shows, in tissue cells proteins are broken down to amino acids, which enter the capillaries. Fats are obtained from adipose cells and also transported to the liver. In liver cells, the complex conversion to glucose molecules takes place.*

Most chylomicrons are removed from the body by the liver and the adipose tissue. Lipase enzymes then break down the triglycerides and release the free fatty acids and glycerol molecules. Some of the molecules are recombined for storage in the adipose, and some are metabolized in the liver, depending on the amount of dietary intake of fat.

For transport to the cells, many fatty acids are bound to albumin molecules in the plasma. Fat molecules are also transported as smaller protein-containing particles called **lipoproteins**. Lipoproteins are divided into three classes according to their density. **Very low density lipoproteins (VLDLs)** contain approximately 60 percent triglycerides and 15 percent cholesterol. **Low density lipoproteins (LDLs)** have almost 50 percent cholesterol, depending on the dietary intake of cholesterol and saturated fat. A high level of LDL in the blood indicates much cholesterol in the blood and the potential for coronary heart disease. **High density lipoproteins (HDLs)** contain about 20 percent cholesterol, about 5 percent triglycerides, and about 50 percent protein. These lipoproteins transport cholesterol to the liver for metabolism and clear triglycerides and cholesterol from the blood. A high concentration of HDLs is associated with a low potential for heart disease.

## Catabolism of Fats

Fats are broken down by cells and used as important sources of energy. In the breakdown process, the **glycerol** molecule is separated from the fatty acids. Within the cell cytoplasm, enzymes convert the glycerol to dihydroxy-acetone-phosphate (DHAP). DHAP is an intermediary compound in the process of glycolysis, and the DHAP continues along the pathway to pyruvic acid. Alternately, the DHAP molecule may reverse its track and follow a pathway that leads backward through glycolysis toward glucose. In this way, glycerol may be used to synthesize glucose molecules.

**Fatty acids** are metabolized in the mitochondria of the cell. Here they are converted into 2-carbon units of acetyl-CoA by a process known as **beta-oxidation**, also known as the **fatty acid spiral**. A single fatty acid molecule containing 16 carbon atoms will result in eight molecules of acetyl-CoA. Each acetyl-CoA molecule then enters the Krebs cycle (as if it had come from pyruvic acid) and is metabolized to release its energy. Thus, the energy yield from a 16-carbon fatty acid is considerable.

During the process of fat catabolism, some of the acetyl-CoA molecules condense with one another to form acetoacetic acid. This substance is then converted to acetone molecules and molecules of beta-hydroxybutyric acid. These molecules are called **ketone bodies** (or ketone molecules) because they contain keto groups ($-C = 0$). As normal products of fat catabolism their level is generally low because they are rapidly converted back to acetyl-CoA molecules. However, when much fat catabolism is taking place as in a person with diabetes mellitus, a high number of ketone bodies is formed, and the condition is called **ketosis**. The excessive ketone bodies raise the body's acidity, a condition that may result in diabetic coma. Ketosis also occurs when glucose is in short supply in the body such as in starvation. A high fat, low carbohydrate diet may also be responsible.

## Anabolism of Fat

The synthesis of fats in fat anabolism occurs from acetyl-CoA molecules (Figure 19.9). Acetyl-CoA molecules used in the synthesis are generally obtained from glucose molecules, thus providing a mechanism for the conversion of glucose molecules to fatty acids. Enzymes in the liver are able to convert one fatty acid to another and form the triglycerides found in fat molecules. Three unsaturated fatty acids, linolenic, linoleic, and arachidonic acids, cannot be synthesized by the body. These fatty acids are obtained from the diet and are known as **essential fatty acids**.

When the diet is high in carbohydrates, glucose is converted to fat in the process of **lipogenesis**. Amino acids can also be converted to fats through the intermediaries of cellular respiration. In this way, much body fat can be derived from carbohydrate and protein.

An important controlling factor of fat metabolism is insulin, which prevents fat breakdown by inhibiting the activity of lipase enzymes. This inhibition is lifted when insulin is deficient, as in

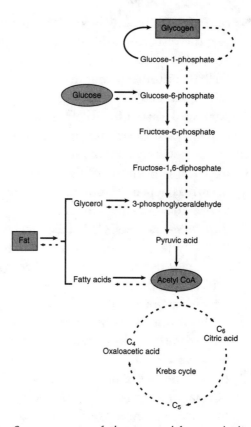

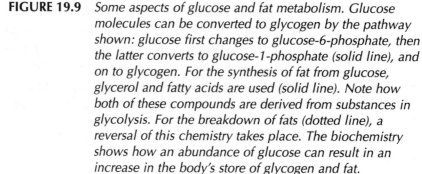

FIGURE 19.9    *Some aspects of glucose and fat metabolism. Glucose molecules can be converted to glycogen by the pathway shown: glucose first changes to glucose-6-phosphate, then the latter converts to glucose-1-phosphate (solid line), and on to glycogen. For the synthesis of fat from glucose, glycerol and fatty acids are used (solid line). Note how both of these compounds are derived from substances in glycolysis. For the breakdown of fats (dotted line), a reversal of this chemistry takes place. The biochemistry shows how an abundance of glucose can result in an increase in the body's store of glycogen and fat.*

persons with diabetes mellitus. Other hormones stimulating the release of fatty acids from the adipose tissue include epinephrine, human growth hormone, glucagon, ACTH, and thyroxine. Nerve impulses of the sympathetic nervous system accelerate fat breakdown in the adipose tissue, while impulses from the parasympathetic stimulation increase fat deposit.

## Protein Metabolism

In the body, proteins are digested in the gastrointestinal tract to their component amino acids. Amino acids are then absorbed from the intestine by active transport and carried to the liver. Here they may be synthesized into protein molecules or released to the circulation for transport to other cells. In the cells, the amino acids are linked together in a sequence that reflects the genetic code in the DNA of the cell. The process of protein synthesis is discussed in Chapter 3.

The body also utilizes some of the amino acids as energy sources. The conversion of amino acids to energy compounds occurs in the liver, especially when the diet is rich in protein. The conversion process begins with a step known as **deamination**. In this chemical reaction, the amino group ($-NH_2$) is removed from the amino acid by the enzyme deaminase and used to form a molecule of ammonia. The latter then passes through a cyclic series of events called the **urea cycle**, where it unites with carbon dioxide molecules to form **urea**. The urea is expelled into the bloodstream and removed by the kidneys.

After the amino group has been removed in deamination, an oxygen atom is added where the amino group once existed (Figure 19.10). The result is a compound normally found somewhere in the metabolic sequence of glycolysis or the Krebs cycle. This molecule may now be utilized for energy through the process of cellular respiration. Alternatively, the molecule can be used to form acetyl-CoA molecules, which can be used to form fatty acids in the process of lipogenesis. Proteins are utilized for energy after carbohydrates and fats have been used up.

There are certain amino acids that the body can synthesize by converting one amino acid into another in the liver by the process of transamination. These are **nonessential amino acids**. Eleven amino acids are nonessential. **Essential amino acids** are those obtained from the diet. They include tryptophan, valine, lysine, leucine, isoleucine, methionine, and histidine. Animal protein is likely to contain these amino acids and is regarded as **complete protein**. Plant protein often lacks some and is known as **incomplete protein**.

Protein metabolism is regulated by a number of hormones including human growth hormone, which stimulates the active transport of amino acids into cells and promotes protein synthesis. The male sex hormone testosterone and the female sex hormone estrogen also stimulate protein synthesis and bring about increased protein deposit in the tissues. Thyroxine increases the rate of cell metabolism and influences protein synthesis, and glucocorticoids encourage protein breakdown in cells.

## Metabolic States

The human body exists in two different metabolic states. Following the consumption of a meal, the body is in an **absorptive (feasting) state**, where food substances are being absorbed from the gastrointestinal tract into the bloodstream. When these absorptive processes are completed, the body enters a **postabsorptive (fasting) state** (Figure 19.11). In the postabsorptive state, body needs are met by materials already present in the body.

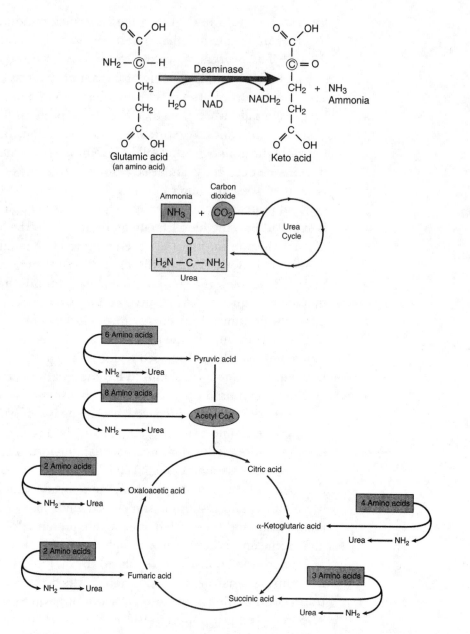

**FIGURE 19.10** *Aspects of protein metabolism. (a) To utilize proteins for energy metabolism, proteins are digested to yield their component amino acids. Certain amino acids are then converted by the enzyme deaminase in the process of deamination to a keto acid that can be used in glycolysis or the Krebs cycle. The remaining ammonia molecule is metabolized through the ornithine cycle to form the waste product urea. (b) Six different amino acids are converted directly or indirectly to pyruvic acid, eight different amino acids are converted to acetyl-CoA by deamination, two are converted to oxaloacetic acetic acid, and so on. Urea results from each conversion.*

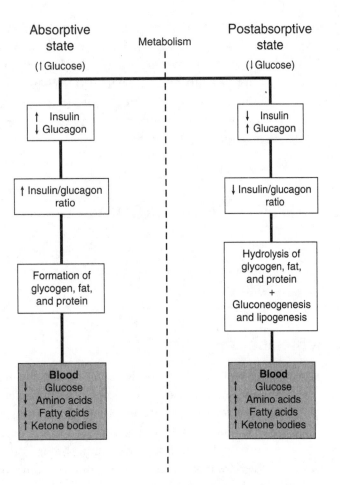

**FIGURE 19.11**   *A comparison of nutrient levels in the blood during the absorptive (feasting) state and the postabsorptive (fasting) state.*

During the absorptive state, the level of insulin is high, and the body transports glucose molecules into cells and utilizes them as a major energy source. It stores the excess carbohydrate as adipose fat and glycogen synthesized in its cells. The liver converts excess carbohydrate into fat or glycogen and releases most fat into the blood for transport to cells of the adipose tissue. Also, the body uses amino acids for protein synthesis and stores much of the protein as fat, while converting some to carbohydrates. A major portion of fat is stored as adipose fat.

During the postabsorptive state, the level of glucagon is high, and the body maintains blood glucose levels in homeostasis. Glucose sources are supplemented by metabolic pathways involving the breakdown of fat and glycogen in the liver. Fat catabolism and amino acid catabolism also supply energy-containing compounds. During periods of prolonged fasting, the body uses muscle protein not essential to cellular functions for energy purposes.

During the postabsorptive state, almost all the tissues and organs depend primarily on fat for energy. This spares glucose for use by the nervous system, which normally requires glucose as its principal energy source. The liver also utilizes fatty acids, thus sparing its amino acids for use in glucose synthesis. As a result of these adjustments, an average person can fast for many days, so long as water is provided to the body.

# OTHER ASPECTS OF METABOLISM

The broad concept of metabolism includes the body's use of minerals and vitamins and the mechanism by which its temperature is regulated.

## Mineral Metabolism

In addition to organic compounds, the body also requires certain inorganic elements also known as **minerals**. Minerals make up about 5 percent of the body's weight, and many perform essential functions such as regulating various body processes and assisting enzyme activity. Minerals also maintain the osmotic pressures of body fluids; they are often found in combination with organic compounds, such as the iron of hemoglobin molecules.

The mineral **calcium** is important for bone and tooth formation. It is the most abundant positively charged ion in the body and is required for blood clotting and normal muscle and nerve activity. **Sodium** is the most abundant positively charged ion of the extracellular fluids. It is important in water balance in the body, and it aids in maintaining the responsiveness of muscle, nerve, and heart tissues. **Potassium** assists the transmission of nerve impulses and the contraction of muscle cells. **Phosphorus** is used for bone and tooth formation and as a component of ATP and nucleic acids.

Among the other minerals, **magnesium** is used for nerve and muscle cell function and in bone formation. It is also a component of many enzymes. **Iron** is a component of myoglobin, hemoglobin, and the cytochromes used in electron transport. **Iodine** is used by the thyroid gland for the production of thyroxine and other hormones involved in metabolic control. **Sulfur** is used to formulate certain amino acids and is found in several vitamins. **Copper** serves in the manufacture of hemoglobin and the pigment melanin. **Zinc** is a constituent of several enzymes and is essential for normal growth. **Manganese** participates in the formation of urea in the urea cycle and serves as an enzyme activator. **Cobalt** is a component of vitamin $B_{12}$ and functions in the maturation of erythrocytes.

# Vitamin Metabolism

In addition to carbohydrates, fats, proteins, and minerals, the body also requires minute amounts of other organic nutrients known as vitamins. **Vitamins** usually act as coenzymes by regulating numerous physiological processes. Normally they cannot be synthesized by the body and must be obtained from external sources such as food.

Vitamins are generally divided into two major groups: water-soluble vitamins and fat-soluble vitamins. **Water-soluble vitamins** are absorbed with water from the gastrointestinal tract, while **fat-soluble vitamins** are absorbed with dietary fats. The body stores minimal amounts of water-soluble vitamins but stores larger amounts of fat-soluble vitamins, especially vitamins A and D, in the liver.

Among the water-soluble vitamins is **vitamin $B_1$**, also known as **thiamine** (Table 19.3). This vitamin is an important coenzyme in carbohydrate metabolism and is used in the synthesis of acetylcholine. A deficiency leads to **beriberi**, a condition characterized by digestive disturbances, weakness, muscle atrophy, and some paralysis.

**Vitamin $B_2$** is a coenzyme used in carbohydrate and protein metabolism. The vitamin is also known as **riboflavin**. It is utilized to synthesize the coenzyme FAD; a deficiency results in skin inflammation. Another water-soluble vitamin, **niacin**, is used to formulate the coenzyme NAD. The vitamin is also known as **nicotinamide**. A deficiency leads to **pellagra**, characterized by muscular weakness, diarrhea, and mental disturbances. **Vitamin $B_6$**, known as **pyridoxine**, is utilized as a coenzyme in amino acid and fat metabolism. A deficiency may result in anemia, nerve problems, dermatitis, and gastrointestinal disturbances. **Vitamin $B_{12}$**, also known as **cyanocobalamin**, is needed for erythrocyte formation and for the entry of certain amino acids into the growth cycle. A deficiency of vitamin $B_{12}$ leads to **pernicious anemia**.

**Pantothenic acid** is a vitamin that serves as a portion of the coenzyme A molecule. As such it is involved in the entry of pyruvic acid into the Krebs cycle. Deficiencies result in fatigue, spasms, and neuromuscular degeneration. **Folic acid** participates in the enzyme system for nucleic acid and hemoglobin synthesis. It is also involved in erythrocyte and leukocyte formation, and a deficiency can lead to anemia. **Biotin** is used as a coenzyme for the addition of organic groups during the synthesis of fatty acids and in nucleic acid metabolism. **Vitamin C**, known as **ascorbic acid**, promotes protein metabolism and the deposit of collagen during connective tissue formation. A deficiency leads to **scurvy** and is related to poor connective tissue growth, poor wound healing, and bone fracture.

Fat-soluble vitamins include **vitamin A**, also known as retinol. This vitamin is used in the formation of the visual pigment rhodopsin, and a deficiency can lead to a condition called nightblindness. It is also used in the growth of bones and teeth and in epithelial cell maintenance. **Vitamin D** (calciferol) is another fat soluble vitamin;

it is used in the absorption of calcium and phosphorous from the gastrointestinal tract. In children, poor bones result from a deficiency of vitamin D, a condition called **rickets**. **Vitamin E**, or tocopherol, is used to form erythrocytes. It is understood that a deficieicy can result in anemia accompanied by erythrocyte lysis. **Vitamin K** (medadione) is used by the body as a coenzyme and is required for the synthesis of prothrombin in the liver (Table 19.4). A deficiency of this vitamin leads to clotting problems and excessive bleeding. Bacteria normally found in the latter portion of the gastrointestinal tract are known to synthesize this vitamin.

**TABLE 19.3**  *Water-Soluble Vitamins and Their Roles*

| Vitamin | Metabolic Role | Deficiency Symptoms |
| --- | --- | --- |
| Thiamine ($B_1$) | Coenzyme in carbohydrate metabolism | Beriberi, loss of appetite, fatigue |
| Riboflavin ($B_2$) | Part of FAD, coenzyme in respiration and protein metabolism | Inflammation and breakdown of skin |
| Pyridoxine ($B_6$) | Coenzyme in amino acid and fat metabolism | Anemia, nerve problems |
| Cyanocobalamin ($B_{12}$) | Coenzyme in formation of erythrocytes and nucleic acids | Pernicious anemia |
| Niacin | Part of NAD, coenzyme in energy metabolism | Pellagra, fatigue |
| Ascorbic acid (C) | Assists synthesis of collagen in connective tissues | Scurvy, anemia, slow wound healing |
| Pantothenic acid | Part of coenzyme A, in carbohydrate and fat metabolism | Similar to other B vitamins |
| Biotin | Coenzyme in addition of carboxyl groups | Rare; tiny amounts required |
| Folic acid | Coenzyme in formation of nucleotides and hemoglobin | Some types of anemia |

**TABLE 19.4**  *Fat-Soluble Vitamins and Their Roles*

| Vitamin | Physiological Role | Deficiency Syndrome |
| --- | --- | --- |
| A (Retinol) | Contributes to visual pigments in eye | Nightblindness, drying of mucous membranes in body |
| D (Calciferol) | Absorption of calcium and phosphorus; construction of teeth and bones | Rickets, especially in children |
| E (Tocopherol) | Protects blood cells from destruction during formation | Lysis of red blood cells, anemia |
| K (Medadione) | Used in synthesis of prothrombin required for blood clotting | Excessive bleeding, especially in newborns; poor blood clotting |

## Metabolic Rates

A measurement of the energy expended by the body over a set period of time represents the **metabolic rate**. The metabolic rate is generally measured twice: when the body is at rest and when it is fasting. During this time, no energy is being stored, and the only work performed is internal work. The energy expenditure by the body is equal to the heat being produced by the body.

Body heat production may be measured directly or indirectly. For a direct measurement, a device called a **calorimeter** is used. This device consists of an insulated chamber or room in which the subject is placed. The heat produced by the subject warms air in the chamber, and the rate of heat gain is measured. An indirect way of measuring body heat production consists of measuring the rate of oxygen consumed by the body.

For comparative purposes, metabolic rates are often measured in a postabsorptive state under standard conditions designed to eliminate most variables. Under these conditions the **basal metabolic rate (BMR)** can be determined as the energy expenditure per unit time under basal conditions. The BMR represents the minimal energy required for the work of respiration, circulation, digestion, and other body activities when the individual is awake. The BMR is influenced by hormones (e.g., thyroid hormones that increase cell metabolism), body size and surface area (larger body size, higher BMR), age (highest BMR during childhood), gender (men have a slightly higher BMR than women), and body temperature (higher BMR for a higher fever).

When a typical meal is ingested, the metabolism increases 10–20 percent. This increase is called the **specific dynamic action (SDA)**. The SDA of protein is greater than for fats or carbohydrates, so a protein-rich meal increases the rate slightly. The increase in metabolic rate is possibly due to the processing of nutrients by the liver. When the caloric value of ingested food equals the energy expended during body activity, the body weight remains the same.

## Temperature Regulation

The human body produces its own heat and maintains a constant body temperature. The normal oral temperature taken in the morning under controlled conditions is approximately 36.7 °C or 98.6 °F. It can vary with a person's activity, the time of day, and the location where the temperature is taken.

The body temperature is a function of heat production during metabolism and of heat loss. Several mechanisms contribute to the loss of heat from the body to the surrounding environment. Once such process is **radiation**, a process in which heat is lost to the environment in the form of infrared rays. A second process is **evaporation** during sweating and through exhalation. A third process, **conduction**, is the process in which energy is transferred from atom to atom during direct contact between two objects. The transfer occurs between the body surface and environmental objects such as the air or water. A fourth process, **convection**, occurs when air and water molecules in contact with the body receive heat from the body by conduction. These molecules then move away and are replaced by other molecules, which in turn receive heat from the body surface. The process constantly brings different air or water molecules in contact with the body surface. Air or water currents such as wind enhance the process.

Temperature regulation in the body depends largely on activity in the thermoregulatory center of the **hypothalamus**. Neurons in the hypothalamus function as a thermostat. When the body temperature falls below an established setpoint, the hypothalamic center initiates impulses to conserve body heat, and when the setpoint rises above this value, the center sends out impulses to increase the loss of heat. The thalamus may also function in temperature regulation.

Inputs to the hypothalamic center are provided from receptors for temperature in the skin and certain mucous membranes. These receptors are called **peripheral thermal receptors**. Other receptors in the hypothalamus called **central thermal receptors** also detect changes in the blood temperature. Central thermal receptors are also located in the spinal cord, abdominal organs, and other internal structures.

**Fever** is an elevation of body temperature above normal resulting from a physiological stress such as allergic reaction or inflammation. Substances called **pyrogens** act on the hypothalamus and set the hypothalamic thermostat to a higher level. Shivering, vasoconstriction, and chills reflect the body's attempts to reach the higher temperature by conserving heat. Hormones called prostaglandins are believed to function in the activity of the hypothalamus. When the fever lowers, sweating and vasodilation reflect a loss of body heat, and the setting of the body temperature to a lower level.

# REVIEW QUESTIONS

**PART A—Completion: Add the word or words that correctly complete each of the following statements.**

1. When a cell needs energy, it utilizes as an immediate energy source a substance called _____ .

2. NAD and FAD are organic substances that work with enzymes and are known as _____ .

3. The cytochromes and coenzymes work together to transfer electrons in a system known as the _____ .

4. ATP production occurs during a process in which protons move across membranes in a cellular structure called the _____ .

5. The principal carbohydrate available to the body for energy is _____ .

6. During the process of cellular respiration, one of the products given off as a waste product is the gas _____ .

7. Also during cellular respiration, the gas that is used as an acceptor of electrons is _____ .

8. During the process of glycolysis, glucose is broken down to form two molecules of _____ .

9. In order to energize the reactions of glycolysis, energy must be supplied from the molecule _____ .

10. Because the process of glycolysis does not involve oxygen, the process is considered to be _____ .

11. The net gain of ATP molecules resulting from glycolysis is _____ .

12. In muscle cells, the pyruvic acid resulting from glycolysis is converted to _____ .

13. The enzymes required for the Krebs cycle reactions and electron transport are located in the mitochondrion along folds of membranes known as _____ .

14. Before entering the Krebs cycle, pyruvic acid is converted into a substance called _____ .

15. An important 6-carbon acid formed during the early stages of the Krebs cycle is _____ .

16. During the reactions of the Krebs cycle, a number of reactions result in the conversion of NAD to _____ .

17. Each of the carbon atoms entering the Krebs cycle is converted to a molecule of _____ .

18. The last chemical compound formed in the Krebs cycle and a starting compound for a new turn of the Krebs cycle is _____ .

19. After an oxygen atom takes up electrons during the electron transport system, it acquires two protons and forms a molecule of _____ .

20. The energy liberated from electron transport is used to pump proteins through the mitochondrial membranes in the process of _____ .

21. The total number of molecules produced through all the reactions of cellular respiration is _____ .

22. When excess glucose is available in the body, it may be stored in the liver as _____ .

23. When the level of glucose is low in the blood, the body breaks down glycogen and releases the glucose in a process called _____ .

24. A small amount of cholesterol is carried in the bloodstream in lipoproteins known as _____ .

25. A high incidence of coronary heart disease is associated with lipoproteins that have almost 50 percent cholesterol and are known as _____ .

26. In the breakdown of fats for energy metabolism, fatty acids are converted into two carbon units of _____ .

27. Also in fat metabolism, the glycerol part of the fat can be utilized for energy after it has been converted to _____ .

28. The processes of fat catabolism result in condensation of acetyl-CoA molecules to yield acetoacetic acid, which is then converted to molecules is called _____ .

29. Among the unsaturated fatty acids that cannot be synthesized by the body are linolenic acid, linoleic acid, and _____ .

30. When the diet contains a large amount of carbohydrate, the glucose is converted to fats in the process of _____ .

31. In the process of deamination, amino acids are converted into compounds that can be used to supply _____ .

32. An important product of the metabolism of amino acids is a waste product expelled by the kidneys and known as _____ .

33. The essential amino acids are those that must be obtained from the _____ .

34. The protein that contains the essential amino acids is known as _____ .

35. The body utilizes carbohydrates as a major energy source and uses amino acids for protein synthesis and stores fat as adipose tissue during the state known as the _____ .

36. The body tissues depend primarily on fat for energy and glycogen is used for energy during the state known as the _____ .

37. The mineral required for blood clotting and normal muscle and nerve activity as well as for bone and tooth formation is _____ .

38. The mineral used as a component of hemoglobin molecules and in the cytochromes of electron transport is _____ .

39. In the extracellular fluid, the most abundant positively charged ion and the mineral used in maintaining water balance in the body and in nerve impulse conduction is _____ .

40. The disease beriberi results from a deficiency of a vitamin known as vitamin $B_1$ or _____ .

41. The vitamin used in the synthesis of FAD is vitamin $B_2$, also known as _____ .

42. A deficiency of vitamin $B_{12}$ leads to a blood disorder known as _____ .

43. The vitamin that promotes protein metabolism and the deposit of collagen during the formation of connective tissue is ascorbic acid, also called _____ .

44. The fat soluble vitamin D promotes the absorption of phosphorus and calcium from the _____ .

45. The energy expenditure of the body per unit of time under basal conditions is the _____ .

46. After a meal has been ingested, the metabolism increases in a phenomenon called _____ .

47. Heat is lost from the body during sweating and through exhalation in the process of _____ .

48. The regulation of temperature in the body is related to the activity in the thermoregulatory center of a brain structure called the _____ .

49. Air and water molecules receive heat from the body through conduction and move away to be replaced by other molecules in the process of _____ .

50. Those substances that increase the body's thermostat and bring about fever are called _____ .

**PART B—Multiple Choice:** Circle the letter of the item that correctly completes each of the following statements.

1. When the body utilizes ATP as an energy source, the molecule releases its energy and breaks down into
   (A) adenine and phosphorus
   (B) adenosine diphosphate and a phosphate ion
   (C) phosphorus and adenosine monophosphate
   (D) phosphorus and adenine molecules

2. The function of coenzymes during cellular metabolism is to
   (A) accept electrons and pass them to other coenzymes
   (B) serve as energy sources
   (C) participate in reactions of chemiosmosis
   (D) replace cytochromes

3. All the following are associated with the process of glycolysis except
   (A) glucose is broken down to pyruvic acid
   (B) ATP must be supplied to the process
   (C) there is a net gain of two ATP molecules
   (D) citric acid is an important component of the process

4. Glycolysis is considered an anaerobic process because
   (A) oxygen is not involved in the process
   (B) no energy is released during the process
   (C) pyruvic acid is not produced during the process
   (D) enzymes are not used during the process

5. Before entering the Krebs cycle, the compound pyruvic acid is converted into
   (A) citric acid
   (B) oxaloacetic acid
   (C) acetyl-CoA
   (D) NAD

6. The chemical reactions of the Krebs cycle occur within the cell
   (A) in the lysosome
   (B) in the Golgi bodies
   (C) along the endoplasmic reticulum
   (D) in the mitochondrion

7. All the reactions of the Krebs cycle are
   (A) catalyzed by enzymes
   (B) accompanied by the release of energy
   (C) dependent on an input of ATP
   (D) accomplished within the lysosome

8. During the Krebs cycle the carbon atoms originally in glucose molecules are released as
   (A) glycogen molecules
   (B) carbon dioxide molecules
   (C) FAD molecules
   (D) electrons

9. In the reactions of the Krebs cycle, electrons are accepted for transfer by molecules of
   (A) protons and neutrons
   (B) NAD and FAD
   (C) ATP and ADP
   (D) MNF and MNG

10. Oxygen atoms serve in the process of cellular respiration as
    (A) producers of carbon dioxide
    (B) coenzymes and cofactors
    (C) sources of NAD molecules
    (D) final electron acceptors

11. The reactions of the Krebs cycle yield enough energy to synthesize
    (A) 36 molecules of ATP
    (B) 10 molecules of water
    (C) 29 molecules of NAD
    (D) 15 molecules of glucose

12. In the process of gluconeogenesis
    (A) glucose molecules are formed from amino acids
    (B) glycogen molecules are formed from glucose
    (C) glycogen molecules are broken down to release glucose
    (D) glycogen molecules are broken down and fats are synthesized

13. The hormone insulin is essential for the proper metabolism of
    (A) amino acids in the cell
    (B) fat molecules in the liver
    (C) glucose molecules in tissue cells
    (D) sodium and potassium ions in nerve cells

14. A high concentration of high density lipoproteins (HDLs) is associated with a
    (A) high incidence of heart disease
    (B) high rate of nerve impulse transfer
    (C) low incidence of heart disease
    (C) low rate of nerve impulse transfer

15. Fatty acids enter the Krebs cycle as molecules of
    (A) glutamic acid
    (B) acetyl-CoA
    (C) acetoacetic acid
    (D) urea

16. A high level of ketone bodies in the bloodstream reflects a high rate of
    (A) glycogenolysis
    (B) fat breakdown
    (C) amino acid utilization
    (D) mineral absorption

17. The hormones epinephrine, human growth hormone, glycogen, and insulin all have an effect on
    (A) fat metabolism
    (B) the transport of vitamin A
    (C) the production of water in the Krebs cycle
    (D) the absorption of sodium in the kidney

18. To be utilized for energy metabolism, an amino acid must be changed by
    (A) adding an additional acid group
    (B) adding an additional calcium atom
    (C) removing an amino group
    (D) removing its coenzyme portion

19. Nonessential amino acids are those that
    (A) are absorbed from the intestine
    (B) are synthesized from glucose molecules
    (C) are conversion products of fat metabolism
    (D) can be synthesized in the body

20. All the following take place during the absorptive state except
    (A) the body uses amino acids for protein synthesis
    (B) most of the fat is stored as adipose tissue
    (C) carbohydrates are used as the main energy source
    (D) glycogen is used for the body's energy needs

21. The minerals calcium and phosphorus are both used
    (A) to assist nerve impulse transmission
    (B) for tooth and bone formation
    (C) to maintain the water balance of the body
    (D) as a component of hemoglobin

22. It is important that iodine be contained in the diet for use by the
    (A) pancreas
    (B) blood cells that contain hemoglobin
    (C) cells that synthesize certain amino acids
    (D) thyroid gland

23. All the following apply to vitamins except
    (A) many vitamins act as coenzymes
    (B) the fat-soluble vitamins include vitamins A and D
    (C) a deficiency of niacin leads to beriberi
    (D) folic acid participates in the synthesis of nucleic acids

24. The minimal amount of energy required for the work of respiration, circulation, digestion, and other bodily activities when the individual is awake is represented by the
    (A) basal metabolic rate
    (B) postabsorptive state
    (C) specific dynamic action
    (D) lower limit metabolism

25. The processes of radiation, evaporation, and convection help to control
    (A) the Krebs cycle
    (B) the mineral requirements of the body
    (C) the body temperature
    (D) the metabolism of vitamins in the body

**PART C—True/False:** For each of the following statements, mark the letter "T" next to the statement if it is true. If the statement is false, change the underlined word to make the statement true.

1. In the cellular reactions of metabolism, every oxidation reaction is accompanied by a reduction reaction.

2. A single molecule of adenosine triphosphate can be broken down with the release of 38 calories of energy.

3. The function of cytochromes in cellular metabolism is to accept and release protons during energy transfer among molecules.

4. The gaseous waste product of cellular respiration is oxygen.

5. During the process of glycolysis, a glucose molecule is metabolized through a series of enzyme reactions with the result of two molecules of acetic acid.

6. The reactions of glycolysis occur in the mitochondria of human tissue cells.

7. Another name for the Krebs cycle is the citric acid cycle, because citric acid is formed during the process.

8. The six molecules of carbon dioxide that form during the processes of cellular respiration contain the six carbon atoms originally in the molecule of glucose that began the process.

9. The energy used to synthesize ATP molecules is that energy released during the flow of protons in the process of glycolysis.

10. During the process of glycogenolysis, glucose molecules are bound to one another for storage in the liver.

11. Microscopic droplets of fat that enter the lacteal vessels and the general circulation are called chylomicrons.

12. A high level of high density lipoproteins indicates that much cholesterol is being transported in the blood, and this level is associated with a high incidence of coronary heart disease.

13. During the breakdown of fats and energy metabolism, the glycerol portion of the fat molecule is converted to acetyl-CoA for metabolism in the Krebs cycle.

14. When fat catabolism is occurring at a high rate the body forms a high number of aldehyde bodies, which accumulate in the bloodstream.

15. Among the unsaturated fatty acids that cannot be synthesized by the body are linolenic, linoleic, and arachidonic.

16. During the process of deamination, the amino groups of amino acids are removed and used to form molecules of glucose, which are then metabolized to form urea.

17. <u>Animal</u> protein is generally regarded as incomplete protein because it lacks several essential amino acids.

18. During the absorptive state, the body uses amino acids for <u>energy metabolism</u>, and stores much of it as fat.

19. During the postabsorptive state, almost all the tissues and organs depend primarily on <u>protein</u> for energy.

20. The mineral <u>iron</u> is required by the body for the formation of hemoglobin and for the synthesis of cytochromes used in electron transport.

21. The vitamin riboflavin is essential in the diet because it is utilized to form <u>NAD</u>.

22. The synthesis of prothrombin in the liver requires the presence of <u>vitamin E</u>.

23. A device called a <u>sphygmomanometer</u> is used to determine the basal metabolic rate of an individual.

24. A large amount of heat is produced in the body during the breakdown of food and during the activity of <u>epithelial cells</u>.

25. The major center for temperature regulation is an area of the brain known as the <u>hypothalamus</u>.

## PART A—Completion

1. adenosine triphosphate
2. coenzymes
3. electron transport system
4. mitochondrion
5. glucose
6. carbon dioxide
7. oxygen
8. pyruvic acid
9. ATP
10. anaerobic
11. two
12. lactic acid
13. cristae
14. acetyl-CoA
15. citric acid
16. NADH
17. carbon dioxide
18. oxaloacetic acid
19. water
20. chemiosmosis
21. 38
22. glycogen
23. glycogenolysis
24. very low density lipoproteins
25. low density lipoproteins
26. acetyl-CoA
27. DHAP
28. ketone bodies
29. arachidonic acid
30. lipogenesis
31. energy
32. urea
33. diet
34. complete protein
35. absorptive state
36. postabsorptive state
37. calcium
38. iron
39. sodium
40. thiamin
41. riboflavin
42. pernicious anemia
43. vitamin C
44. gastrointestinal tract
45. basal metabolic rate
46. specific dynamic action
47. evaporation
48. hypothalamus
49. convection
50. pyrogens

## PART B—Multiple Choice

| | | | | |
|---|---|---|---|---|
| 1. B | 6. D | 11. A | 16. B | 21. B |
| 2. A | 7. A | 12. A | 17. A | 22. D |
| 3. D | 8. B | 13. C | 18. C | 23. C |
| 4. A | 9. B | 14. A | 19. D | 24. A |
| 5. C | 10. D | 15. B | 20. D | 25. C |

## PART C—True/False

1. true
2. 7.3 kilocalories
3. electrons
4. carbon dioxide
5. pyruvic acid
6. cytoplasm
7. true
8. true
9. chemiosmosis
10. glycogenesis
11. true
12. low
13. fatty acid
14. ketone
15. true
16. ammonia
17. plant
18. protein synthesis
19. fat
20. true
21. FAD
22. vitamin $B_{12}$
23. calorimeter
24. muscle
25. true

# CHAPTER 20

# THE URINARY SYSTEM

The primary function of the urinary system is to regulate the composition and concentration of the extracellular fluids surrounding the body cells. These extracellular fluids, known as **interstitial fluids**, are the plasma and the tissue fluids. The urinary system accomplishes its function by forming urine in the kidneys and associated ducts and organs.

In the process of urine formation, the kidneys perform numerous functions: they regulate the volume of blood plasma and thereby contribute to the blood pressure; they control the concentration of waste products in the blood; they regulate the concentration of the plasma's electrolytes (including ions such as sodium, potassium, carbonate, and bicarbonate); and they contribute to the acid/base level (the pH) of the plasma.

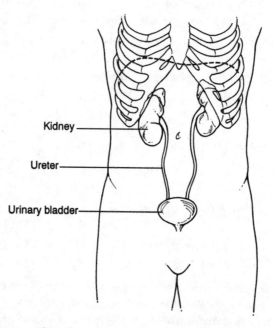

Kidney

Ureter

Urinary bladder

**FIGURE 20.1**   *The urinary system in place in the human body. Note the position of the kidneys relative to the ribs and the location of the urinary bladder. The dotted line represents the position of the diaphragm.*

# THE KIDNEYS

The **kidneys** are attached to the posterior abdominal wall of the body outside the peritoneum (and are said to be retroperitoneal). They lie lateral to the vertebral column and are held in place by adipose and connective tissues (Figure 20.1). In an adult, each kidney weighs about 6 oz. (175 g) and is about the size of a fist.

The external structures of the kidney include a concave notch on the medial surface called the **hilium** and an enveloping capsule of white, fibrous tissue. Each kidney delivers its urine into a cavity known as the **renal pelvis**. From the pelvis, the urine is channeled into a long duct called the **ureter**. The ureter delivers urine to the **urinary bladder**.

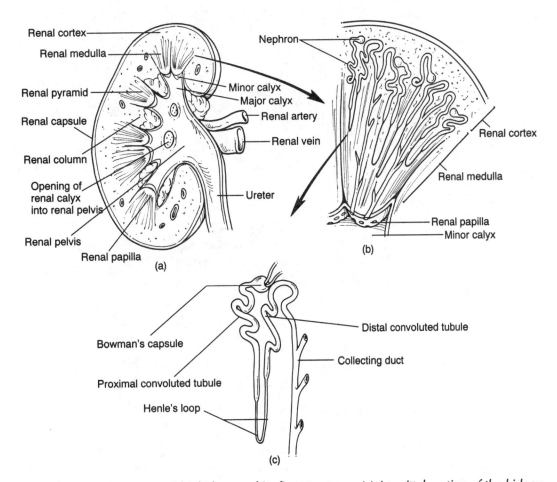

**FIGURE 20.2**    *Structure of the kidney and its fine structures. (a) A sagittal section of the kidney showing the entry and exit of the renal artery and renal vein, respectively. Note the minor calyx leading to the major calyx then to the renal calyx and the renal pelvis and ureter for the elimination of urine. (b) The renal cortex and renal medulla are enlarged and the nephron can be seen. The swollen portion is Bowman's capsule. Tubes lead down into the renal medulla then turn up back toward the cortex. Here they join with ducts leading down to the minor calyx. (c) A nephron expanded to show its details. Note the location of the capsule and how it leads to the proximal tubule, Henle's loop, the distal tubule, and the collecting duct. The peritubular capillary is not shown.*

A sagittal section through the kidney reveals two distinct regions: an outer region called the **cortex**, and a deeper region called the **medulla** (Figure 20.2). The medulla is composed of numerous triangular wedges of tissue called **renal pyramids** and inward extensions of the cortex between the pyramids known as **renal columns**. The apex of each of the renal pyramids extends into a tiny cavity known as the **minor calyx**. Several minor calyces unite to form a **major calyx**. The major calyces then combine to a funnel that delivers urine into the renal pelvis.

## The Nephron

The **nephron** is the functional unit of the kidney and the structure in which urine forms. Each kidney contains over a million nephrons. A nephron consists of small vessels carrying blood and a set of tubules carrying fluid derived from the blood. The fluid results from capillary filtration. It enters the tubules and is then modified by a number of different processes as the blood reabsorbs many materials of value to the body (Figure 20.3). The fluid left over after these modifications is the **urine**.

Arterial blood enters the kidney through the **renal artery**. This artery then divides into smaller arteries passing into the renal

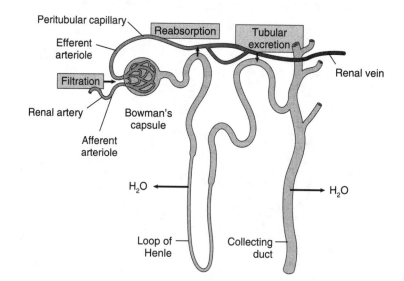

**FIGURE 20.3**  *A simplified view of activity at the nephron. Fluid passes out of the bloodstream into the wall of Bowman's capsule during the process of filtration. The efferent arteriole becomes the peritubular capillary, and materials pass back into the capillary during reabsorption. The fluid passes through Henle's loop, then receives more materials from the peritubular capillary by tubular excretion. The blood then passes out a vein, while the fluid (now urine) moves to the collecting duct.*

medulla. The arteries give rise to still smaller arteries that enter the kidney's cortex. Finally, the small arteries subdivide into numerous microscopic **afferent arterioles.**

### Structures of the Nephron

The microscopic afferent arterioles deliver blood into a tuft of capillaries called a **glomerulus.** Each nephron has one glomerulus. In the glomerulus, blood plasma passes into the hollow walls of a surrounding capsule (Table 20.1). The remaining blood then flows from the glomerulus into an **efferent arteriole.** The efferent arteriole then forms a capillary network called the **peritubular capillaries.** These surround the nephron tubules, which will be discussed in the following paragraph. Later, the peritubular capillaries drain into small veins, which then unite to form larger veins. The larger veins eventually form the **renal vein,** which conducts blood out of the kidney.

The tubular portion of a nephron consists of several structures: the hollow capsule surrounding the glomerulus, the proximal convoluted tubule, the descending limb of the loop of Henle, the loop of Henle, the ascending limb of the loop of Henle, and the distal convoluted tubule.

The capsule surrounding the glomerulus is called the **Bowman's capsule,** also known as the **glomerular capsule.** It is somewhat analogous to a long balloon pushed in at one end by a fist so the balloon surrounds the fist. The fist represents the glomerulus; the balloon represents Bowman's capsule.

**TABLE 20.1** *The Nephron and Its Physiology*

| Nephron Structure | Physiology |
| --- | --- |
| Glomerulus | Filtration of blood plasma; removal of water and solutes with exception of proteins |
| Proximal tubules | Reabsorption of sodium ions, other ions, glucose, and amino acids by active transport; reabsorption of chloride ions by diffusion; reabsorption of water by osmosis |
| Loop of Henle<br>  Descending limb | Reabsorption of sodium ions by diffusion |
|   Ascending limb | Reabsorption of sodium chloride by active transport |
| Distal collecting tubules | Reabsorption of selective ions by active transport; reabsorption of water by osmosis under influence of ADH; secretion of ammonia, certain ions, drugs, hormones, and other substances |

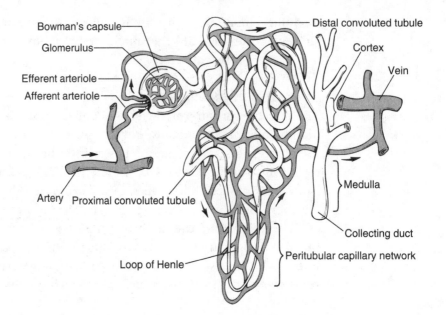

Bowman's capsule
Glomerulus
Efferent arteriole
Afferent arteriole
Artery   Proximal convoluted tubule
Loop of Henle
Distal convoluted tubule
Cortex
Vein
Medulla
Collecting duct
Peritubular capillary network

**FIGURE 20.4**   *A detailed view of activity at the nephron. Note the arrow indicating the flow of blood from the artery to the afferent arteriole, through the glomerulus, and out the efferent arteriole. The blood then flows through the peritubular capillary network before leaving at the vein. Meanwhile, plasma enters the wall of Bowman's capsule, then flows through the proximal convoluted tubule, loop of Henle, and distal convoluted tubule before entering the collecting duct. Modifications of the plasma in the tubules results in urine. Note the intertwined peritubular capillary and nephron tubules that make the transfers possible. Also note that the collecting duct receives urine from other nephrons in the area. Portions of the nephron are in the cortex of the kidney, as indicated, while other portions are in the medulla.*

## Filtration

Blood fluid (plasma) enters the tissue of Bowman's capsule through microscopic slits (Figure 20.4). This transfer of fluids and dissolved substances (solutes) from the glomerular capillaries to the Bowman's capsule is called **filtration**. Filtration occurs because the permeability of glomerular capillaries is greater than other body capillaries and because the blood pressure in the glomerulus is higher than in other capillaries. The higher blood pressure develops because the efferent arteriole has a narrower diameter than the afferent arteriole.

Approximately 7.5 liters of blood plasma pass into the glomeruli of the kidney nephrons each hour. The fluid entering the walls of Bowman's capsule is called the **glomerular filtrate**. For a male, the glomerular filtration rate is about 125 milliliter (mL) per minute (min) and about 110 mL/min for females.

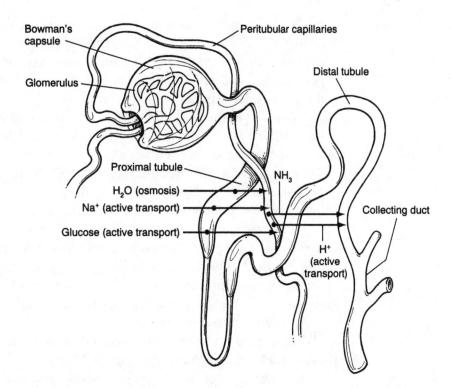

**FIGURE 20.5**   *A summary of material movements among the tubules and peritubular capillaries of the nephron. Various activities occur at the glomerulus, at the proximal tubule, and at the distal tubule.*

## General Reabsorption

The glomerular filtrate leaves the Bowman's capsule and passes into the lumen of the **proximal convoluted tubule**. The wall of the tubule contains millions of microvilli, which increase the surface area. **Reabsorption** takes place here. During reabsorption, varying amounts of water, salts, and other molecules are transported from the cavity through the epithelial tubular cells and into the peritubular capillaries surrounding the tubules.

The transport of molecules is generally the function of specific membrane carriers, and the transport is therefore selective. The reabsorption of glucose and amino acids occurs by **active transport**, a process in which ATP is utilized as an energy source. Specific carrier proteins in the cell membranes transport the substances out of the tubular cells and deposit them in the blood of the peritubular capillaries (Figure 20.5).

## Salt and Water Reabsorption

The reabsorption of salts and water from the proximal tubule occurs by a different mechanism. First, sodium ions are transported by active transport from the fluid in the proximal tubule to the blood in the peritubular capillary. The transport of sodium ions creates a difference in electrical charges across the wall of the tubule, since positively charged sodium ions have accumulated in the peritubular capillary. This electrical gradient stimulates the passive transport of chloride ions toward the higher concentration of sodium ions in the peritubular capillary. Chloride ions follow the sodium ions out of the glomerular filtrate.

As a result of the sodium chloride concentration in the peritubular capillary, an osmotic gradient is established. Water flows in the direction of the higher sodium chloride concentration and dilutes the sodium chloride ions until their concentrations are equal in the proximal tubule and peritubular capillary. Sodium chloride essentially exerts a "pulling power" on the water molecules.

Through these mechanisms, over 65 percent of the salt and water in the glomerular filtrate is reabsorbed through the proximal tubule and returned to blood in the peritubular capillary. Due to the active transport of sodium ions, as much as 5 percent of all calories consumed by the body while at rest may be expended in this process.

As a result of passive and selective reabsorption from the proximal tubules, most water, nutrients, and required salts and ions are taken back into the blood. Among the filtrate components not reabsorbed are much of the nitrogenous waste produced by the body, as well as some water, ions, and salts.

The tubular fluid (glomerular filtrate) then enters the descending limb of the loop of Henle. The descending limb extends down into the medulla where the loop itself is found. Then the fluid enters the ascending limb leading out of the medulla back toward the cortex. In the ascending limb, sodium and chloride ions pass out of the tubule, and they accumulate in the surrounding tissues of the medulla.

The salt accumulation in the medulla tissues creates an osmotic gradient, and as sodium chloride exerts its pulling power, water flows out of the glomerular filtrate in the descending limb into the surrounding tissues (Figure 20.6). The water eventually makes its way back into the blood through nearby capillaries and lymphatic vessels. This mechanism is called the **countercurrent mechanism**. The increasing concentration of sodium chloride ensures that water molecules continue to leave the descending limb from the top to bottom of the limb. They also leave the tubule at the loop of Henle and from the ascending limb.

Some nitrogenous waste products, especially a compound called **urea**, are also believed to leave the lower portion of the nearby collecting duct. The accumulation of urea contributes to the high concentration of organic materials in the medulla. The urea eventually passes back into the loop of Henle and flows back into the collecting duct. The activity taking place in the loop of Henle helps remove water from the tubules and send the water back to the circulation by the capillaries and lymphatic vessels. It also ensures that the fluid remaining in the nephron will become concentrated as urine.

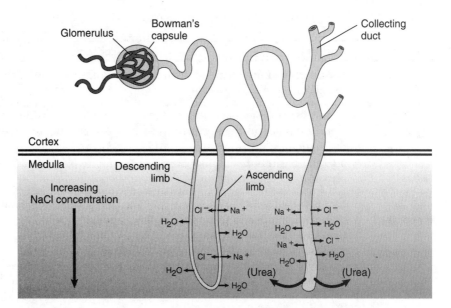

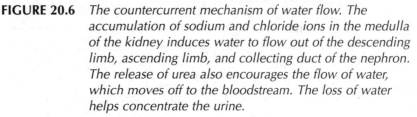

**FIGURE 20.6**  *The countercurrent mechanism of water flow. The accumulation of sodium and chloride ions in the medulla of the kidney induces water to flow out of the descending limb, ascending limb, and collecting duct of the nephron. The release of urea also encourages the flow of water, which moves off to the bloodstream. The loss of water helps concentrate the urine.*

The glomerular filtrate then enters the distal convoluted tubule, where the work begun in the proximal tubule continues. Once again salt and water are reabsorbed into the bloodstream. As before, sodium ions are taken from the fluid by active transport, chloride ions follow, and water follows passively.

### Tubular Excretion

In the distal tubules, another process called **tubular excretion** takes place (Table 20.2). Tubular excretion is an active process in which chemical compounds move from the peritubular capillary (from the blood) into the distal convoluted tubule (into the glomerular filtrate). Active transport accounts for the movement. Some molecules excreted this way include uric acid, creatinine, hydrogen ions, ammonia, and antibiotics such as penicillin.

After processing in the distal convoluted tubule, the glomerular filtrate exists as **urine**. The urine now drips into the collecting duct. The collecting duct descends into the medulla of the kidney on its way to the renal pelvis. Once again, the duct encounters the salty environment of the medulla fluid and a high concentration of urea and other organic compounds (as noted previously). Since the surrounding fluid is rich in salts and organic compounds, water is once again drawn out of the collecting ducts by osmosis. The water is then transported by capillaries and lymphatic vessels back to the general blood circulation. This activity completes the return of water to the blood.

**TABLE 20.2**  *Physiological Activities in the Kidney*

| Activity | Outcome | Location |
|---|---|---|
| Filtration | Forces plasma out of the glomerular vessels into the nephron tubule | Glomerulus and Bowman's capsule |
| Selective reabsorption | Recovers nutrients, salts, and water from the liquid in the proximal and distal tubules; transports materials into the peritubular capillary for return to the bloodstream | Proximal convoluted tubules; loop of Henle; distal convoluted tubule |
| Tubular secretion | Excretes molecules from the peritubular capillary into the nephron tubule; refines ion amounts to maintain blood homeostasis | Distal convoluted tubule; collecting duct |
| Excretion | Eliminates urine from the collecting duct into the renal pelvis; transports urine to the ureters, then to the urinary bladder | Collecting duct, renal pelvis, and accessory organs of excretion |

### Hormone Activity

The rate of reabsorption of water from the distal convoluted tubule and collecting duct is determined by the permeability of the cell membranes of cells of the duct. The permeability of the membranes of the duct, in turn, is controlled by a hormone called **antidiuretic hormone (ADH)**. By a complex chemical mechanism, ADH opens the membrane pores and allows water to pass (Figure 20.7). The hormone is produced by neurons in the hypothalamus and secreted by the posterior lobe of the pituitary gland (Chapter 13).

The secretion of ADH is stimulated when chemical receptors in the hypothalamus respond to an increase or decrease in sodium and other ions in the blood. During dehydration, for example, the ion concentration increases and the receptors signal the hypothalamus to release ADH. The ADH increases cell membrane permeability in the duct cells, and more water is reabsorbed from the glomerular filtrate. Conversely, when an excessive amount of water exists in the body, the ions become more diluted and their concentration drops. The receptors detect this drop and inhibit the secretion of ADH, with the result that less water is reabsorbed in the tubules and more remains to dilute the urine. The process also involves a substance called renin and a system known as the renin-angiotensin system, a topic considered in Chapter 21.

Another hormone functioning in kidney regulation is **aldosterone**, which is secreted by the cortex of the adrenal glands. The hormone acts primarily on the distal convoluted tubules, with three important effects: it stimulates the reabsorption of sodium ions from the distal convoluted tubules; it stimulates the reabsorption of water, since water molecules "follow" the sodium ions in the process of

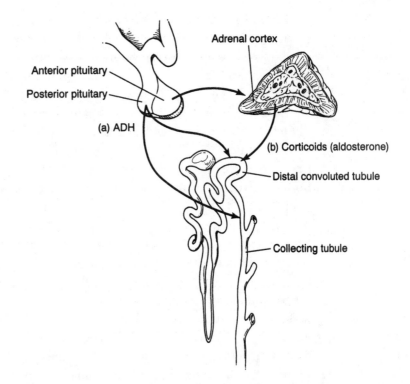

**FIGURE 20.7**    *Two hormones that control water reabsorption in the nephron. (a) Antidiuretic hormone (ADH) from the hypothalamus and posterior pituitary gland. (b) Aldosterone from the adrenal cortex.*

osmosis; and it stimulates the secretion of potassium from blood into the fluid of the distal convoluted tubule. The release of potassium by tubular excretion is the principal method for potassium's elimination by the body. Without aldosterone, all potassium would be reabsorbed. Excessive potassium in the body can lead to heart failure. Another condition called **Addison's disease** may also result from poor secretion of aldosterone. President John F. Kennedy is believed to have suffered from Addison's disease.

## Urine

The product of nephron activity is **urine**. Approximately 95 percent of the urine is water, and 5 percent is solids such as organic wastes, ions, and salts. Among the organic wastes is **urea**. Urea is a product of the liver metabolism. It is produced during the conversion of amino acids to energy-supplying compounds. In the conversion, the amino group is removed and combined with another amino group and a carbon and oxygen atom to form urea in a process known as the ornithine cycle (Chapter 18). Urea is toxic to the cells of the body and must be removed in the urine.

Other organic substances in the urine include ammonia, uric acid (resulting from nucleic acid breakdown), and creatinine (a product of creatine phosphate utilization in the muscle cells). Ions in the urine include cations (positively charged ions), such as sodium, potassium, magnesium, and calcium; and anions (negatively charged ions), such as chloride, sulfate, and phosphate.

Other substances found in the urine include hippuric acid and ketone bodies. **Hippuric acid** is an organic substance derived from benzoic acid, which is often found in fruits and vegetables. **Ketone bodies** (actually ketone molecules) are organic substances resulting from the breakdown of fat molecules. Persons with diabetes mellitus have large numbers of ketone bodies in their urine since the body is using fat molecules as a primary source of energy rather than glucose molecules. Other substances in the urine depend on the diet and may include pigments, fatty acids, mucus, hormones, and various carbohydrates and therapeutic drugs.

The urine usually has a yellow or amber color (Table 20.3). Its color is due to pigments derived from substances present in the diet, or from bilirubin derived from the breakdown of hemoglobin of red blood cells. One of the important urine pigments is **urobilinogen**. This pigment is produced by bacterial action on bilirubin in the intestine. The hepatic portal system delivers urobilinogen to the liver, then the circulation brings the pigment to the kidney. If the urine is red due to red blood cells, this generally indicates bleeding in the urinary system since red blood cells are passing across the walls of the glomerulus.

The urine is usually cloudy, and it has an ammonialike odor on standing. Its pH varies from 4.6 to 8.0, with an average of 6.0, depending on the diet. Approximately 1 to 2 liters of urine are produced per day, a number that can vary considerably.

**TABLE 20.3** *Characteristics of Human Urine*

| Characteristic | Description |
|---|---|
| Clarity | Transparent or clear; becomes cloudy on standing |
| Specific gravity | 1.015 to 1.020; highest in morning |
| Color | Amber or straw-colored; varies according to diet and amount voided |
| Amount in 24 hours | About three pints (1500 ml); varies according to fluid intake, amount of perspiration, and other factors |
| pH | Acidic, may be alkaline if diet contains large amounts of vegetables; high protein diet increases acidity; stale urine has alkaline reaction from decomposition of urea to ammonium compounds; normal pH range is 4.6 to 8.0; average, about 6. |
| Odor | Characteristic urine odor; develops ammonia on standing from formation of ammonium compounds |

## ACCESSORY STRUCTURES

In addition to the kidneys, certain accessory structures are parts of the urinary system. The **ureters**, for example, are tubular organs extending from the kidneys to the urinary bladder (Figure 20.8). They are about 10 to 12 inches long and carry urine to the bladder by waves of peristalsis taking place in their muscles. Urine enters the bladder in jets occurring at a rate of five per minute. Where they meet the kidney, the ureters swell to form the renal pelvis.

Another accessory organ is the **urinary bladder** lying posterior to the pubic symphysis. The bladder is a collapsible sac consisting of a mucous coating and walls of smooth muscle fibers. It is capable of considerable expansion and has three openings: two into the ureters and one into the urethra. The urinary bladder stores up to 600 mL of urine, and, with the urethra, expels urine from the body. The process of urine expulsion is called **micturition**. Involuntary micturition is referred to as **incontinence**.

The tube leading from the floor of the bladder to the exterior is the **urethra**. In females, the urethra lies ventral to the vagina and is about 2.5 cm in length. In males, the urethra passes through the penis and is approximately 15 cm in length when the penis is relaxed. In males the urethra also serves as a passageway for semen. It is surrounded by the prostate gland in males, and enlargement of the gland can impede urinary flow. The opening of the urethra to the external environment is called the **urinary meatus**.

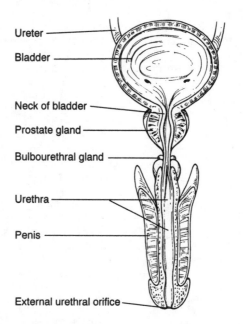

Ureter

Bladder

Neck of bladder

Prostate gland

Bulbourethral gland

Urethra

Penis

External urethral orifice

**FIGURE 20.8**    *Some of the accessory structures functioning in the male urinary system. The prostate and bulbourethral glands function in the reproductive system.*

# OTHER EXCRETORY ORGANS

In addition to the urinary system, there are other excretory organs in the body. One such organ is the **liver**, which metabolizes the products of hemoglobin of red blood cells and excretes the products of hemoglobin breakdown as bile pigments, as noted earlier. A portion of the pigments that color the urine are derived from the liver. Other excretory organs are the **lungs**. These organs excrete carbon dioxide and give off some water.

The intestine and skin are also considered excretory organs (Figure 20.9). Epithelial cells lining the **intestine** excrete certain salts, such as iron and calcium salts, and release water. (The process of defecation is not considered excretion because excreted substances are waste products of metabolism. In defecation undigested food is eliminated from the body.) The **skin** is an excretory organ because it excretes perspiration during sweating. Perspiration contains water as well as salts, and certain amounts of ammonia, urea, and uric acid. Sweating is actually a cooling process because it allows the body to give off excessive amounts of heat.

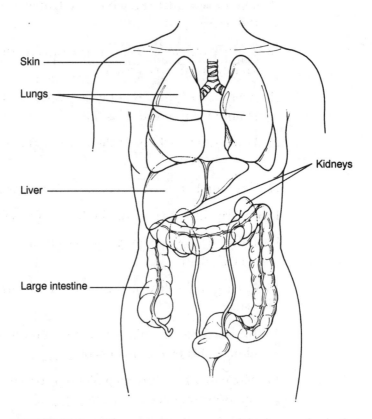

**FIGURE 20.9** *Other excretory organs of the human body that supplement the urinary system.*

# REVIEW QUESTIONS

**PART A—Completion: Add the word or words that correctly complete each of the following statements.**

1. The extracellular fluids surrounding the cells of the body are called _____ .

2. The kidneys regulate the volume of blood plasma in the body and thereby contribute to the regulation of _____ .

3. In the body, the kidneys are found lateral to the _____ .

4. The adult kidney is about the size of a _____ .

5. The concave notch on the medial surface of the kidney is known as the _____ .

6. Each kidney is enclosed in white, fibrous tissue that forms a _____ .

7. The long duct leading away from the kidney and carrying urine is the _____ .

8. The two distinct regions of the kidney are the outer cortex and the inner _____ .

9. The triangular wedges of tissue composing the medulla of the kidney are referred to as _____ .

10. A branch of the renal pelvis located at the apex of each renal pyramid is referred to as the _____ .

11. Urine is formed within the functional unit of the kidney, a structure called the _____ .

12. The number of nephrons in each kidney is over one _____ .

13. Arterial blood entering the kidney flows through the _____ .

14. Blood flows into the glomerulus by means of a microscopic vessel called an _____ .

15. Each glomerulus of the nephron is surrounded by a capsule called _____ .

16. Blood fluid, or plasma, enters Bowman's capsule from the glomerulus by the process of _____ .

17. The major force pushing blood plasma into Bowman's capsule is the pressure exerted by the _____ .

18. In a single hour, the amount of blood plasma passing through the glomeruli is approximately _____ .

19. The tubule of the nephron leading away from Bowman's capsule is the _____ .

20. The transport of molecules from the proximal convoluted tubule into the peritubular capillary occurs by the process of _____ .

21. The surface area for reabsorption is increased in the wall of the proximal tubule by the presence of _____ .

22. Active transport, which accounts for the reabsorption of amino acids and glucose, requires the expenditure of energy in the form of _____ .

23. The passage of chloride ions out of the proximal convoluted tubule is encouraged by the preliminary passing out of _____ .

24. The accumulation of sodium chloride molecules in the peritubular capillary creates an _____ .

25. The sodium chloride molecules accumulating in the peritubular capillary exert a pulling power on molecules of _____ .

26. The passage of water molecules from the proximal convoluted tubule to the peritubular capillary occurs by the process of _____ .

27. The passage of sodium ions into the peritubular capillary occurs by active transport, a process that requires much _____ .

28. The proximal convoluted tubule leads to the descending limb of the _____ .

29. The descending limb extends down into the portion of the kidney known as the _____ .

30. In the ascending limb of the loop of Henle, active transport brings about the passage of _____ .

31. Water flows out of the descending limb of the loop of Henle into the surrounding tissues as a result of the accumulation of _____ .

32. The water exiting the descending loop of Henle does so by the process of _____ .

33. The mechanism of accounting for the passage of water out of the descending loop of Henle is known as the _____ .

34. The accumulation of organic materials in the medulla includes a high concentration of the nitrogenous waste product known as _____ .

35. The water released at the loop of Henle flows back into the bloodstream by means of capillaries and _____ .

36. After leaving the loop of Henle, the fluid flows into the _____ .

37. In the process of tubular excretion, compounds move from the blood fluid into the _____ .

38. Among the molecules entering the nephron fluid by tubular excretion are hydrogen ions, ammonia, uric acid, and _____ .

39. After leaving the distal convoluted tubule, the newly formed urine drips into the _____ .

40. The reabsorption of water in the nephron of the kidney is controlled in part by a hormone known as _____ .

41. The secretion of ADH is controlled by chemical receptors that respond to an increase in _____ .

42. The hormone ADH, which is involved in water reabsorption, is stored in the posterior lobe of the _____ .

43. The adrenal hormone that stimulates the reabsorption of sodium ions from the distal convoluted tubules is _____ .

44. In stimulating the reabsorption of sodium ions, the adrenal hormone also stimulates the reabsorption of _____ .

45. The secretion of potassium in the nephron of the kidney is regulated by the hormone _____ .

46. The waste product urea, which is present in the urine, is a product of amino acid metabolism taking place in the _____ .

47. Chloride, sulfate, and phosphate ions are all found in the urine and all carry a charge that is _____ .

48. Large amounts of ketone bodies in the urine are often a symptom of _____ .

49. The pigments that give urine its color are derived from substances in the diet or from the pigment in red blood cells known as _____ .

50. The process of expelling urine from the body is referred to as _____ .

**PART B—Multiple Choice:** Circle the letter of the item that correctly completes each of the following statements.

1. All the following are functions of the kidney except
   (A) regulating the volume of blood plasma
   (B) regulating the concentration of waste products in the blood
   (C) regulating the digestion of carbohydrates and proteins
   (D) regulating the concentration of ions in the plasma

2. The kidneys are located at the
   (A) posterior abdominal wall
   (B) ventral thoracic wall
   (C) thoracic wall
   (D) ventral abdominal wall

3. The two major regions of the kidney are the
   (A) major and minor calyx
   (B) renal and nephritic pyramids
   (C) medulla and cortex
   (D) jejunum and ileum

4. Urine flows to the urinary bladder from the kidney by means of the
   (A) urethra
   (B) proximal tubule
   (C) peritubular capillary
   (D) ureter

5. The kidney structure in which urine is formed is known as the
   (A) calyx
   (B) nephron
   (C) neuron
   (D) nephridium

6. The fluid that enters the glomerulus is essentially similar to
   (A) serum
   (B) blood
   (C) sea water
   (D) fresh water

7. The fluid that enters Bowman's capsule is essentially similar to
   (A) sea water
   (B) plasma
   (C) lymph
   (D) sterilized water

8. The structures leading away from Bowman's capsule include the
   (A) renal artery and renal vein
   (B) peritubular capillary
   (C) proximal convoluted tubule and loop of Henle
   (D) glomerulus

9. The driving force that pushes fluid from the blood into Bowman's capsule is exerted
   (A) by the heart
   (B) by the muscles lining the abdominal cavity
   (C) by the urinary bladder
   (D) by the urethra

10. In the proximal convoluted tubule, the reabsorption of amino acids and glucose takes place by
    (A) osmosis
    (B) diffusion
    (C) facilitated diffusion
    (D) active transport

11. In the nephron, the accumulation of sodium and chloride ions accounts for the movement of
    (A) ATP molecules
    (B) protein carriers in the cell membranes
    (C) urine
    (D) water molecules

12. In the activities of the nephron, much of the energy requirement is fulfilled by the utilization of
    (A) NAD molecules
    (B) glucose molecules
    (C) ATP molecules
    (D) electrons transported in the membranes

13. The loop of Henle exists between the
    (A) renal artery and renal vein
    (B) peritubular capillary and collecting duct
    (C) proximal convoluted tubule and distal convoluted tubule
    (D) glomerulus and peritubular capillary

14. In the process of tubular excretion, materials move from
    (A) the peritubular capillary to the proximal convoluted tubule
    (B) the glomerulus to the Bowman's capsule
    (C) the Bowman's capsule to the glomerulus
    (D) the peritubular capillary to the distal convoluted tubule

15. Which of the following describes the flow of urine through the kidney?
    (A) Bowman's capsule to renal vein to collecting duct
    (B) distal convoluted tubule to collecting duct to renal pelvis
    (C) collecting duct to glomerulus to peritubular capillary
    (D) renal artery to peritubular capillary to renal vein

16. The rate of reabsorption of water from the collecting duct is determined by the hormone
    (A) oxytocin
    (B) cortisone
    (C) antidiuretic hormone
    (D) lactogenic hormone

17. All the following are functions of the hormone aldosterone except
    (A) it stimulates the reabsorption of water in the nephron
    (B) it stimulates the secretion of potassium from the blood
    (C) it stimulates the reabsorption of sodium ions from the distal convoluted tubules
    (D) it regulates the excretion of calcium from the peritubular capillary

18. The inadequate secretion of aldosterone may result in a condition known as
    (A) Addison's disease
    (B) Grave's disease
    (C) Hashimoto's syndrome
    (D) Job's syndrome

19. All the following apply to the organic substance urea except
    (A) it is harmless to the body cells
    (B) it is a product of amino acid metabolism
    (C) it is a major component of the urine
    (D) it is a waste product of body metabolism

20. The pigments that give urine its yellow or amber color are derived from
    (A) nerve cells
    (B) plant cells
    (C) bile and hemoglobin molecules
    (D) ketone bodies

21. The urinary bladder has openings to the
    (A) nephron and kidney
    (B) urethra and ureters
    (C) urinary meatus and glomerulus
    (D) renal vein and loop of Henle

22. The flow of urine in the ureters is assisted by the actions of
    (A) blood pressure arising from the heart
    (B) emptying of the bladder
    (C) peristalsis in muscles of the ureter
    (D) movement of the diaphragm muscle

23. The urinary bladder lies
    (A) anterior to the rectum
    (B) anterior to the transverse colon
    (C) posterior to the pubic symphysis
    (D) lateral to the kidneys

24. The term micturition refers to
    (A) the process of urine formation
    (B) a disease of the kidney
    (C) the process of urine expulsion
    (D) activities taking place in the renal calyx

25. All the following are considered accessory organs of excretion except
    (A) the pancreas
    (B) the lungs
    (C) the skin
    (D) the intestines

**PART C—True/False:** For each of the following statements, mark the letter "T" next to the statement if it is true. If the statement is false, change the <u>underlined</u> word to make the statement true.

1. In the human body, the kidneys are found lateral to the <u>linea alba</u>.

2. The renal pyramids are found primarily in the <u>cortex</u> of the kidney.

3. Just before leaving the kidney, the urine flows through a funnel-shaped structure known as the <u>major calyx</u>.

4. The renal artery delivers blood to the glomerulus by means of the <u>efferent</u> arteriole.

5. During the process of <u>active transport</u>, plasma from the blood leaves the glomerulus and enters Bowman's capsule.

6. The fluid in the glomerular filtrate normally contains no <u>amino acids</u>.

7. Blood leaving the glomerulus normally has the same number of <u>ions</u> as it had when it entered the glomerulus.

8. Immediately after leaving the Bowman's capsule, the glomerular filtrate enters the <u>loop of Henle</u>.

9. Immediately after leaving the glomerulus, the blood enters the <u>peritubular capillary</u>.

10. The process of active transport accounts for the reabsorption of amino acids and <u>starch</u> molecules taking place in the proximal convoluted tubule.

11. In the proximal convoluted tubule, the release of water is related to the process of <u>phagocytosis</u>.

12. The "pulling power" exerted on water molecules is due to an accumulation of <u>calcium phosphate</u> molecules.

13. The descending and ascending limbs of the loop of Henle are found in the <u>cortex</u> of the kidney.

14. Certain organic compounds such as <u>urea</u> leave the lower portion of the collecting duct and exert a pulling power on water.

15. The process by which water flows from the tubules in response to the high sodium chloride concentration in the surrounding area is an <u>active</u> process.

16. The process by which compounds move from the peritubular capillary into the distal convoluted tubule is known as <u>tubular excretion</u>.

17. The antidiuretic hormone regulates the rate of reabsorption of water from the <u>glomerulus</u>.

18. The secretion of antidiuretic hormone is stimulated when chemical receptors in the <u>thalamus</u> respond to an increase in blood pressure.

19. The hormone aldosterone has a substantial impact on the reabsorption of <u>calcium</u> ions from the distal tubules of the nephron.

20. The condition known as <u>Wellington's</u> disease may result from an inadequate secretion of aldosterone.

21. Urea, a major component of the urine, is produced by the <u>brain</u> during its metabolism.

22. A common acid found in the urine is <u>hippuric</u> acid, an organic acid derived from benzoic acid in fruits and vegetables.

23. The urethra is the longer tubule in <u>females</u>.

24. Pigments that give color to the urine are derived from substances produced primarily in the <u>pancreas</u>.

25. The skin is considered an excretory organ because it excretes water and <u>salts</u> in the perspiration.

**Answers**

## PART A—Completion

1. interstitial fluids
2. blood pressure
3. vertebral column
4. fist
5. hilium
6. capsule
7. ureter
8. medulla
9. renal pyramids
10. minor calyx
11. nephron
12. million
13. renal artery
14. afferent arteriole
15. Bowman's capsule
16. filtration
17. heart
18. 7.5 liters
19. proximal convoluted tubule
20. active transport
21. microvilli
22. ATP
23. sodium ions
24. osmotic gradient
25. water
26. osmosis
27. energy
28. loop of Henle
29. medulla
30. sodium and chloride ions
31. sodium chloride
32. osmosis
33. countercurrent mechanism
34. urea
35. lymph vessels
36. distal convoluted tubule
37. urine
38. creatinine
39. collecting duct
40. antidiuretic hormone
41. sodium and other ions
42. pituitary gland
43. aldosterone
44. water
45. aldosterone
46. liver
47. negative
48. diabetes mellitus
49. hemoglobin
50. micturition

## PART B—Multiple Choice

| | | | | |
|---|---|---|---|---|
| 1. C | 6. B | 11. D | 16. C | 21. B |
| 2. A | 7. B | 12. C | 17. D | 22. C |
| 3. C | 8. C | 13. C | 18. A | 23. C |
| 4. D | 9. A | 14. D | 19. A | 24. C |
| 5. B | 10. D | 15. B | 20. C | 25. A |

## PART C—True/False

1. vertebral column
2. medulla
3. true
4. afferent
5. filtration
6. red blood cells
7. red blood cells
8. proximal convoluted tubule
9. true
10. glucose
11. osmosis
12. sodium chloride
13. medulla
14. true
15. passive
16. true
17. collecting tubule
18. hypothalamus
19. sodium
20. Addison's
21. liver
22. true
23. males
24. liver
25. true

# CHAPTER 21

# FLUID, ELECTROLYTE, AND ACID/BASE BALANCE

The homeostasis, or constancy, of the body environment depends on a balance of fluids, electrolytes, and acids and bases. In the normal body, the levels of fluids and electrolytes remain constant, and the input of water and electrolytes is balanced by selective elimination through the excretory system. If serious depletion of fluids or electrolytes occur, immediate replacement is critical.

**Electrolytes** are the products of substances that break up, or dissociate, into electrically charged components when they dissolve in water. For example, the compound sodium chloride ($NaCl$) dissociates into sodium ions ($Na^+$) and chloride ions ($Cl^-$), which are electrolytes carrying electrical charges.

An **acid** is a chemical compound that liberates hydrogen ions into a solution. A **base**, by contrast, is a chemical compound that takes up hydrogen ions from a solution, leaving the solution with an excess of hydroxyl ($-OH$) ions. Examples of acids are sulfuric acid, hydrochloric acid, and lactic acid. Bases are represented by sodium hydroxide and potassium hydroxide.

## BODY FLUIDS

The fluid content of the body refers to the amount of water in the body. The amount of body water can vary, depending upon the weight, sex, age, and fat content of the individual. For instance, women have a relatively lower water content than men because a woman's body contains a higher percentage of fat, and fat tissue contains very little cellular water. Water makes up about 60 percent of the body weight of an adult male and about 50 percent of a female's body weight.

## Fluid Compartments

The body's total water content may be conveniently subdivided into two fluid compartments known as the intracellular fluid compartment and the extracellular fluid compartment. The **intracellular fluid compartment** refers to the water existing in all body cells.

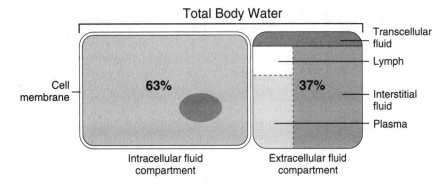

Total Body Water

Cell membrane — 63% — Intracellular fluid compartment

37% — Extracellular fluid compartment

Transcellular fluid — Lymph — Interstitial fluid — Plasma

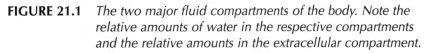

**FIGURE 21.1**  *The two major fluid compartments of the body. Note the relative amounts of water in the respective compartments and the relative amounts in the extracellular compartment.*

About two-thirds of the body's water exists in this compartment. The **extracellular fluid compartment** consists of the body area outside the cells (Figure 21.1). Extracellular fluid provides a constant environment for the cells and transports substances to and from the cells. Approximately one-third of the body water exists in this compartment: about 25 percent is the blood plasma, and about 75 percent is interstitial fluid and lymph. **Interstitial fluid** surrounds the body cells, while **lymph** is found in lymphatic vessels (Chapter 16).

About 1 percent of the extracellular fluid is **transcellular fluid**. This fluid is separated from other body fluids by epithelial cell layers. Transcellular fluid includes cerebrospinal fluid, synovial fluid, sweat, and fluid in the pleural, pericardial, and peritoneal cavities, as well as fluid in the eye chambers and the digestive, respiratory, and urinary tracts.

The extracellular fluid contains large amounts of sodium, chloride, and bicarbonate ions and small amounts of potassium, magnesium, calcium, phosphate, sulfate, and organic ions. Plasma is protein-rich, while interstitial fluid is protein-poor. The intracellular fluid contains small amounts of sodium and large amounts of potassium ions, due mainly to the sodium-potassium pump existing in cells such as neurons. Within body cells, the protein concentration is generally high, since protein is constantly being synthesized and used for cellular functions.

Water enters the body through the digestive tract and as a result of chemical reactions that produce water as end products (Chapter 2). Water leaves the body by several exits: the kidneys excrete urine; the lungs give off water in expired air; the skin gives off sweat; and the intestines eliminate water in feces. In general terms, the amount of water exiting the body equates to the amount entering the body (Figure 21.2).

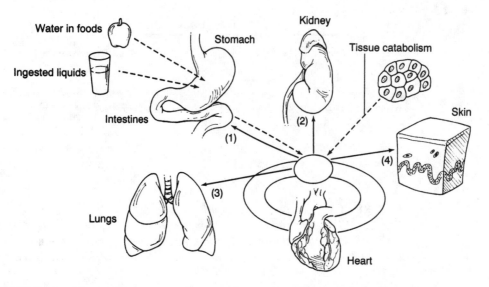

**FIGURE 21.2** *The various routes of entry of water to the blood (dotted lines) and four routes of exit from the blood to the exterior (solid lines).*

## Water Movement

Water moves into and out of cells as a result of **osmosis**. Osmosis is the diffusion of water from a region of high water concentration through a semipermeable membrane (the cell membrane) to a region of low water concentration. Put another way, osmosis is the flow of water from a region of low solute concentration (for example, low sodium ions) to a region of high solute concentration (high sodium ions). The ions are said to exert a "water pull."

Normally there is an equal concentration of solute ions inside and outside cells. The **osmotic pressure** is therefore said to be zero, and water does not flow in either direction under these conditions. However, if ions are lost from the extracellular environment, the equilibrium is disturbed and a higher concentration of ions then exists inside the cell. An osmotic pressure is established, and water molecules move through the membrane into the cell in the direction of the higher solute concentration until equilibrium is established.

A constant movement of water occurs between the plasma and interstitial fluid across the membranes of capillary cells. Osmotic pressures, usually due to proteins, account for much of the water movement. Excess fluid and excess protein existing in the interstitial fluid is removed by the lymphatic system to establish homeostasis.

## Regulation of Fluid Balance

Several devices exist in the body to maintain the fluid balance and ensure that the intake of fluid equals the outflow. One device is **thirst**, the conscious desire for water. The thirst mechanism is controlled by a nerve center in the **thalamus**, where neurons known as **osmoreceptors** are found. When a person fails to take in water, salivary secretion decreases, and a dry mouth and thirst sensations develop. In addition, decreased blood volume and increased amounts of solute (for example, salt) in the plasma signal a need for water. The osmoreceptors shrink, depolarize, and send impulses to the cerebral cortex to stimulate thirst sensations and the desire to drink. Baroreceptors in the cardiovascular system also respond to decreased blood volume (Chapter 15) and transmit their impulses to the osmoreceptors.

Water intake is also related to hormone activity. For example, urine volume is determined by the filtration rate in the glomerulus and the amount of water reabsorption in the nephron tubules. The rate of tubular reabsorption is controlled by **antidiuretic hormone (ADH)** released in response to the same stimuli as the thirst sensation. ADH increases the absorption of water from the tubules and results in a decreased amount of urine, or more concentrated urine. In addition, the kidneys release a hormone called **renin**. Renin activates the production of angiotensin II (to be discussed presently), which stimulates the brain to increase thirst sensations. **Aldosterone**, an adrenal cortex hormone, regulates water reabsorption in response to the renin-angiotensin system. Angiotensin stimulates aldosterone to act on the tubules to increase sodium ion reabsorption. Water molecules then follow the sodium ions.

## Fluid Movement Across the Capillary Membrane

The amount of blood passing through a blood vessel at a given time period is determined by two important factors: blood pressure and resistance, which is the opposition to the flow of blood measured as the force of friction in the blood vessels. The **resistance** to blood flow is related to three factors: blood viscosity measured as the ratio of red blood cells and solutes to blood fluid; the length of the blood vessel (since resistance is directly proportional to the length of the vessel); and blood vessel radius (the smaller the blood vessel's radius, the greater resistance to blood flow).

As blood flows through the capillary, water moves between the plasma and the interstitial fluid. The mechanism controlling the water movement is known as **Starling's law of the capillaries**: Starling's law indicates in which directions the fluids flow between capillary and interstitial fluid (Figure 21.3). Two major controlling factors are the blood hydrostatic pressure and blood colloid osmotic pressure. **Hydrostatic pressure** (or "blood pressure") refers to the pressure of water in the blood. This pressure accounts for the movement of fluid out of the capillaries into the interstitial fluid. The **colloid osmotic pressure**, by contrast, depends upon the presence of plasma proteins such as albumin. Plasma proteins draw water

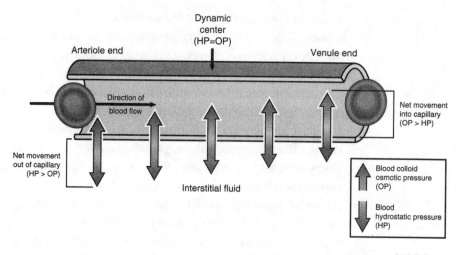
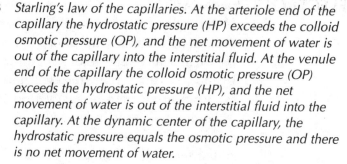

**FIGURE 21.3**  *Starling's law of the capillaries. At the arteriole end of the capillary the hydrostatic pressure (HP) exceeds the colloid osmotic pressure (OP), and the net movement of water is out of the capillary into the interstitial fluid. At the venule end of the capillary the colloid osmotic pressure (OP) exceeds the hydrostatic pressure (HP), and the net movement of water is out of the interstitial fluid into the capillary. At the dynamic center of the capillary, the hydrostatic pressure equals the osmotic pressure and there is no net movement of water.*

through the capillary wall by osmosis. The plasma proteins exist in a colloid state within the plasma and are in higher concentration in the capillary than in the interstitial space outside the capillary.

According to Starling's law, at the arteriole end of the capillary, water leaves the capillary and enters the interstitial space because the hydrostatic pressure (blood pressure) exceeds any colloid osmotic pressure present. The process is known as **filtration** and is similar to what takes place in the glomerulus of the kidney. At the venule end of the capillary, water leaves the interstitial space and enters the capillary because the colloid osmotic pressure exceeds any hydrostatic pressure (blood pressure) present. This process is known as **absorption**. Fluid also leaves the interstitial spaces and enters the lymph capillaries because of osmotic pressures developing within the lymph capillaries.

The movement of fluid between the interstitial space and the intracellular environment is controlled by similar pressures. Hydrostatic pressure tends to be stable, however, because the solutes present within the cell and outside the cell are usually equivalent. Any fluid movement occurring between the cell's interior and exterior environments is likely due to changes in osmotic pressure, not hydrostatic pressure. For example, if the sodium ion concentration is high outside the cells, water will flow through the cell membrane out of the cell in response to the osmotic pressure. The cell will therefore tend to shrink.

**Edema** is the presence of abnormally large amounts of fluid in the interstitial tissue spaces caused by various circumstances. For example, blood hydrostatic pressure in the capillaries may increase due to an obstruction in the veins (due to cardiac failure or blood clots), and the pressure then forces fluid out of the capillaries. Another cause may be a decreased amount of plasma proteins (resulting from malnutrition, liver disease, kidney disease, or increased capillary permeability) and thus a lower osmotic pressure. This lower pressure results in less drawing power to bring water back into the capillary. A final cause may be increased extracellular fluid resulting from fluid retention in the kidney.

## ELECTROLYTE BALANCE

The quantities of various electrolytes taken into the body must equal the quantities lost by the body. In the extracellular fluid, two types of ions can be found: **cations** (such as sodium ions), which are positively charged; and **anions** (such as chloride ions), which are negatively charged. Electrolytes are obtained from foods, drinking water, and metabolic reactions, and they are lost from the body in feces and perspiration, and primarily as a result of kidney function and urine production.

### Sodium Ions

Sodium ions account for about 90 percent of cations in extracellular fluids. When the input of sodium ions exceeds the output, water is also retained in the body, and the volume of plasma and extracellular fluid increases. Edema and weight gain may result. When the output exceeds the input, a decrease occurs in the volume of plasma and extracellular fluid, and the blood pressure lowers.

Regulation of sodium levels in the body reflect excretion processes in the kidney, especially the glomerulus. When the blood pressure drops, the blood flow to the glomerulus is reduced, and the filtration rate diminishes, resulting in less sodium filtered. As the concentration of salt and water increases, the blood pressure elevates and the sodium excretion increases.

Sodium ion's reabsorption from the kidney is regulated by the hormone aldosterone (Chapter 20) operating in the distal and collecting tubules of the kidney as well as in the sweat glands and gastrointestinal tract. The secretion of aldosterone is a part of the renin-angiotensin-aldosterone system. **Renin** is a substance released from the juxtaglomerular apparatus in the nephron when there is a low blood volume or low concentration of sodium in the blood, or when excessive amounts of water have been lost (Figure 21.4). In the plasma, renin reacts with the liver protein **angiotensinogen** and

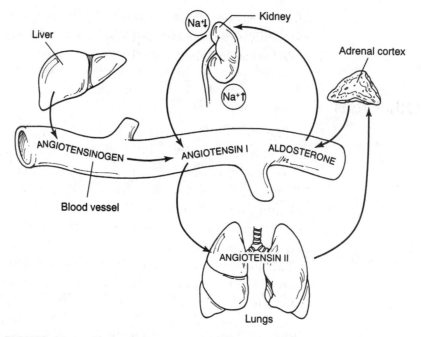

**FIGURE 21.4** *The angiotensin system for sodium regulation. Renin from the kidney reacts with the liver protein angiotensinogen and converts it to angiotensin I. In the lungs angiotensin I converts to angiotensin II, which stimulates the adrenal cortex to release aldosterone. Aldosterone regulates sodium ion reabsorption in the kidney tubules.*

converts it to **angiotensin I**. In the lung, angiotensin I is converted to **angiotensin II**. Angiotensin II stimulates the release of aldosterone from the adrenal cortex. It also stimulates thirst and stimulates the secretion of antidiuretic hormone and ACTH. Moreover, it constricts the blood vessels and increases the blood pressure. All these activities directly or indirectly affect the reabsorption of sodium ions in the kidney.

Aldosterone also regulates the level of potassium ions. An important stimulus for aldosterone secretion is an increase in potassium ion concentration, and the potassium ions appear to stimulate the adrenal cortex directly. Aldosterone then causes the secretion of potassium ions into the urine while it is encouraging the reabsorption of sodium ions.

## Potassium

Potassium is the major intracellular cation. It functions in electrical activity of the muscle and nerve tissue, and like sodium, it maintains the osmotic balance inside the cells. In the extracellular fluid, potassium influences the balance of acid and base. Aldosterone, the adrenal hormone, regulates the amount of potassium in the blood. Deficits of potassium may arise through diarrhea, kidney disease, or edema. Excesses result from inadequate renal excretion. Excessive potassium presence can cause fibrillations of the heart, while deficits can cause arrhythmia of the heart.

## Other Ions

Among the other ions balanced in the body are calcium, magnesium, sulfate, chloride, phosphate, and bicarbonate. **Calcium** ions are regulated by the hormones of the parathyroid gland (Chapter 13) and by the hormone calcitonin secreted by the thyroid gland. Calcium ions have roles in blood clot formation, muscle contraction, hormonal activity, nerve conduction, and as structural components of teeth and bones.

The most common extracellular anion is **chloride**, which is almost always linked to sodium. Chloride ions provide an isotonic environment for the cells and contribute to the stable osmotic pressure inside and outside cells (Figure 21.5). Chloride ions are generally excreted as potassium salts (e.g. potassium chloride), and potassium deficiencies are normally accompanied by chloride deficiencies. The concentration of chloride and other anions is generally due to secondary regulatory mechanisms. For example, chloride reabsorption takes place when chloride ions are electrically attracted to cations during reabsorption.

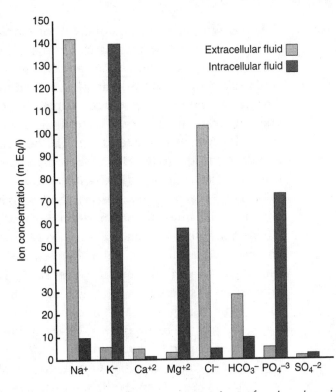

**FIGURE 21.5** *The relative concentrations of various ions in the extracellular and intracellular fluids of the body.*

## ACID/BASE BALANCE

The balance of acids and bases in the body is associated with the regulation of the hydrogen ion concentration in the body fluids. This concentration, expressed as pH, influences the activity of cell enzymes as well as the maintenance of cell structural and cell membrane permeability.

As noted previously, acids (such as hydrochloric and lactic acids) are chemical compounds that release hydrogen ions into a solution, while bases (such as sodium hydroxide and ammonia) are chemical compounds that remove hydrogen ions from a solution. A **strong acid** produces the maximum number of hydrogen ions possible and forms ions more completely than weak acids. Hydrochloric acid is a strong acid, while carbonic acid is a weak acid.

There are many sources of hydrogen ions in the metabolism of the body. For example, during the respiration of glucose, carbon dioxide reacts with water to form carbonic acid, which dissociates to release bicarbonate ions and hydrogen ions. The breakdown of fatty acids and amino acids also results in acid compounds.

The regulation of the concentration of hydrogen ions in the body occurs primarily by acid/base buffer systems, by the activity of the brain's respiratory center, and by the action of the kidneys.

## Acid/Base Buffer Systems

A **buffer** is a solution containing two or more chemical compounds, which prevent substantial changes in the pH when either an acid or base enters the system. Usually a buffer system contain a weak acid and a salt of that acid. An example is the **carbonic acid–sodium bicarbonate system**: that is, a solution containing both carbonic acid and sodium bicarbonate. Should a strong acid such as hydrochloric acid enter the system, it will react with the bicarbonate ions ($HCO_3^-$) of sodium bicarbonate ($NaHCO_3$), producing carbonic acid and sodium chloride. Carbonic acid ($H_2CO_3$) is a weaker acid than hydrochloric acid, and therefore, the increase in hydrogen ions in the solution will be minimal. The carbonic acid then dissociates to water and carbon dioxide, and the water molecules help remove the hydrogen ions of the acid, while the carbon dioxide increases the respiratory rate to eliminate the carbon atoms (Figure 21.6). If a strong base such as sodium hydroxide enters the solution, it will react with the carbonic acid of the buffer system to produce sodium bicarbonate, a weaker base than sodium hydroxide, and there will be a minimal basic shift.

Another important buffer system is the **phosphate buffer system**. This system consists of disodium-hydrogen-phosphate ($Na_2HPO_4$), which is a weak base, and sodium dihydrogen-phosphate ($NaH_2PO_4$), which is a weak acid.

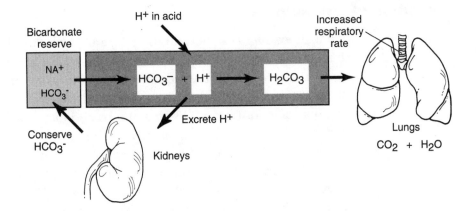

**FIGURE 21.6** *How the bicarbonate ions (HCO$_3$) of sodium bicarbonate (Na$_2$CO$_3$) act as a buffer when hydrogen ions from an acid enter the system. The hydrogen reacts with the bicarbonate ions to produce carbonic acid, which dissociates to carbon dioxide and water. The water is lost from the body by various means, and the CO$_2$ stimulates an increased inspiratory rate in the lungs to remove the excess carbon atoms. Hydrogen ions are also removed by the kidney and used to form a reserve of bicarbonate ions.*

The most powerful buffer system in the body is the **protein buffer system**, which consists of intercellular proteins (such as hemoglobin) and extracellular plasma proteins (such as albumin). Proteins have both amino groups and carboxyl groups in their amino acids. The amino groups function as bases, and the carboxyl groups function as acids. By releasing hydrogen ions from the carboxyl groups or by accepting hydrogen ions into the amino groups, proteins act as acids or bases and comprise an acid/base buffer system.

## Respiratory Regulation

The brain stem contains the respiratory center that helps regulate the hydrogen ion concentration by controlling the rate and depth of breathing (Chapter 17). For example, when cells increase their carbon dioxide production such as during physical exercise, the carbonic acid content of the blood increases. The carbonic acid dissociates and releases hydrogen ions, which cause the acidity of body fluids to increase. The acidity stimulates chemical receptors of the respiratory center, and the center increases the depth and rate of breathing to release more carbon dioxide from the lungs. As carbon dioxide is lost, the hydrogen ion concentration drops in the body fluids because less carbonic acid is present.

## Regulation by the Kidneys

The kidneys regulate the acid/base balance at various levels during the excretion of hydrogen ions in the urine. For instance, carbon dioxide molecules diffuse from the plasma into the epithelial cells of the tubules and form carbonic acid with water molecules. The carbonic acid ionizes into hydrogen ions and bicarbonate ions, and the hydrogen ions are transported into the lumen of the tubules for excretion in the urine. This action reduces the plasma acidity.

In addition, as the hydrogen ions are secreted, sodium ions are taken into the epithelial cells lining the tubules. The sodium ions and the bicarbonate ions are then transported from the epithelial cells into the interstitial fluid and the blood. When the blood is alkaline, the secretion of hydrogen ions decreases and fewer appear in the urine. Now, the bicarbonate ions are poorly reabsorbed, and they remain in the fluid to be excreted into the urine. The loss of bicarbonate ions relieves the alkaline condition in the blood (since bicarbonate ions act as bases).

Regulation can also be accomplished by a phosphate system. Phosphate buffers concentrate in the fluid of the tubules and remove hydrogen ions from the tubular fluid for removal in the urine. In this way, hydrogen ions can be effectively removed without making the urine too acidic, a condition that might damage the urinary tract. Buffering is also accomplished by ammonia molecules and ammonium ions present in the tubule lumen. Ammonia molecules react with hydrogen ions to form ammonium ions, thereby removing the hydrogen ions from the system.

The acid/base status of the body is evaluated in systemic arterial blood. The normal pH of arterial blood is 7.4, while the normal pH of venous blood and interstitial fluid is slightly more acidic. The decrease of arterial pH below 7.35 is a condition known as **acidosis**, while the increase above 7.45 is called **alkalosis**.

# REVIEW QUESTIONS

*PART A—Completion:* **Add the word or words that correctly complete each of the following statements.**

1. Substances that break up or dissociate into separate compounds when they dissolve in water are called _____ .

2. A base is a chemical compound that takes up hydrogen ions from a solution and leaves the solution with an excess of _____ .

3. Two examples of bases are potassium hydroxide and _____ .

4. In a solution, an acid liberates _____ .

5. Water encompasses about 50 percent of the body weight of a normal adult _____ .

6. The water existing in all cells of the body occupies a compartment known as the _____ .

7. The amount of body water that exists in the extracellular fluid compartment is approximately _____ .

8. About 75 percent of the extracellular fluid of the body exists in interstitial fluid and _____ .

9. Sinovial fluid, sweat, and cerebrospinal fluid are different types of _____ .

10. The concentration of protein in the interstitial fluid is usually low, but the concentration of protein is high in the _____ .

11. Water leaves the body by means of activities occurring in the lungs, skin, intestines, and _____ .

12. The principal mechanism by which water moves into and out of cells is _____ .

13. Much of the water movement between the plasma and the interstitial fluid is due to the presence of _____ .

14. The body's thirst mechanism is controlled by a nerve center in the _____ .

15. Tubular reabsorption of water taking place in the kidneys is controlled by the hormone known as _____ .

16. The hormone aldosterone regulates water reabsorption in the kidneys through its activity on the reabsorption of _____ .

17. Such things as blood viscosity and the length of the blood vessel influence the fluid movement factor called _____ .

18. The mechanism for controlling the flow of water between the plasma and interstitial fluid is known as _____ .

19. The pressure of water in the blood is referred to as the hydrostatic pressure, or the _____ .

20. The colloid osmotic pressure present in the capillaries depends upon the presence of plasma _____ .

21. According to Starling's law, at the arteriole end of the capillary water leaves the capillary and enters the interstitial fluid because the hydrostatic pressure exceeds the _____ .

22. The loss of water at the arteriole end of the capillary is similar to that taking place in the glomerulus of the kidney, and the process is called _____ .

23. At the venule end of the capillary, water leaves the interstitial space and enters the capillary because the colloid osmotic pressure exceeds the _____ .

24. The movement of water into the capillary at the venule end is known as _____ .

25. The movement of fluids between the interstitial space and the intracellular environment is controlled by pressure known as _____ .

26. The presence of abnormally large amounts of water in the interstitial tissue causes the condition _____ .

27. Those ions that are negatively charged are known as _____ .

28. Electrolytes are obtained for the body from drinking water, metabolic reactions, and _____ .

29. Approximately 90 percent of the cations in the extracellular fluids consist of _____ .

30. The juxtaglomerular apparatus of the nephron secretes a sodium-regulating substance called _____ .

31. The release of aldosterone to regulate sodium reabsorption in the tubules is stimulated by the substance _____ .

32. In addition to regulating the sodium ion concentration, aldosterone also regulates the body's level of _____ .

33. Excessive amounts of potassium ions in the body can lead to fibrillations of the _____ .

34. The ion that plays roles in blood clot formation, muscle contraction, nerve conduction, and as the structural component of teeth and bones is _____ .

35. The regulation of calcium ions in the body is governed by hormones produced by the parathyroid gland and the _____ .

36. The maximum number of hydrogen ions possible in a solution is provided by a _____ .

37. An example of a weak acid is _____ .

38. The concentration of hydrogen ions in the body fluid is expressed as _____ .

39. A buffer system is used to prevent substantial changes in a solution's _____ .

40. A buffer solution generally contains a weak acid and a _____ .

41. Proteins act as powerful buffers in the body because they contain carboxyl groups and _____ .

42. The respiratory center helps regulate the acid/base balance of the body by controlling the rate of _____ .

43. During periods of intense physical exercise, the cells increase carbon dioxide production, which leads to an increase of the blood content of _____ .

44. One of the important proteins that provides a protein buffer system is _____ .

45. The kidneys help regulate the acid/base balance by excreting hydrogen ions in the _____ .

46. In the kidneys, the alkaline condition of the blood can be regulated by the excretion of _____ .

47. Hydrogen ions can be removed by a reaction between hydrogen and ammonia molecules to produce _____ .

48. To determine the acid/base status of the body, samples are removed from the _____ .

49. Normally, the pH of arterial blood is _____ .

50. Should the pH increase significantly above the set level in the body, the condition is known as _____ .

**PART B—Multiple Choice: Circle the letter of the item that correctly completes each of the following statements.**

1. All the following are important electrolytes in the body except
   (A) potassium ions
   (B) carbon ions
   (C) chloride ions
   (D) sodium ions

2. A base may be defined as a chemical compound that
   (A) removes hydrogen ions from a solution
   (B) adds sodium chloride to a solution
   (C) adds hydrogen ions to a solution
   (D) eliminates sodium ions from a solution

3. The intracellular fluid compartment refers to all the water existing
   (A) in the bones of the body
   (B) in areas outside the body cells
   (C) in areas within the gastrointestinal tract
   (D) in all cells of the body

4. Approximately one-third of the body water exists in the
   (A) kidneys and urinary bladder
   (B) blood
   (C) extracellular fluid compartment
   (D) transcellular fluid compartment

5. The interstitial fluid is generally poor while the plasma is generally rich in
   (A) hydrogen ions
   (B) sodium and chloride ions
   (C) protein
   (D) carbohydrates

6. Water leaves the body by all the following mechanisms except
   (A) through air expired from the lungs
   (B) through metabolic reactions taking place in the cells
   (C) through sweat given off at the skin
   (D) from feces eliminated from the intestine

7. In the process of osmosis
   (A) water moves from a region of high solute concentration to a region of low solute concentration
   (B) water moves from a region of low solute concentration to a region of high solute concentration
   (C) sodium ions move through a semipermeable membrane
   (D) chloride ions follow the movement of sodium ions to a region of low concentration

8. When the concentration of solute ions such as sodium ions is the same on both inside and outside cells, then
   (A) chloride ions leave the cell
   (B) potassium ions rush into the cell
   (C) water flows out of the cell into the extracellular environment
   (D) the osmotic pressure is zero

9. Osmoreceptors detect a decreased blood volume of increased blood concentration of salt and stimulate
   (A) increased kidney activity
   (B) increased salivary secretions
   (C) thirst
   (D) increased secretion of progesterone

10. The hormones aldosterone and ADH both have an important function in
    (A) fluid balance in the body
    (B) the regulation of acid concentration in the body
    (C) stimulation of a conscious desire for water
    (D) the activity of buffer systems

11. Starling's law of the capillaries indicates
    (A) the directions fluid flows between interstitial fluid and capillary
    (B) how much protein will be found in the plasma of the blood
    (C) the effectiveness of the substance renin
    (D) the flow of sodium and chloride ions through the capillary walls

12. At the venule end of the capillary, water enters the capillary
    (A) because the hydrostatic pressure exceeds the colloid osmotic pressure
    (B) because potassium ions are located within the capillary
    (C) because the higher acid concentration is found in the interstitial fluid
    (D) because the colloid osmotic pressure exceeds the hydrostatic pressure

13. Which of the following may be a possible cause of edema in the tissues?
    (A) decreased blood pressure
    (B) a decreased level of proteins in the plasma
    (C) the ingestion of a large amount of carbohydrates in the diet
    (D) reduced temperature at the skin surface

14. Ninety percent of the cations in the extracellular fluids consist of
    (A) hydroxyl ions
    (B) calcium ions
    (C) sodium ions
    (D) bicarbonate ions

15. The release of aldosterone from the adrenal cortex is related to the presence of
    (A) angiotensin II produced in the lung
    (B) renin produced in the bone marrow
    (C) parathormone produced in the parathyroid gland
    (D) estrogen and progesterone, the sex hormones

16. A low concentration of sodium in the blood stimulates the production of
    (A) potassium ions from potassium hydroxide
    (B) hydrogen ions from strong acids
    (C) renin from the nephron of the kidney
    (D) lymph from the lymphatic channels

17. The concentration of potassium in the blood is regulated by
    (A) the enzyme potassium
    (B) the hormone aldosterone
    (C) the level of plasma proteins
    (D) the acidic content of the blood

18. Excessive amounts of potassium in the body may lead to
    (A) arrhythmia of the heart
    (B) poor muscle contraction
    (C) accumulation of acid in the body
    (D) fibrillations of the heart

19. Deficiencies of chloride are often accompanied by deficiency of
    (A) calcium
    (B) albumin proteins
    (C) angiotensin
    (D) potassium ions

20. An example of a strong acid found in the body is
    (A) carbonic acid
    (B) propionic acid
    (C) acetic acid
    (D) hydrochloric acid

21. A buffer system generally contains
    (A) a strong acid and its accompanying base
    (B) a strong acid and its accompanying salt
    (C) a weak acid and a strong base
    (D) a weak acid and a salt of that acid

22. The carboxyl groups of amino acids in proteins
    (A) function as bases
    (B) take up hydrogen ions from the surrounding environment
    (C) increase the acidity of the surrounding environments
    (D) react with amino acids in the protein

23. The rate and depth of breathing has a regulatory influence on the
    (A) acid/base balance of the body
    (B) protein metabolism of the body
    (C) amount of water taken into the body
    (D) rate at which fats are broken down in the body

24. The loss of bicarbonate ions from the body through urine excretion
    (A) relieves the alkaline condition of the blood
    (B) increases the excretion of protein from the blood
    (C) has an effect on the nerve conduction system
    (D) changes the temperature at the body surface

25. The normal pH of the venous blood and interstitial fluid is slightly
    (A) more basic than arterial blood
    (B) neutral as compared to arterial blood
    (C) more acidic than arterial blood
    (D) the same as arterial blood

**PART C—True/False:** For each of the following statements, mark the letter "T" next to the statement if it is true. If the statement is false, change the underlined word to make the statement true.

1. An acid is a chemical compound that liberates hydrogen ions into a solution.

2. A normal adult male has about <u>forty</u> percent body weight of water.

3. The extracellular fluid compartment of the body consists of all the body area <u>outside</u> the cells.

4. About one-quarter of the <u>intracellular</u> fluid compartment of the body is blood plasma.

5. The transcellular fluid is separated from other body fluids by layers of <u>muscle</u> cells.

6. Interstitial fluid is generally <u>rich</u> in protein.

7. In general terms, the amount of water leaving the body <u>exceeds</u> the amount of water entering the body.

8. When ions are lost from the extracellular environment, water tends to flow through the cell membrane <u>out of</u> the cells.

9. Excess fluid and protein may be removed from the interstitial fluid by the <u>circulatory</u> system to establish homeostasis.

10. A nerve center in the <u>thalamus</u> in the brain is the location of osmoreceptors involved in fluid regulation.

11. An increase in the amount of ADH in the tubules of the kidney results in an <u>increase</u> in the amount of urine expelled by the kidney.

12. <u>Boyle's</u> law states the directions that fluids flow between the capillary and interstitial fluid.

13. The pressure exerted by water in the blood is the <u>colloid osmotic</u> pressure of the blood.

14. At the arteriole end of the capillary, water leaves the capillary because the hydrostatic pressure is <u>higher</u> than the colloid osmotic pressure.

15. Edema is the presence of abnormally large amounts of fluid in the <u>interstitial tissue spaces</u> caused by high blood pressure or other circumstances.

16. <u>Anions</u> are ions such as sodium ions that carry a positive charge.

17. When the volume of blood is low, the substance <u>angiotensin</u> is released from the nephron of the kidney.

18. The substance angiotensin II stimulates thirst and <u>inhibits</u> the secretion of antidiuretic hormone and ACTH.

19. The hormone aldosterone regulates the body's level of sodium ions and <u>calcium</u> ions.

20. <u>Chloride</u> ions have roles in blood clot formation, muscle contraction, nerve conduction, and the construction of bones and teeth.

21. In a buffer system, a strong base will react with a <u>strong</u> acid to minimize a basic shift.

22. The most powerful buffer system in the body is the <u>sodium chloride</u> buffer system.

23. Increased metabolism of the body cells leads to a <u>decrease</u> in the concentration of carbonic acid in the blood.

24. The acid/base status of the body is evaluated in system <u>venous</u> blood.

25. When the pH of the blood drops below 7.35, a condition known as <u>acidosis</u> results.

**Answers**

## PART A—Completion

1. electrolytes
2. hydroxyl ions
3. sodium hydroxide
4. hydrogen ions
5. female
6. intracellular fluid compartment
7. one-third
8. lymph
9. transcellular fluid
10. plasma
11. kidney
12. osmosis
13. proteins
14. thalamus
15. antidiuretic hormone
16. sodium ions
17. resistance
18. Starling's law of the capillaries
19. blood pressure
20. proteins
21. colloid osmotic pressure
22. filtration
23. hydrostatic pressure
24. absorption
25. osmotic pressure
26. edema
27. anions
28. foods
29. sodium ions
30. renin
31. angiotensin II
32. potassium ions
33. heart
34. calcium
35. thyroid gland
36. strong acid
37. carbonic acid
38. pH
39. pH
40. salt
41. amino groups
42. breathing
43. carbonic acid
44. albumin
45. urine
46. bicarbonate ions
47. ammonium ions
48. systemic arterial blood
49. 7.4
50. alkalosis

## PART B—Multiple Choice

| | | | | |
|---|---|---|---|---|
| 1. B | 6. B | 11. A | 16. C | 21. D |
| 2. A | 7. B | 12. D | 17. B | 22. C |
| 3. D | 8. D | 13. B | 18. D | 23. A |
| 4. C | 9. C | 14. C | 19. D | 24. A |
| 5. C | 10. A | 15. A | 20. D | 25. C |

## PART C—True/False

1. true
2. sixty
3. true
4. extracellular
5. epithelial
6. poor
7. is the same as
8. into
9. lymphatic
10. true
11. decrease
12. Starling's
13. hydrostatic
14. true
15. true
16. cations
17. renin
18. stimulates
19. potassium
20. calcium
21. weak
22. protein
23. increase
24. arterial
25. true

# CHAPTER 22

# THE MALE REPRODUCTIVE SYSTEM

The human reproductive system is responsible for producing, storing, nourishing, and transporting reproductive cells. These reproductive cells are called **gametes**. The reproductive systems of both males and females contain numerous analogous structures: two reproductive organs known as **gonads**, which produce gametes and hormones; **ducts** to receive and transport the gametes; **accessory glands and organs** that secrete fluids into the ducts; and a number of external structures associated with the reproductive process and collectively known as **external genitalia**. The male reproductive system is considered in this chapter; the female system is the topic of Chapter 23.

## THE TESTES

The **testes** are the male organs of reproduction (Figure 22.1). Their functions are to produce sperm cells and hormones associated with the male reproductive process. Each testis resembles a flattened oval, and is about 5 cm in length and about 2.5 cm in width.

The testes are contained within the **scrotum**, a fleshy pouch suspended below the perineum and anterior to the anus. The scrotum is divided into two separate chambers, one chamber for each testis. The boundary between the two chambers is marked by a raised thickening within the scrotal surface known as the **raphe**. A thin layer of smooth muscle called the **dartos** lies in the dermis of the scrotum and contracts to give the characteristic wrinkled appearance of the scrotal surface.

## Development of the Testes

The testes develop during the fetal stage within the abdominal cavity near the kidneys. By the end of the seventh month of pregnancy, the testes have moved through the abdominal musculature into the scrotum, where it is a few degrees cooler than in the abdominal cavity. A fibrous cord of tissue called the **gubernaculum** is responsible

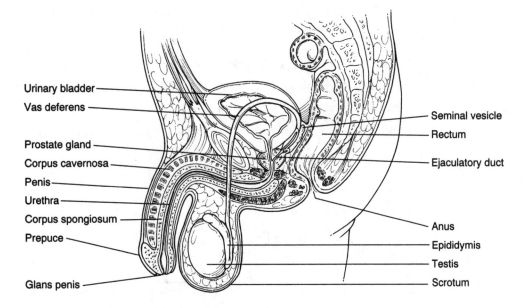

**FIGURE 22.1**  *A view of the male reproductive system observed from the left lateral position.*

for pulling the testes downward. If this descent does not occur, the condition is called **cryptorchidism**, and surgery may be required to bring the testes into the scrotum. Cryptorchidism may lead to infertility because the temperature in the abdominal cavity is too high for sperm cell production.

As the testes move through the body wall, each testis is accompanied by blood vessels, nerves, and the vas deferens, the tube leading from the testis (Table 22.1). Together, these structures form the **spermatic cord**. The canal through the peritoneum linking the scrotal chambers with the peritoneal cavity is called the **inguinal canal**. It is a weak spot in the abdominal wall and peritoneum, and it may be the site of inguinal hernias. **Hernias** are protrusions of an abdominal structure through the abdominal wall.

Each testis is composed of small lobes of cells (separated by connective tissues) surrounded by a series of tightly coiled tubules called **seminiferous tubules**. The lining of the seminiferous tubules contains two types of cells: germinal epithelial cells that produce sperm cells; and nourishing cells called **Sertoli cells**. Between the seminiferous tubules, there exist a series of **interstitial cells** (also called the **cells of Leydig**). The interstitial cells produce male sex hormones called androgens. A capsule of connective tissue encloses each testis.

The seminiferous tubules unite to form a plexus called the **rete testis**. A series of efferent ducts drain the rete testis, and the efferent ducts emerge at the top of the testis to enter the head of coiled tube known as the epididymis.

**TABLE 22.1** *Major Organs of the Male Reproductive System*

| Organ | Function |
|---|---|
| Testes | Produces male sex hormones |
|   Seminiferous tubules | Environment where sperm cells are produced |
|   Interstitial cells | Produces male sex hormones |
| Epididymis | Environment where sperm cells mature; conveys sperm cells to vas deferens |
| Vas deferens (Ductus deferens) | Conveys sperm cells to ejaculatory duct |
| Seminal vesicle | Secretes alkaline fluid containing nutrients and prostaglandins |
| Prostate gland | Secretes alkaline fluid to neutralize acidic seminal fluid and enhances sperm motility |
| Bulbourethral gland | Secretes mucus that lubricates end of penis and alkaline substances to neutralize acidity |
| Scrotum | Encloses and protects testes |
| Penis | Conveys urine and semen to outside of body; organ of sexual intercourse; glans penis is associated with feelings of pleasure during sexual stimulation |

**TABLE 22.2** *The Phases of the Two Stages of Meiosis*

| Phase | Activity |
|---|---|
| | **Stage I** |
| Prophase I | Homologous chromosomes form a tetrad during synapsis. Chromatids of homologous chromosomes overlap, forming chiasmata. Chromatid segments may cross over, leading to increased genetic variability |
| Metaphase I | Paired homologues assort independently as they line up on the metaphase (equatorial) plate |
| Anaphase I | Paired homologues separate and move to opposite poles. Centromeres of duplicated chromosomes do not split |
| Telophase and Cytokinesis | Separated homologues are partitioned into two daughter cells |
| | **Stage II** |
| Interphase | Intermediate period between chromosome movements |
| Prophase II | Chromosomes condense and move to center of cell |
| Metaphase II | Centromeres line up on the metaphase (equatorial) plate |
| Anaphase II | Centromeres split, and sister chromosomes move to opposite poles |
| Telophase II and Cytokinesis | Four haploid daughter cells form, each with exactly half the number of chromosomes (23) as the mother cell |

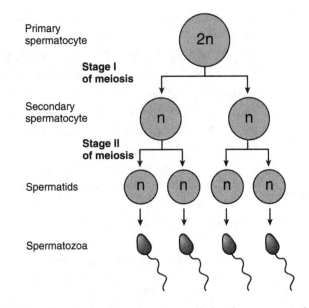

Primary spermatocyte

**Stage I of meiosis**

Secondary spermatocyte

**Stage II of meiosis**

Spermatids

Spermatozoa

**FIGURE 22.2**   *The formation of sperm cells by the process of meiosis. Spermatogonia form primary spermatocytes, which have 46 chromosomes (the diploid or 2N number). Primary spermatocytes enter the Stage I meiosis, where they undergo a reduction division to produce secondary spermatocytes with 23 chromosomes (the haploid or N number). In Stage II of meiosis, the chromosome number remains the same as the cells duplicate to produce 4 spermatids, each with 23 chromosomes. The spermatids develop to sperm cells.*

## Spermatogenesis

Sperm cells are produced through the process of **spermatogenesis**. Spermatogenesis begins at the outermost layer of germinal cells in the seminiferous tubules. Primordial (stem) cells called **spermatogonia** undergo duplications to produce cells, which are gradually pushed toward the openings to the tubule. These cells are called **primary spermatocytes** (Figure 22.2). The spermatocytes then undergo **meiosis**, a cellular process in which spermatocytes with 46 chromosomes per cell experience a reduction division to yield cells with 23 chromosomes per cell (Table 22.2). The cells that result from meiosis are called **spermatids**. The spermatids mature to form **sperm cells**, or **spermatozoa**, contained in fluid.

## Sperm Cells

The goal of the testes is to produce sperm cells. Each sperm cell has three distinct regions: the head, the middle piece, and the tail.

The **head** of a sperm cell is a flattened oval body densely packed with the nucleus of 23 chromosomes. The tip of the head forms an area called the **acrosome**, or **acrosomal cap**, which contains an enzyme that functions in fertilization. The **neck** of the sperm cell is very short, and the **middle piece** contains microtubules surrounded by an inner layer of dense fibers and an outer layer filled with mitochondria. The mitochondria provide sites for energy metabolism required to move the tail. The **tail** of the sperm cell contains dense fibers surrounded by a fibrous sheath. It acts as a **flagellum** and beats about, thereby pushing the sperm cell forward (Figure 22.3).

Sertoli cells have several functions in sperm cell production. They provide mechanical support for sperm cells and prevent female structures from developing in the embryo. Also, they assist hormonal activity in the testis by maintaining a high concentration of the hormones, and they interact with the anterior pituitary gland, from which reproductive hormones are released.

## ACCESSORY DUCTS AND ORGANS

A number of accessory ducts and organs assist the physiology of the testes by removing sperm cells from the body and activating them.

## Male Reproductive Ducts

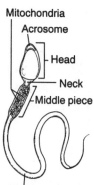

Mitochondria
Acrosome
Head
Neck
Middle piece
Flagellum (tail)

**FIGURE 22.3**
*The human sperm cell displaying the major anatomical parts. The nucleus containing 23 chromosomes is the key feature of the sperm cell.*

To reach the external environment, the mature sperm cells pass through a duct system having several subdivisions. The first subdivision of the duct system is the **epididymis**. Mature sperm cells reach the epididymis from the efferent ducts arising from the rete testis. The epididymis lies along the posterior border of the testis and consists of an elongated tubule twisted and coiled.

Cells lining the epididymis adjust the composition of the sperm fluid by adding secretions. The overall pH of the fluid is acidic due to the waste products produced by the stored sperm cells. The epididymis is also where damaged sperm cells and debris are absorbed. Moreover, it is the site of sperm cell maturation occurring over a period of about two weeks.

After leaving the epididymis, sperm cells enter the next duct, the **vas deferens**, also called the **ductus deferens**. The vas deferens is a tubular extension of the epididymis extending through the inguinal canal into the abdominal cavity. In the abdominal cavity, the vas deferens passes over the top and posterior surface of the urinary bladder toward the superior and posterior margin of the prostate gland. Just before reaching the prostate gland, the vas deferens enlarges to a portion called the **ampulla**.

The function of the vas deferens is to propel and conduct seminal fluid from the epididymis of each testis. At the ampulla the two vas deferens ducts join with the duct leading from the seminal vesicle. The merge forms the **ejaculatory duct**. This relatively short duct

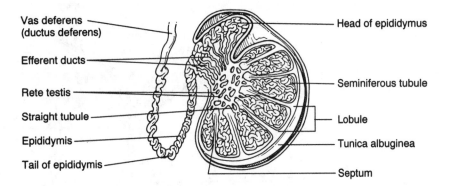

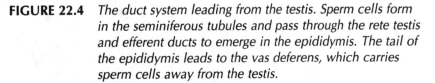

**FIGURE 22.4**    *The duct system leading from the testis. Sperm cells form in the seminiferous tubules and pass through the rete testis and efferent ducts to emerge in the epididymis. The tail of the epididymis leads to the vas deferens, which carries sperm cells away from the testis.*

penetrates the wall of the prostate gland and unites with the urethra (Figure 22.4).

The **urethra** extends from the urinary bladder to the tip of the penis. It is divided into three portions: the prostatic, membranous, and penile portions. The **prostatic urethra** passes through the center of the prostate gland where it receives secretions from the gland. The **membranous urethra** is a short segment penetrating the muscular floor of the pelvic cavity. The **penile urethra** extends through the penis to the **external urethral meatus** at the tip of the penis. The penile urethra receives secretions from the bulbourethral glands.

## Accessory Organs

There are several accessory organs that contribute fluid to the sperm cells or serve as organs for delivery of sperm cells during fertilization. One such organ is the **seminal vesicle**. The seminal vesicle consists of paired saclike structures drained by ducts merging with the vas deferens. The seminal vesicle (as well as other body cell) secretes hormones known as prostaglandins and adds nutrient fluids (especially fructose) to support the sperm cells during the ejaculation process. The fluid produced is alkaline to neutralize the acidity developing in the epididymis, and the vesicle fluid represents about 60 percent of the total seminal fluid known as semen.

Another important accessory organ is the **prostate gland**. This is a single gland that secretes a slightly alkaline fluid, which contributes to sperm motility by neutralizing the natural acidity of the vagina. The prostate gland contains muscle for support and encircles the urethra. Its enlargement in older males may interfere with urination. The prostate gland contributes approximately 30 percent of the volume of semen (seminal fluid).

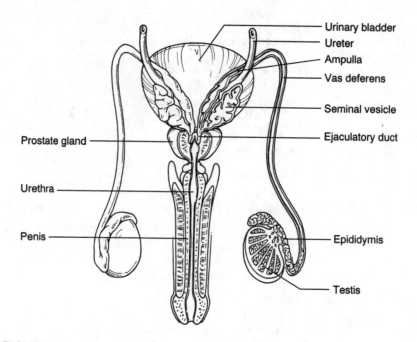

**FIGURE 22.5**   *Accessory ducts and organs of the male reproductive system. The vas deferens carrying sperm cells extends through the inguinal canal, passes near the urinary bladder, and joins with the duct from the seminal vesicle to form the ejaculatory duct. The latter penetrates the wall of the prostate gland and joins with the urethra. The urethra extends through the penis to the exterior.*

The **bulbourethral glands** are also called **Cowper's glands**. They are two small glands situated near the base of the penis. The glands secrete lubricating mucus and alkaline substances that neutralize vaginal acid and activate the sperm cells. Secretions of the bulbourethral glands, prostate gland, seminal vesicle, and testes combine with the sperm cells to form the **semen** (Figure 22.5).

Another accessory organ is the **penis**. The penis is the male organ of urination and copulation. It consists of a root, body (shaft), and glans. The glans is the portion that surrounds the external urethral meatus. A fold of skin called the **prepuce** (or foreskin) surrounds the tip of the penis.

Most of the body of the penis consists of three masses of erectile tissue. Erectile tissue contains a maze of vascular channels separated by partitions of connective tissue and smooth muscle fibers. Two of the erectile tissue masses are called the **corpora cavernosae** (singular **corpus cavernosa**). The third mass is called the **corpus spongiosum**. During sexual excitation, impulses from the parasympathetic branch of the nervous system cause arterioles in the erectile tissues to dilate, and blood flow to these tissues increases substantially. The vascular network becomes engorged with blood, and erection occurs. The semen then passes through the urethra to the external urethral meatus. After ejaculation of semen, impulses from the sympathetic branch of the nervous system produce arteriole constriction and the blood supply diminishes. As veins carry blood away from the erectile tissues, the penis becomes flaccid. Continued sympathetic stimuli maintain the vasoconstriction and the flaccid condition.

A typical ejaculation of semen measures about two to five milliliters. This volume, called the **ejaculate**, contains sperm cells, seminal fluid, and enzymes. The sperm cell count is approximately 20 to 100 million sperm cells per milliliter of semen. The fluid is a mixture of glandular secretions from the accessory organs. Enzymes in the fluid include a protease and other enzymes (some unidentified) to assist fertilization. Peristaltic contractions in the reproductive ducts move semen during ejaculation. Contractions in the respective glands augment the sperm cells with fluids.

## MALE HORMONES

Several hormones contribute to the male reproductive process. The anterior pituitary gland releases **follicle-stimulating hormone (FSH)** as well as **interstitial cell stimulating hormone (ICSH)**. FSH induces spermatogenesis to occur in the seminiferous tubules, while ICSH assists spermatogenesis and stimulates the production of testosterone by the interstitial cells. ICSH is analogous to LH, a hormone of the female system and is also known by that name (Table 22.3).

**TABLE 22.3**   *Hormones of the Male Reproductive System*

| Hormone | Origin | Principal Effects | Control |
|---|---|---|---|
| Gonadotropin releasing hormone (GnRH) | Hypothalamus | Causes pituitary to release FSH and LH; increases blood levels of FSH and LH | — |
| Follicle stimulating hormone (FSH) | Pituitary | Stimulates maturation of seminiferous tubules and sperm production | Hypothalamus |
| Luteinizing hormone (LH) | Pituitary | Stimulates maturation of interstitial cells | Hypothalamus |
| Testosterone | Testes | Produces and maintains male sex characteristics; stimulates sperm production; inhibits LH production | LH |

**Testosterone** is one of the **androgens**, which are male hormones produced by the interstitial cells of the testis. In the fetus, testosterone regulates the differentiation of male tissues and is involved in descent of the testes to the scrotum. Little testosterone is then produced until puberty. After puberty, testosterone has numerous functions, such as promoting the maturation of sperm cells; maintaining the accessory organs of the male reproductive tract; influencing the development of secondary male sexual characteristics; stimulating metabolic processes that concern protein synthesis and muscle growth; and influencing sexual behavior and sexual drive. Other androgens accelerate puberty and initiate sexual maturation and the appearance of secondary male characteristics.

The release of FSH and ICSH (LH) is apparently regulated by a hormone of the hypothalamus referred to as gonadotropin releasing hormone (GnRH). Together with FSH and ICSH, the releasing hormone works in a negative feedback mechanism to control testosterone synthesis and secretion.

# REVIEW QUESTIONS

**PART A—Completion: Add the word or words that correctly complete each of the following statements.**

1. Another name for reproductive cells is _____ .

2. The name given to the male organs of reproduction is the _____ .

3. The two products of the testes are sperm cells and _____ .

4. Suspended below the perineum is a fleshy pouch that contains the testes and is called the _____ .

5. The boundary between the two chambers of the scrotum is a raised thickening known as the _____ .

6. In the dermis of the scrotum there is a thin layer of muscle called the _____ .

7. The smooth muscle in the dermis of the scrotum contracts to give the scrotum a characteristic appearance of _____ .

8. Within the fetus, the testes have moved into the scrotum by the end of month number _____ .

9. The failure of the testes to descend into the scrotum is called _____ .

10. The nerves, ducts, and blood vessels emerging from the testes together form the _____ .

11. The canal through which the spermatic cord passes into the peritoneal cavity is the _____ .

12. The tightly coiled tubules of the testes are known as _____ .

13. Nourishing cells lining the seminiferous tubules are the _____ .

14. The cells lying between the seminiferous tubules are _____ .

15. The interstitial cells of the testes are responsible for producing _____ .

16. The plexus formed by the union of the seminiferous tubules is the _____ .

17. Efferent ducts arising from the rete testes enter the tubule called the _____ .

18. The process through which sperm cells are produced is _____ .

19. Sperm cells are formed from primordial cells known as _____ .

20. The process of spermatogenesis takes place in the _____ .

21. The cells produced by duplication of the spermatogonia are _____ .

22. The cellular process by which spermatocytes form spermatids is called _____ .

23. Spermatids will mature to form sperm cells, also known as _____ .

24. During the process of meiosis a spermatocyte with 46 chromosomes will produce a spermatid having chromosomes that number _____ .

25. The chromosomes of the sperm cell are packed into the region of the cell known as the _____ .

26. Enzymes important in fertilization are contained in the tip of the head of the sperm cell called the _____ .

27. The mitochondria of the sperm cell are contained in a portion of the cell known as the _____ .

28. The tail of the sperm cell provides motion by acting as a _____ .

29. The site of sperm cell maturation over a period of about two weeks is a duct called the _____ .

30. The duct where damaged sperm cells and debris are absorbed is the _____ .

31. Another name for the vas deferens is the _____ .

32. The tubular extension of the epididymis that extends through the inguinal canal is the _____ .

33. Just before reaching the prostate gland, the vas deferens enlarges to a portion called the _____ .

34. Within the abdominal cavity, the vas deferens passes over the top and posterior surface of the _____ .

35. At the ampulla, the vas deferens joins with the duct leading from the _____ .

36. The duct passing from the urinary bladder to the tip of the penis is the _____ .

37. The three portions of the urethra are the membranous urethra, the penile urethra, and the _____ .

38. The penile portion of the urethra receives secretions from the _____ .

39. Prostaglandins and nutrient fluids are added to the sperms cells by _____ .

40. An alkaline fluid contributing to sperm motility is produced by a gland that encircles the urethra and is called the _____ .

41. Older males may experience interference with urination if enlargement takes place in the _____ .

42. Two small glands situated at the base of the penis are called bulbourethral glands, or _____ .

43. The root, shaft, and glans are three portions of the _____ .

44. Most of the body of the penis consists of tissue that is _____ .

45. Erection of the penis takes place after the accumulation of _____ .

46. The opening of the penis to the exterior is the _____ .

47. The ejaculate contains sperm cells that number approximately _____ .

48. Spermatogenesis is stimulated by the pituitary hormone known as _____ .

49. The male hormone that promotes sperm cell maturation and maintains the accessory organs is called _____ .

50. The general name for male hormones that accelerate puberty and initiate sexual maturation is _____ .

*PART B—Multiple Choice:* **Circle the letter of the item that correctly completes each of the following statements.**

1. The alternative term for the human reproductive organs is
   (A) genitalia
   (B) gametes
   (C) gonads
   (D) spermatogonia

2. The anatomical structures associated with the scrotum include
   (A) the corpus cavernosa and corpus spongiosum
   (B) the dartos and raphe
   (C) the neck and flagellum
   (D) Cowper's glands and the prostate gland

3. Cryptorchidism is the condition in which
   (A) the vas deferens fails to develop
   (B) the scrotum is too small to accommodate the testes
   (C) the epididymis is blocked
   (D) the testes fail to descend into the scrotum

4. The spermatic cord consists of all the following structures except
   (A) nerves
   (B) the epididymis
   (C) blood vessels
   (D) the vas deferens

5. The epithelial cells that produce sperm cells are located in the testes within the
   (A) seminiferous tubules
   (B) vas deferens
   (C) corpus spongiosum
   (D) ampulla

6. The male sex hormones are produced in the testes by the
   (A) interstitial cells
   (B) blood cells
   (C) penile cells
   (D) prostate cells

7. The process of spermatogenesis is one in which
   (A) sperm cells mature
   (B) sperm cells gather in the vas deferens
   (C) sperm cells are ejaculated from the body
   (D) sperm cells are produced

8. Spermatocytes differ from spermatids in that
   (A) spermatocytes are produced in the interstitial cells
   (B) spermatocytes are produced by the prostate gland
   (C) spermatocytes have more chromosomes
   (D) spermatocytes are found in the corpus cavernosa

9. The chromosomes of a sperm cell are contained
   (A) within the flagellum
   (B) packed in the head of the cell
   (C) within the acrosome of the cell
   (D) in the neck of the sperm cell

10. To service the energy needs of the sperm cell, the cell has
    (A) many ribosomes
    (B) a large number of chromosomes
    (C) many mitochondria
    (D) a large number of fibrils

11. The epididymis receives mature sperm cells
    (A) from the rete testes
    (B) from the vas deferens
    (C) from the prostate gland
    (D) from the penis

12. The vas deferens is a tubular extension of the
    (A) prostatic urethra
    (B) membranous urethra
    (C) ejaculatory duct
    (D) epididymis

13. The vas deferens passes into the abdominal cavity through the
    (A) ejaculatory duct
    (B) inguinal canal
    (C) corpus spongiosum
    (D) membranous urethra

14. The tube passing from the urinary bladder to the tip of the penis is the
    (A) seminal tube
    (B) ejaculatory duct
    (C) urethra
    (D) prostate tubule

15. Secretions of the seminal vesicle
    (A) are alkaline
    (B) pass into the testes
    (C) are acidic
    (D) contain sperm cells

16. All the following apply to the prostate gland except
    (A) the gland encircles the urethra
    (B) the gland contributes about 30 percent of the volume of semen
    (C) the secretion of the gland is acidic
    (D) the prostate is a single gland

17. The semen contains secretions from all the following except
    (A) the testes
    (B) the penis
    (C) the bulbourethral glands
    (D) prostate gland

18. Enlargement of the prostate gland in older males
    (A) may result in sterility
    (B) never happens
    (C) interferes with secretions from the seminal vesicle
    (D) may interfere with urination

19. During erection, the erectile tissues of the penis
    (A) fill with blood
    (B) contract into the scrotum
    (C) secrete alkaline fluid
    (D) absorb sperm cells

20. The corpus cavernosa and the corpus spongiosum are both
    (A) masses of erectile tissue
    (B) glands that add secretions to the semen
    (C) sites where sperm cells are stored
    (D) areas where urine is held prior to expulsion

21. Dilation of the blood vessels in the erectile tissues in the penis is stimulated
    (A) by the sensory somatic nervous system
    (B) by the central nervous system
    (C) by the sympathetic and parasympathetic nervous systems
    (D) by reflexes in the spinal cord

22. An ejaculate of semen contains approximately
    (A) 100,000 sperm cells
    (B) 50 million sperm cells
    (C) 75–100 sperm cells
    (D) less than one million sperm cells

23. The process of sperm formation in the testes is influenced by
    (A) the hormone calcitonin
    (B) the lactogenic hormone
    (C) the follicle stimulating hormone
    (D) thyroxin

24. All the following characteristics are associated with testosterone except
    (A) it maintains the accessory organs of the male reproductive tract
    (B) it influences the secondary male characteristics
    (C) it promotes protein synthesis and muscle growth
    (D) it is produced by the seminiferous tubules of the testes

25. The general term given to male sex hormones is
    (A) estrogens
    (B) thymosins
    (C) androgens
    (D) prostaglandins

**PART C—*True/False*: For each of the following statements, mark the letter "T" next to the statement if it is true. If the statement is false, change the <u>underlined</u> word to make the statement true.**

1. An alternate name for the reproductive organs is <u>gametes</u>.

2. The two main functions of the testes are to produce <u>sperm cells</u> and hormones associated with the reproductive process.

3. The boundary between the two chambers of the testes is marked by a raised thickening called the <u>dartos</u>.

4. Thin layers of <u>striated</u> muscle are located within the dermis of the scrotum, and they contract to give it a wrinkled appearance.

5. The testes develop in a male <u>after</u> birth has taken place.

6. If the testes fail to descend into the scrotum, a condition called <u>accommodation</u> exists.

7. The canal through which the spermatic cord passes into the peritoneal cavity is the <u>pleural</u> canal.

8. The tightly coiled tubules known as <u>interstitial</u> tubules are the places where sperm cells are produced.

9. The main function of the interstitial cells is to produce <u>sex hormones</u>.

10. <u>Spermatogenesis</u> is the process by which sperm cells are produced in the male.

11. In the process of meiosis, a <u>spermatid</u> undergoes a change in which cells with 23 chromosomes are produced.

12. Another name for <u>spermatozoa</u> is sperm cells.

13. The acrosome is located in the <u>middle piece</u> of the mature sperm cell.

14. The number of chromosomes present in a sperm cells is <u>46</u>.

15. The sperm cell is pushed forward by a beating tail known as the <u>flagellum</u>.

16. The plexus of seminiferous tubules that unite to enter the epididymis is known as the <u>raphe</u>.

17. The site of maturation of sperm cells is the <u>vas deferens</u>.

18. The duct leading from the testes that passes through the inguinal canal and enters the abdominal cavity is the <u>epididymis</u>.

19. The duct extending from the urinary bladder to the tip of the penis is the <u>ejaculatory duct</u>.

20. The fluid produced by both the seminal vesicles and the prostate gland is <u>acidic</u>.

21. The enlargement of the <u>seminal vesicles</u> in older males may interfere with urination.

22. The secretions of the bulbourethral glands, prostate gland, and seminal vesicles when combined with sperm cells form the <u>semen</u>.

23. The foreskin that surrounds the tip of the penis and is removed in circumcision is known as the <u>prepuce</u>.

24. One of the masses of erectile tissue found in the penis is known as the <u>corpus callosum</u>.

25. An example of an androgen that promotes sperm cell maturation is <u>FSH</u>.

**Answers**

## PART A—Completion

1. gametes
2. testes
3. hormones
4. scrotum
5. raphe
6. dartos
7. wrinkling
8. seven
9. cryptorchidism
10. spermatic cord
11. inguinal canal
12. seminiferous tubules
13. Sertoli cells
14. interstitial cells
15. sex hormones
16. rete testis
17. epididymis
18. spermatogenesis
19. spermatogonia
20. seminiferous tubules
21. spermatocytes
22. meiosis
23. spermatozoa
24. twenty-three
25. head
26. acrosome
27. middle piece
28. flagellum
29. epididymis
30. epididymis
31. ductus deferens
32. vas deferens
33. ampulla
34. urinary bladder
35. seminal vesicles
36. urethra
37. prostatic urethra
38. bulbourethral glands
39. seminal vesicles
40. prostate gland
41. prostate gland
42. Cowper's glands
43. penis
44. erectile
45. blood
46. external urethral meatus
47. 20 to 100 million per ml
48. follicle stimulating hormone
49. testosterone
50. androgens

## Part B—Multiple Choice

| | | | | |
|---|---|---|---|---|
| 1. C | 6. A | 11. A | 16. C | 21. C |
| 2. B | 7. D | 12. D | 17. B | 22. B |
| 3. D | 8. C | 13. B | 18. D | 23. C |
| 4. B | 9. B | 14. C | 19. A | 24. D |
| 5. A | 10. C | 15. A | 20. A | 25. C |

## Part C—True/False

1. gonads
2. true
3. raphe
4. smooth
5. before
6. cryptorchidism
7. inguinal
8. seminiferous
9. true
10. true
11. spermatocyte
12. true
13. head
14. 23
15. true
16. rete testes
17. epididymis
18. vas deferens
19. urethra
20. alkaline
21. prostate gland
22. true
23. true
24. corpus cavernosa
25. testosterone

# CHAPTER 23

# THE FEMALE REPRODUCTIVE SYSTEM

The female reproductive system produces, stores, nourishes, and transports reproductive cells called **gametes**. Gametes include sperm cells from the male reproductive system (Chapter 22) and egg cells, or ova (singular ovum) from the female system. These cells unite in fertilization to form the single fertilized egg cell. The female reproductive system includes the reproductive organs (the ovaries, also known as the **gonads**) that are responsible for producing gametes and hormones; several **ducts** that receive and transport the gametes, **accessory glands and organs** that secrete fluids; and structures of the **external genitalia** associated with the reproductive system (Figure 23.1).

The major organs of the female reproductive tract are enclosed within an extensive mesentery (fold of peritoneum) called the **broad ligament**. The Fallopian tubes run along the superior border of the broad ligament and open into the pelvic cavity lateral to the ovaries. The broad ligament attaches to the sides and floor of the pelvic cavity, and its epithelium is continuous with the epithelium of the parietal peritoneum.

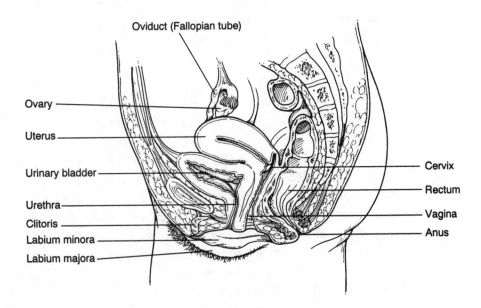

**FIGURE 23.1** *A view of the female reproductive tract from the left lateral aspect. The major organs and structures of the system are illustrated.*

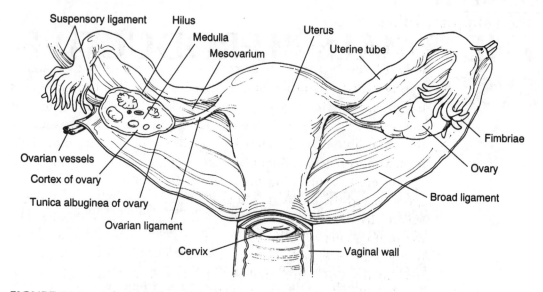

**FIGURE 23.2**  *A frontal view of the female reproductive tract. The ovaries in place in the female body enclosed within the broad ligament. Each ovary is supported by the ovarian ligament and the suspensory ligament. Various anatomical aspects of the ovary are shown. The uterine tubes (Fallopian tubes) are seen along the superior border of the broad ligament leading to the uterus covered by the ligament. The vagina has been sectioned to show the opening from the uterus into the vagina.*

## THE OVARIES AND ACCESSORY ORGANS

Egg cells are produced in the female reproductive system by ovaries, then transported to the exterior by a number of accessory organs, which also provide support for the egg cell.

## The Ovaries

The female organs where egg cells are produced are the paired **ovaries**. The ovaries are small, almond shaped organs located near the lateral walls of the pelvic cavity. The ovaries are retroperitoneal (behind the peritoneum). Each is approximately 5 cm across and approximately 2.5 cm wide. In addition to producing egg cells, these organs produce female sex hormones called **estrogens**. Each ovary is stabilized by a pair of supporting ligaments called the **ovarian ligament** and the **suspensory ligament** (Figure 23.2).

Internally, the ovary contains many clusters of cells called **follicles**. Each follicle contains layers of cells enclosing an immature egg cell called an **oocyte**. Oocytes mature within the follicle and are released in the process of **ovulation**. Ovulation occurs approximately every 28 days, and after ovulation the follicle reorganizes and becomes a structure known as the **corpus luteum** (yellow body). The corpus luteum is a hormone-secreting organ.

## The Fallopian Tubes

Each **Fallopian tube** (also known as the oviduct) is about 12.5 cm long. Closest to the ovary, the end of the tube forms an expanded funnel called the **infundibulum**. The infundibulum has numerous irregular, branched projections called **fimbriae** extending toward the ovary. The elongated segment of the Fallopian tube proximal to the infundibulum is the **ampulla**, and the ampulla leads to the **isthmus**, a short segment opening to and joining with the uterine wall.

The epithelium lining the ampulla has numerous pockets and grooves, and the exposed epithelial surface has hairlike appendages called **cilia**. Ovum transport through the Fallopian tube is assisted by the beating cilia as well as peristaltic contractions of smooth muscle in the walls. Normally, it takes about three to four days for an ovum to travel from the infundibulum to the chamber of the uterus. Fertilization will take place if sperm cells are encountered during the first one to two days of passage through the Fallopian tube. Unfertilized egg cells degenerate in the terminal portion of the Fallopian tubes.

No actual contact exists between the Fallopian tubes and the ovaries. When egg cells are released from ovarian follicles, they are swept into the Fallopian tubes by the beating fimbriae. It is possible for egg cells to miss the opening to the Fallopian tubes and enter the pelvic portion of the abdominopelvic cavity.

## The Uterus

The **uterus** (or **womb**) is a hollow organ normally about the size and shape of a pear, except during pregnancy when it expands considerably (Table 23.1). It is located medially within the anterior portion of the pelvic cavity, above the vagina and over the urinary bladder. The uterus is supported by the broad ligament.

**TABLE 23.1**  *Major Organs of the Female Reproductive System*

| Organ | Function |
| --- | --- |
| Ovary | Produces egg cells and female sex hormones |
| Fallopian tube | Delivers egg cell or developing embryo toward uterus; site of fertilization; also called uterine tube |
| Uterus | Protects and nourishes fetus during development |
| Vagina | Female organ of sexual intercourse; transports fetus during birth process; delivers endometrial lining to the exterior during menstruation |
| Labia major | Encloses and protects external reproductive organs |
| Labia minor | Forms margins of vestibule; protects opening of vagina and urethra |
| Vestibule | Space between labia minora that includes vaginal and urethral openings |

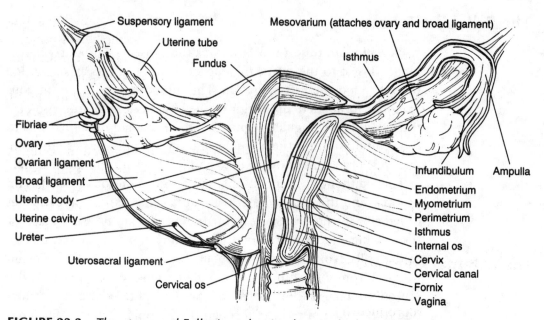

**FIGURE 23.3** *The uterus and Fallopian tubes in place in the broad ligament. The illustration shows the details of the Fallopian tubes and the entry of the tubes to the uterus. Note the thick muscular structure of the uterus and its anatomical parts. The entry to the vagina at the cervix can be seen.*

The uterus provides mechanical protection and nutritional support to the developing embryo and fetus. Its superior, thick-walled portion is called the **body**. The **fundus** is the bulging upper surface of the body of the uterus where the Fallopian tubes enter, and the area where the fundus joins with the Fallopian tubes is the isthmus. The term isthmus also applies to the region where the body of the uterus joins with the neck of the uterus called the **cervix**. The cervix projects a short distance into the vagina, and the uterine cavity opens into the vagina at the external orifice of the cervix, also called the **cervical os**.

Three tissue layers make up the relatively thick uterine wall. The inner tissue layer is called the **endometrium**. This is an inner layer of mucosa where the embryo implants. When fertilization does not occur, the endometrium sloughs off during the process of **menstruation**. The middle layer of the uterine wall is composed of a thick layer of smooth muscle called **myometrium** (Figure 23.3). The muscles contract rhythmically during the process of delivery in birth. The outer tissue layer of the uterine wall is called the **perimetrium**, or **serosa**. It is continuous with the mesothelium of the broad ligament.

## The Vagina

The **vagina** is a fibromuscular tube about 9 cm in length and extending from the cervix to the **vaginal orifice** at the vestibule. It is highly distensible in length and width, and it extends upward and back into the pelvic cavity. The vagina is the female organ of copulation and is often called the **birth canal**. At the point where the cervix projects into the vagina, there is a shallow recess known as the **fornix**. The walls of the vagina contain a network of blood vessels and layers of smooth muscle. Prior to the onset of sexual activity, a thin fold of epithelium known as the **hymen** partially or completely blocks the entrance to the vagina.

The vagina serves as a passageway for the elimination of fluids during menstruation. It also receives the penis during sexual intercourse and is the site of sperm cell deposit. During childbirth, the vagina serves as the passageway for the delivery of the newborn.

## The External Genitalia

The external genitalia are among the accessory organs of the female reproductive tract. The general name for the external genitalia is **vulva**, or **pudendum**.

The vagina opens at the vaginal orifice into a region of the vulva called the **vestibule** (Figure 23.4). This region contains several anatomical structures. One structure, the **clitoris**, is a small mass of erectile tissue that projects into the vestibule and enlarges during sexual arousal. Another structure, the paired **greater vestibular glands**, produce lubricants at the distal end of the vagina during sexual intercourse. They are also called **Bartholin's glands**. The vestibule also contains the opening from the urethra of the urinary tract known as the **urethral orifice**.

The vestibule is bounded by the **labia minora** (singular labium minorum, also called minor lips), two elongate, delicate folds of skin containing sebaceous glands. In the posterior aspect, the labia minora meet at the **perineum**, the skin area between the anus and

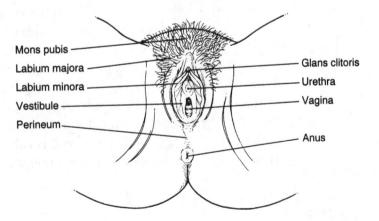

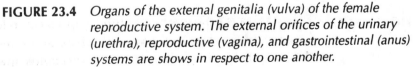

**FIGURE 23.4**  *Organs of the external genitalia (vulva) of the female reproductive system. The external orifices of the urinary (urethra), reproductive (vagina), and gastrointestinal (anus) systems are shows in respect to one another.*

the vulva. The labia minora are smaller than the labia majora (major lips). The outer limits of the vulva are established by the mons pubis and the labia majora. The **mons pubis** is a fatty, prominent region formed by adipose tissue beneath the skin anterior to the pubic symphysis. After puberty, it is covered with pubic hair. The **labia majora** (singular labium majorum, also called major lips) are two elongated folds of skin that encircle and partially conceal the labia minora and structures of the vestibule. Fluid-secreting glands provide secretions into the inner surface of the labia majora and lubricate them.

## The Mammary Glands

The newborn infant gains nourishment after birth from the milk secreted from the mother's **mammary glands**, which are a type of **alveolar glands** (Figure 23.5). The glands are located within the subcutaneous tissue of the anterior thorax within the breasts. The production of milk is called **lactation**.

The mammary gland is composed of lobes containing milk-secreting apocrine glands (alveolar glands) drained by mammary ducts (alveolar ducts). The lobes are separated by connective and fat tissues, and they are connected to the conical projection of each breast called the **nipple**. The skin area surrounding each nipple has a reddish-brown coloration and is called the **areola**. The areola has both sebaceous and sweat glands.

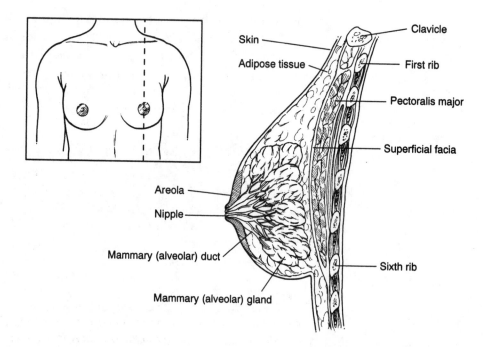

**FIGURE 23.5**   *A sagittal section of the human breast showing the anatomical relationship of the breast to the underlying bones and muscles. The mammary (alveolar) glands of the breast are drained by the mammary (alveolar) ducts.*

The secretion of milk is mediated by the hormone **prolactin**, or **lactogenic hormone (LH)**. The ejection of milk is mediated by the hormone **oxytocin**, a hormone whose release from the posterior pituitary gland is a reflex response to the mechanical stimulation of the infant's sucking at the nipple. Milk production continues as long as milk continues to be removed from the mammary glands.

## FEMALE REPRODUCTIVE PHYSIOLOGY

The physiology of female reproduction includes the changes in the system that relate to the production and fertilization of the egg cell.

## The Menstrual Cycle

The **menstrual cycle** is composed of the physiological and structural changes in the female reproductive tract as it responds to changes in the levels of ovarian hormones. The menstrual cycle is generally about 28 days long, and ovulation usually occurs about midway through the cycle.

During days 1–5 of the menstrual cycle, the thick endometrial lining of the uterus detaches (sloughs off) from the uterine wall. Bleeding (menses) generally occurs for three to five days, and the material

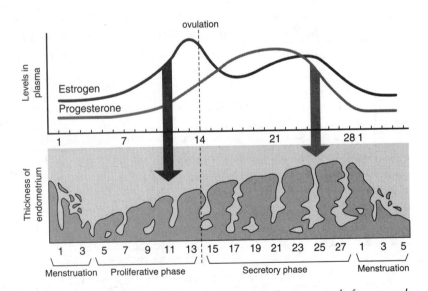

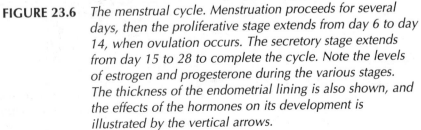

**FIGURE 23.6**  *The menstrual cycle. Menstruation proceeds for several days, then the proliferative stage extends from day 6 to day 14, when ovulation occurs. The secretory stage extends from day 15 to 28 to complete the cycle. Note the levels of estrogen and progesterone during the various stages. The thickness of the endometrial lining is also shown, and the effects of the hormones on its development is illustrated by the vertical arrows.*

contains endometrial tissue, glandular secretions, mucus, and a small amount of blood. This stage is referred to as the **menstrual stage**, or **menstruation**. During this time, the levels of estrogen and progesterone are low, and repair of the endometrium has begun.

During days 6–14 of the cycle, the follicles in the ovaries develop, and the endometrium is completely repaired. This stage is called the **proliferative stage** of the cycle. Tubular glands form in the endometrium, and the endometrial blood supply increases. The levels of the hormones estrogen and progesterone increase during this stage and influence the repair. **Ovulation** occurs at about day 12–14, accompanied by a surge of luteinizing hormone (LH) and an increase of estrogen and progesterone. After release of the egg cell, the follicle converts to an enlarged structure, the corpus luteum, as discussed below.

During days 15–28 of the cycle, progesterone is produced by the corpus luteum in the ovary, and the estrogen content of blood in the endometrium also increases. The endometrial glands begin secreting nutrients into the uterus to sustain an embryo, if present. This stage is called the **secretory stage** of the cycle (Figure 23.6).

If an embryo is not present, the corpus luteum degenerates, and the levels of estrogen and progesterone drop off. The corpus luteum degenerates because it lacks the stimulation to remain viable. The stimulation is given by the human chorionic gonadotropin, a hormone secreted by cells of the blastocyst resulting after fertilization. The lack of ovarian hormones accompanies blood vessel constriction, and without blood the endometrial cells begin to die. Menstruation begins on about day 28, which corresponds to day one of a new menstrual cycle. The first menstruation is known as **menarche**, and the cessation of menstrual cycles is **menopause**.

**TABLE 23.2** *Hormones of the Female Reproductive Systems*

| Hormone | Origin | Principal Effects | Control |
|---|---|---|---|
| Gonadotropin-releasing hormone (GnRH) | Hypothalamus | Induces pituitary to release FSH and LH | |
| Follicle stimulating hormone (FSH) | Pituitary | Stimulates growth of ovarian follicle and production of estrogen | Hypothalamus |
| Luteinizing hormone (LH) | Pituitary | Stimulates progesterone production and ovulation | Hypothalamus |
| Estrogen | Ovary (follicle) | Stimulates development of female sex characteristics; thickens lining of uterus; inhibits FSH production | FSH |
| Prolactin | Pituitary | Stimulates production of milk | Hypothalamus |
| Oxytocin | Pituitary | Stimulates milk release; stimulates uterine contractions in menstruation and birth | |

# Oogenesis

The process by which egg cells are formed in the ovary is called **oogenesis**. Oogenesis begins during the fetal stage of the female, that is, before the woman is born. At this stage, primitive egg cells called **oogonia** enter the process of meiosis and begin the first phase (Prophase I), but do not complete the entire first stage. They convert into cells called **primary oocytes**. About 2 million primary oocytes are surrounded by layers of cells to form the **primary follicles**. The primary follicles are present in the woman's ovary at her birth, and no further primary follicles form during her lifetime. About 75,000 remain at the age of puberty.

At the age of puberty, the hormone **gonadotropin-releasing hormone (GnRH)** is secreted by the hypothalamus (Table 23.2). This hormone stimulates the anterior pituitary gland to release **follicle stimulating hormone (FSH)**. The FSH stimulates the follicles to

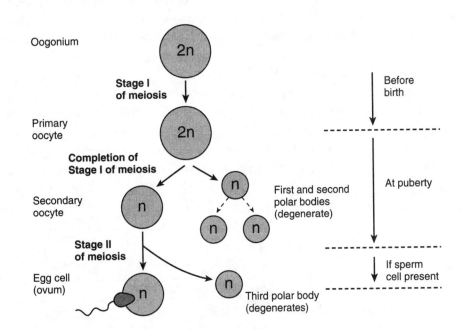

**FIGURE 23.7**   *Egg cell production. The oogonium is the primitive egg cell having 46 chromosomes, the 2N number. Before birth, the oogonium enters the process of meiosis and passes through stage I, developing to the primary oocyte, which is present in the woman's ovary at birth. At puberty, the process continues as the primary oocyte completes the first phase of meiosis to form the secondary oocyte. This cell has 23 chromosomes, the N number of chromosomes. (The other cell formed in Stage I divides to the first and second polar bodies, which degenerate.) The secondary oocyte begins the second stage of meiosis. If a sperm cell is present, it completes the stage and forms the mature egg cell (ovum), having 23 chromosomes, as well as the third polar body. The egg cell can now unite with the sperm cell in fertilization.*

begin growing and maturing at a rate of one per month. Maturations occur in alternating ovaries. All maturations occur from the primary oocytes already present in the ovary. Another hormone from the pituitary called **luteinizing hormone (LH)** stimulates the developing follicle to produce estrogens.

The maturation of a follicle involves the development of follicle cells and changes in the primary oocyte. The primary oocyte having 46 chromosomes completes Stage I of **meiosis** to form two cells, each with 23 chromosomes (Figure 23.7). One of these cells, the **secondary oocyte**, will develop to the mature egg cell (ovum) with 23 chromosomes; the remaining cell forms a functionless **polar body** that divides to two cells, both of which degenerate. The secondary oocyte enters Stage II of meiosis and stops at metaphase. The luteinizing hormone stimulates this development. At this point, the secondary oocyte is known as the **egg cell (ovum)**.

The maturation of the follicle requires about 14 days, and the mature follicle is called the **Graafian follicle**. The egg cell exists within a cavity called the **antrum** and is surrounded by supporting cells called the **corona radiata** (Figure 23.8). After maturation, a surge of LH stimulates release of the egg cell from the follicle, a process called **ovulation**. A prominent bulge in the surface of the ovary signals that ovulation is about to occur.

The mature egg cell (ovum) is released into the pelvic cavity, where fluid currents from the beating fimbriae of the Fallopian tube immediately sweep the ovum into the Fallopian tubes for transport to the uterus. In the ovary, the residual follicle cells undergo structural

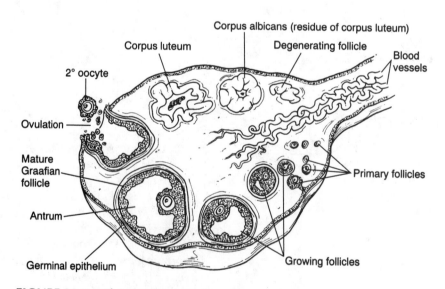

**FIGURE 23.8** *The progression of events occurring monthly in the ovary. The development of the Graafian follicle is illustrated, beginning with the primary follicle and continuing with the release of the secondary oocyte, which becomes the mature egg cell (ovum). The corpus luteum forms from the remaining follicle.*

and biochemical changes to form the enlarged glandular yellow body, or **corpus luteum**. (The luteinizing hormone (LH) was once believed to regulate this change, but that involvement is now uncertain.) The corpus luteum remains active for about 12 days and produces large quantities of progesterone and estrogens, then begins to degenerate if fertilization has not occured. If fertilization occurs, the corpus luteum continues to secrete hormones.

## Fertilization

Fertilization is the union of gametes during sexual reproduction. During **fertilization**, a sperm cell containing 23 chromosomes unites with an egg cell containing 23 chromosomes to form a fertilized egg cell having 46 chromosomes. In order for the sperm cell to unite with the egg cell, the sperm cells must be capacitated. This means that the sperm cell membrane must become fragile in order to allow enzymes in its acrosome to be released. The capacitation happens as the sperm cells swim through the mucus of the female reproductive organs and deplete their cholesterol. As the cholesterol is depleted, the enzymes from the acrosome are released to permit union with the egg cell. The fertilized egg cell is called the **zygote**. The sex of the zygote is determined by which chromosomes are present. If two X chromosomes are present, the individual will be female, while if an X and a Y chromosome are present, then the individual will be a male.

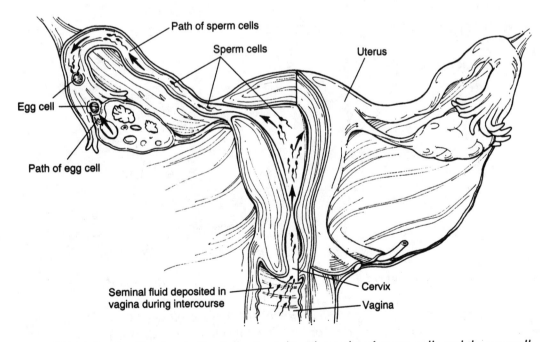

**FIGURE 23.9**   *Fertilization in the Fallopian tube. The paths of sperm cells and the egg cell are shown before union occurs in the tube.*

Fertilization usually occurs in the Fallopian tubes after the sperm cells deposited in the vagina have swum through the cervix and, uterus, and into the tubes (Figure 23.9). Enzymes from the sperms acrosomes digest the outer cell layers of the egg cell, and one sperm cell penetrates. Subsequent changes in the egg cell membrane make other sperm cell penetrations impossible. Now, the egg cell continues its changes in meiosis and completes Stage II, forming a mature egg cell with 23 chromosomes in its nucleus, and a third polar body (which degenerates). The sperm cell nucleus having 23 chromosomes joins with the egg cell nucleus, and the zygote nucleus is formed with 46 chromosomes.

After fertilization, the zygote undergoes mitosis to form two cells, then each of the two cells duplicates to form a four-cell cluster called the **morula**. Then eight cells form, and so on. Soon, there develops a hollow ball of cells called the **blastocyst**. The blastocyst moves down the Fallopian tube and reaches the uterus about three days after ovulation. In the uterine cavity, the blastocyst obtains nutrients by absorbing the secretions of the uterine glands. Within a few days, it contacts the wall of the endometrium, erodes the epithelium, and buries itself in the endometrium (Figure 23.10). This process, known as **implantation**, is usually complete about 11 days after fertilization.

After fertilization has taken place, the corpus luteum in the ovary continues to produce progesterone, which prevents the release of the

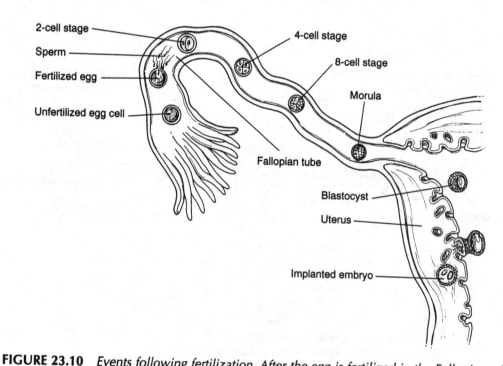

**FIGURE 23.10**   *Events following fertilization. After the egg is fertilized in the Fallopian tube, a number of cell divisions result in the morula and blastocyst, which implants in the uterine wall. The embryo eventually develops from the blastocyst.*

endometrial lining. Projections of the blastocyst called **chorionic villi** unite with the uterine tissues and develop into an organ called the **placenta**. The placenta becomes an endocrine organ and produces estrogen and progesterone to maintain the pregnancy. The placenta also provides a medium for the transfer of dissolved gases, nutrients, and waste products between the embryonic and maternal bloodstreams.

Soon after implantation has occurred, the hormone **human chorionic gonadotropin (HCG)** appears in the mother's bloodstream, having been produced by embryonic cells. The presence of HCG in blood or urine provides a reliable indication that fertilization and pregnancy have taken place. In the presence of HCG, the corpus luteum does not degenerate. Rather, it remains functional and continues to produce progesterone and estrogen. After about three months, however, the corpus luteum undergoes degeneration. At that time, the placenta is actively secreting estrogen and progesterone.

## EMBRYONIC AND FETAL DEVELOPMENT

The first two months of development are considered the embryonic period, and the developing individual is referred to as an **embryo**.

After it has implanted in the endometrium, the blastocyst develops an inner cell mass, which then differentiates into three germ layers in the process of **gastrulation**. The structure containing the three germ layers is called the **gastrula**. The layers of the gastrula are the ectoderm, mesoderm, and endoderm.

The germ layers give rise to all the organ systems of the individual (Table 23.3). The **ectoderm** is the outer germ layer. It will develop into the nervous system and the epidermis, as well as parts of the eye and ear. The middle germ layer, the **mesoderm**, will give rise to the

**TABLE 23.3**  *Embryonic Germ Layers and Structures Derived*

| Germ Layer | Tissue or Organ Derived |
| --- | --- |
| Ectoderm | Nervous system<br>Outer layer of skin (epidermis) and associated structures (nails, hair, etc.)<br>Pituitary gland |
| Mesoderm | Skeleton (bone and cartilage)<br>Muscles<br>Circulatory system<br>Excretory system<br>Reproductive system<br>Inner layer of skin (dermis)<br>Outer layers of digestive tube |
| Endoderm | Lining of digestive tube and associated structures<br>Respiratory system |

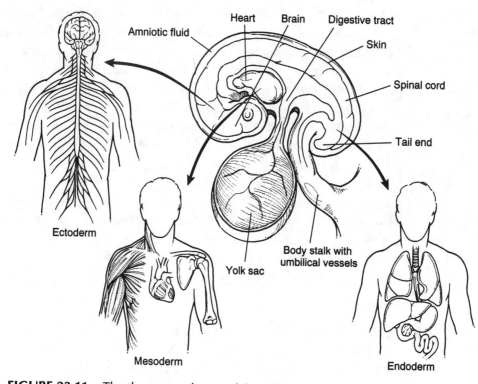

Heart   Brain   Digestive tract

Amniotic fluid

Skin

Spinal cord

Tail end

Ectoderm

Body stalk with
umbilical vessels

Yolk sac

Mesoderm

Endoderm

**FIGURE 23.11**   *The three germ layers of the embryo and the organs they give rise to.*

skeletal muscles, most smooth muscles, and the cardiac muscles. It will also give rise to the blood, dermis, bone, certain epithelial tissues, and parts of the eye and ear. The inner germ layer, the **endoderm**, develops into the gastrointestinal system, structures of the urinary and respiratory tracts, and many of the glands (Figure 23.11).

During its development, the embryo is surrounded by a number of membranes. The sac that entirely encloses the embryo is called the **amnion**, and the enclosing membrane is the **amnionic membrane**. Outside the amnion, the area is called the **chorion** enclosed by the **chorionic membrane**, which is the basis for the chorionic villi. The chorionic villi and allantoic membranes later fuse to form the sac that encloses the fetus. Between the amnion and chorion is a highly vascularized membrane called the **allantoic membrane**, which forms the basis for the umbilical cord. For the first six weeks, the embryo also has a **yolk sac** with an enclosing **yolk sac membrane** (Figure 23.12).

The **umbilical cord** forms from the membranes of the amnion and the endometrium. It is a long ropelike structure containing two umbilical arteries and one umbilical vein. Extending from the placenta to the embryo, and, later, to the fetus, the umbilical cord is the intermediary organ for gas, nutrient, and waste product exchanges between the mother and the child.

The final seven months of development are considered the fetal period, and the developing individual is called a **fetus**. The organ systems have been established in the embryonic period, and they

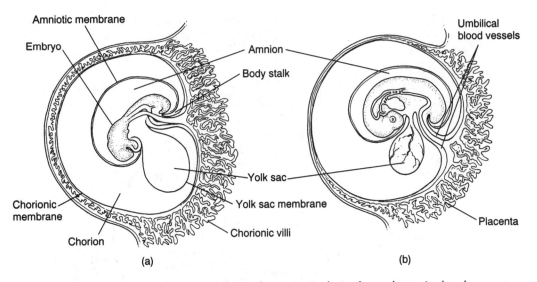

**FIGURE 23.12**   *The membranes of the embryo. (a) Early in the embryonic development, the amnionic, chorionic, allantoic, and yolk sac membranes have formed. (b) After five weeks, the membranes have taken a different structure. The placenta forms from the chorionic membrane and chorionic villi of the endometrial tissues. The allantois contributes to the formation of the umbilical cord and umbilical blood vessels.*

develop during the fetal period. Also, the human appearance becomes apparent. The amnion encloses the fetus during its development.

During the first month of fetal development, which is the **third month** of pregnancy, the body systems have appeared and are developing (Figure 23.13). Now, the growth in body length accelerates and growth of the head slows. The eyes develop, and the arm and leg buds are present. The heartbeat begins, and ossification takes place in most bones. The external reproductive organs become visible.

During the **fourth month**, there is rapid body growth. The limbs become distinct, and facial features become apparent. The major blood vessels are formed, the arms and legs lengthen, and the skeleton continues to ossify. Nails and eyelashes can be observed.

The skeletal muscles are active and movement is first felt during the **fifth month**. The skin becomes covered with fine, downy hair called **lanugo**. The arms and legs achieve their final proportions relative to the body.

During the **sixth month**, there is substantial weight gain. The head of the fetus becomes more proportional to the remainder of the body than before. The skin develops blood vessels and appears wrinkled.

At the **seventh month** there is more fat deposit, and the skin smoothes out. The eyes open. During the **eighth month**, the baby is "full term" and has a good chance of surviving outside the body. During the **ninth month**, the deposit of subcutaneous fat gives the skin a smooth appearance and the baby is ready for birth.

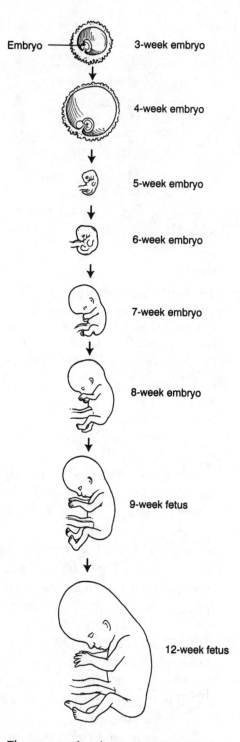

Embryo — 3-week embryo

4-week embryo

5-week embryo

6-week embryo

7-week embryo

8-week embryo

9-week fetus

12-week fetus

**FIGURE 23.13** *The stages of embryonic and fetal development from the 3-week to the 12-week stage. By the 12th week (three months), all the body systems have formed, and development occurs in the succeeding months.*

The **birth process** occurs 266 days (9 calendar months) after fertilization. It involves a number of processes and chemical factors as well as numerous feedback mechanisms (Figure 23.14). The term given to the birth process is called **parturition**. In the early stages of birth, the secretion of progesterone from the placenta decreases. This decrease removes the inhibitory effect of progesterone on the endometrium. Changes in the level of progesterone and estrogen also stimulate the synthesis of **prostaglandins**. These hormones play a role in the birth process by stimulating the smooth uterine muscles and dilating the cervix to open the cervical os. In response to the uterine muscle contractions, the posterior pituitary gland releases the hormone **oxytocin**.

Vigorous uterine contractions are stimulated by oxytocin. The amnion breaks, releasing the amnionic fluid. As the fetal head passes through the cervix, the stretching of cervical tissues stimulates the release of additional quantities of oxytocin. The stretching also initiates waves of contractions over the body of the uterus. Uterine contractions induce contractions of the abdominal wall via reflexes through the spinal cord.

Uterine and abdominal contractions force the baby through the cervix and vagina to the exterior. The baby's head normally presents first (**vertex presentation**), but in about 5 percent of cases, the buttocks presents first, a condition called **breech birth**. Some minutes after the baby's birth, the placenta (afterbirth) is expelled from the reproductive tract.

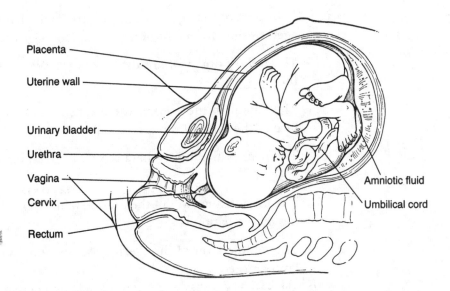

**FIGURE 23.14**   *The fully developed child in place immediately before birth.*

# REVIEW QUESTIONS

**PART A—Completion:** Add the word or words that correctly complete each of the following statements.

1. The major organs of the female reproductive system are enclosed within an extensive mesentery called the _____.

2. The organs where egg cells are produced in the female reproductive tract are the _____.

3. Hormones produced by cells of the female reproductive tract are the _____.

4. The ligaments that support the ovary are the ovarian ligament and the _____.

5. Immature egg cells contained within the ovary are known as _____.

6. The expanded funnel of the Fallopian tube is the _____.

7. The projections that extend from the Fallopian tube into the pelvic cavity are _____.

8. Fertilization of the egg cell by the sperm cell takes place in the _____.

9. Transport of the egg cell through the Fallopian tube is encouraged by the movement of hairlike appendages called _____.

10. The hollow organ where the embryo develops in the female is the _____.

11. The Fallopian tubes enter the uterus at the bulging upper surface of the body of the uterus called the _____.

12. The neck of the uterus that projects into the vagina is called the _____.

13. The inner tissue layer of the uterine wall that is sloughed off during menstruation is the _____.

14. The middle layer of the uterine wall composed of thick muscle layers is the _____.

15. The muscular tube extending between the uterus and the external genitalia is the _____.

16. The fold of epithelium that blocks the entrance to the vagina prior to sexual activity is the _____.

17. The muscular tube where sperm cells are deposited during sexual intercourse is the _____.

18. The vulva is an alternate term for the _____.

19. The small mass of erectile tissue that enlarges during sexual arousal is the _____.

20. During sexual intercourse, lubricants are produced by the gland known as the greater vestibular gland, also known as _____.

21. Two elongated folds of skin that encircle and partially conceal the labia minora and structures of the vestibule are the _____.

22. After birth, the newborn gains nourishment from milk secreted from the _____.

23. The production of milk is called _____.

24. The secretion of milk is mediated by the hormone _____.

25. The physiological and structural changes in the female reproductive tract that occur in response to changes of ovarian hormones is known as the _____.

26. The length of time for a complete menstrual cycle to occur is about _____.

27. The stage of the menstrual cycle during which follicles develop in the ovaries and the endometrium is repaired is the _____.

28. A surge of estrogen and progesterone accompany the release of the egg cell from the follicle, a process known as _____.

29. The process by which egg cells are formed in the ovary is known as _____.

30. The primitive egg cells that will pass through oogenesis to form mature egg cells are called _____.

31. Layers of cell that surround the primary oocytes form a structure called the _____.

32. Within the ovary, the follicle grows and matures under the influence of the hormone _____.

33. The mature egg cell is also known as an _____.

34. The number of chromosomes present in the mature egg cell is _____.

35. After the egg cell is released from the follicle, the follicle cells form a structure called the _____.

36. When an egg cell unites with a sperm cell, the resulting fertilized egg cell is called the _____.

37. The fertilized egg cell forms two cells, then four cells, then eight cells, by the process of _____.

38. The hollow ball of cells resulting from multiple divisions of the fertilized egg cell is the _____.

39. The process in which the blastocyst contacts the wall of the endometrium and buries itself therein is called the _____.

40. Within the uterus, the organ that produces hormones and provides a medium for the transfer of nutrients, gases, and waste products between embryonic and maternal blood streams is the _____.

41. A reliable indication that fertilization has taken place is the presence in the bloodstream of the hormone _____.

42. The main hormone produced by the corpus luteum prevents the contractions of the uterus and is known as _____.

43. During the first two months of life, the developing individual is referred to as an _____.

44. The germ layer that develops into the skeletal muscles, cardiac muscles, blood, bone, and other organs is the _____.

45. The germ layer that develops into the gastrointestinal system, many of the glands, and structures of the urinary and respiratory tract is the _____.

46. The sac that entirely encloses the embryo during development is the _____.

47. The long ropelike structure extending from the placenta to the embryo during development is known as the _____.

48. During its final seven months, the developing individual is known as a _____.

49. The heartbeat begins, ossification takes place in bones, and the body systems are developing during pregnancy in month number _____.

50. During the early stages of birth, contractions of the uterus are stimulated by a hormone released from the pituitary gland, known as _____.

**PART B—Multiple Choice:** Circle the letter of the item that correctly completes each of the following statements.

1. The female reproductive organs are located in the pelvic cavity within a mesentery known as the
   (A) small ligament
   (B) ovarian ligament
   (C) broad ligament
   (D) narrow ligament

2. The process of ovulation occurs every
   (A) 2–3 days
   (B) 14–15 days
   (C) 28–30 days
   (D) 8–9 months

3. The ampulla, fimbriae, and infundibulum are anatomical features of the
   (A) uterus
   (B) cervix
   (C) vagina
   (D) Fallopian tubes

4. When egg cells are released from follicles in the ovaries, they are swept into the Fallopian tubes
   (A) by the flagella they possess
   (B) by currents established by suspensory cells
   (C) by the ciliary action of fimbriae
   (D) by the sucking action of the cervix

5. All the following are associated with the uterus except
   (A) it is a hollow organ normally about the size and shape of a pear
   (B) it expands considerably during pregnancy
   (C) it is also called the womb
   (D) it is composed of one tissue layer called the myometrium

6. Development of the embryo takes place in the tissue layer of the uterus known as the
   (A) perimetrium
   (B) endometrium
   (C) myometrium
   (D) neurometrium

7. In order to reach egg cells present in the Fallopian tubes, sperm cells must swim through the
   (A) vagina, cervix, and uterus
   (B) vagina, ovaries, and vulva
   (C) vulva, vagina, and ovaries
   (D) uterus, cervix, and ovaries

8. The hymen is a fold of epithelium that generally blocks the entrance to the vagina
   (A) before the first act of sexual intercourse
   (B) after the onset of menopause
   (C) after fertilization of the egg cell has taken place
   (D) only until the birth process has taken place

9. The point where the cervix projects into the vagina has a hollow recess known as the
   (A) infundibulum
   (B) fornix
   (C) labia minora
   (D) areola

10. One of the functions of Bartholin's gland is to
    (A) nourish the developing embryo
    (B) provide hormones to stimulate follicle production
    (C) provide progesterone to maintain pregnancy
    (D) lubricate the vagina during sexual intercourse

11. Lactation is the process of
    (A) implantation in the uterine lining
    (B) production of egg cells within the ovary
    (C) production of milk by the mammary glands
    (D) birth

12. During the first few days of the menstrual cycle
    (A) ovulation takes place
    (B) the endometrium is sloughed off
    (C) the endometrial lining builds up with tissue
    (D) fertilization of the egg cell takes place

13. During days 6–14 of the menstrual cycle
    (A) fertilization occurs
    (B) the endometrium is released from the uterus
    (C) follicles develop in the ovaries
    (D) high levels of progesterone are produced by the corpus luteum

14. During days 15–28 of the menstrual cycle
    (A) the level of progesterone is lowest
    (B) the uterine lining is removed from the body
    (C) ovulation takes place
    (D) the corpus luteum secretes progesterone

15. In the process of meiosis
    (A) an egg cell with 23 chromosomes is produced
    (B) sperm cells are deposited in the vagina
    (C) menstruation takes place
    (D) the developing embryo implants in the uterine lining

16. The corpus luteum forms from the
    (A) residual cells of the follicle
    (B) ectoderm and endoderm cells
    (C) ciliated cells of the Fallopian tube
    (D) greater vestibular gland

17. To form the zygote,
    (A) the hormone HCG must be present
    (B) the corpus luteum must secrete progesterone
    (C) the nuclei of sperm and egg cells must unite
    (D) the umbilical cord must form

18. The endometrial lining of the uterus remains in place during pregnancy in large measure because of the
    (A) presence of the hormone FSH
    (B) development of the umbilical cord
    (C) presence of the hormone progesterone
    (D) maturation of the Graafian follicle

19. The endoderm is the germ layer that will develop into the
    (A) skeletal muscles and most smooth muscles
    (B) cardiac muscle and blood
    (C) nervous system and epidermis
    (D) gastrointestinal and respiratory tracts

20. All the following apply to the umbilical cord except
    (A) it a long, ropelike structure
    (B) it contains one umbilical artery and two umbilical veins
    (C) it extends from the placenta to the embryo
    (D) it is the intermediary for gas, nutrient, and waste product exchanges

21. During pregnancy, the corpus luteum remains functional
    (A) throughout the pregnancy
    (B) for about three days
    (C) for about three months
    (D) for six months

22. The sac that entirely encloses the embryo is the
    (A) placenta
    (B) lanugo
    (C) amnion
    (D) mesoderm

23. The developing individual is considered a fetus
    (A) throughout the pregnancy
    (B) only during the last two weeks
    (C) for the first two months
    (D) for the final seven months

24. The birth process is marked by the
    (A) decrease of progesterone and increase of oxytocin
    (B) increase of progesterone and decrease of oxytocin
    (C) increase of FSH and decrease of LH
    (D) increase of LH and decrease of FSH

25. A breach birth is one in which
    (A) The head of the baby presents first
    (B) the buttocks of the baby presents first
    (C) birth takes place by abdominal surgery
    (D) the feet of the baby present first

**PART C—True/False:** **For each of the following statements, mark the letter "T" next to the statement if it is true. If the statement is false, change the <u>underlined</u> word to make the statement true.**

1. The Fallopian tubes run along the superior border of the broad ligament and enter into the <u>pectoral</u> cavity.

2. The ovaries produce egg cells as well as female sex hormones known as <u>androgens</u>.

3. Egg cells mature within the <u>corpus luteum</u> and are released in the process of ovulation.

4. In the female, ovulation occurs once every <u>14</u> days.

5. The fimbriae, ampulla, and isthmus are structures found in the <u>Fallopian tubes</u>.

6. Hairlike appendages called <u>flagella</u> beat and help move egg cells through the Fallopian tubes.

7. Egg cells are fertilized by sperm cells in the <u>uterus</u>.

8. The uterus, also called the <u>womb</u>, is the place where the embryo and fetus are nourished.

9. The neck of the uterus, where the uterine cavity opens into the vagina, is known as the <u>fundus</u>.

10. If fertilization of the egg cell does not occur, the <u>myometrium</u> of the uterus is removed in the process of menstruation.

11. In order for sperm cells to reach the egg cells in the Fallopian tubes, they must swim through the <u>uterus</u>.

12. The muscular tube where sperm cells are deposited is the birth canal—also known as the <u>cervix</u>.

13. Before sexual intercourse begins, the entrance to the vagina is usually blocked by a fold of epithelium known as the <u>infundibulum</u>.

14. The region of the vulva where the vagina opens is known as the <u>isthmus</u>.

15. The labia minora and the <u>labia majora</u> are both parts of the female external genitalia.

16. The newborn gains nourishment from the milk secreted by the mother's <u>pituitary</u> glands.

17. The skin area surrounding each nipple of the breast having a reddish-brown coloration is called the <u>areola</u>.

18. The first few days of the menstrual cycle are those in which the endometrium detaches from the wall of the <u>cervix</u>.

19. During the last two weeks of the menstrual cycle, the corpus luteum produces large amounts of the hormone <u>testosterone</u>.

20. The primitive egg cells that will form mature egg cells are known as <u>oogonia</u>.

21. Development of the follicle within the ovary is stimulated by the hormone <u>lactogenic</u> hormone.

22. The process in which a primary oocyte undergoes a reduction in its chromosome numbers to form a mature egg cell is called <u>mitosis</u>.

23. The mature egg cell is released into the pelvic cavity and swept into the <u>vagina</u> for transport to the uterus.

24. Implantation occurs when the <u>fetus</u> implants itself in the wall of the uterus.

25. The presence of HCG in the bloodstream of the mother is a reliable indication that <u>menstruation</u> has taken place.

**Answers**

## PART A—Completion

1. broad ligament
2. ovaries
3. estrogens
4. suspensory ligament
5. oocytes
6. infundibulum
7. fimbriae
8. Fallopian tube
9. cilia
10. uterus
11. fundus
12. cervix
13. endometrium
14. myometrium
15. vagina
16. hymen
17. vagina
18. external genitalia
19. clitoris
20. Bartholin's gland
21. labia majora
22. mammary glands
23. lactation
24. prolactin
25. menstrual cycle
26. 28 days
27. proliferative stage
28. ovulation
29. oogenesis
30. oogonia
31. follicle
32. follicle stimulating hormone
33. ovum
34. 23 chromosomes
35. corpus luteum
36. zygote
37. mitosis
38. blastocyst
39. implantation
40. placenta
41. human chorionic gonadotropin
42. progesterone
43. embryo
44. mesoderm
45. endoderm
46. amnion
47. umbilical cord
48. fetus
49. three
50. oxytocin

## Part B—Multiple Choice

| | | | | |
|---|---|---|---|---|
| 1. C | 6. B | 11. C | 16. A | 21. C |
| 2. C | 7. A | 12. B | 17. C | 22. C |
| 3. D | 8. A | 13. C | 18. C | 23. D |
| 4. C | 9. B | 14. D | 19. D | 24. A |
| 5. D | 10. D | 15. A | 20. B | 25. B |

## Part C—True/False

1. pelvic
2. estrogens
3. follicle
4. 28
5. true
6. cilia
7. Fallopian tube
8. true
9. cervix
10. endometrium
11. true
12. vagina
13. hymen
14. vestibule
15. true
16. mammary
17. true
18. uterus
19. progesterone
20. true
21. follicle stimulating
22. meiosis
23. Fallopian tube
24. blastocyst
25. fertilization

# GLOSSARY

**abdomen** The area between the diaphragm and the pelvis.

**abduct** To move away from the midline of the body.

**absorption** The process by which the products of digestion pass through the intestinal wall into capillaries or lymphatic vessels.

**accommodation** A change in the curvature of the lens of the eye to adjust for various distances.

**acetabulum** The cuplike cavity on the lateral surface of the hipbone; it receives the femur.

**acetylcholine** A chemical transmitter substance released at the ends of certain nerve cells.

**Achilles tendon** The tendon that attaches the soleus, gastrocnemius, and plantaris muscles to the calcaneus, a tarsal.

**acidosis** The state of abnormally high acidity in the extracellular fluid and blood.

**acini** Groups of cells that secrete digestive enzymes in the pancreas; also, cell groups in other exocrine glands.

**actin** One of the essential contractile proteins of muscle cells.

**action potential** A depolarization event conducted along the membrane of a muscle cell or a nerve fiber; the nerve impulse.

**adduct** To move a body part toward the midline of the body.

**adenohypophysis** The anterior pituitary gland.

**adenoids** The pharyngeal tonsil.

**adenosine triphosphate (ATP)** The organic molecule that stores and releases chemical energy for use in body cells.

**adrenal glands** Hormone-producing glands superior to the kidneys; they produce glucocorticoids and mineralocorticoids from different regions; also called suprarenal glands.

**adrenaline** *See* Epinephrine.

**adrenergic fibers** Nerve fibers that release norepinephrine at the synapse.

**adrenocorticotropic hormone (ACTH)** The hormone produced by the anterior lobe of the pituitary gland that influences the production and secretion of certain hormones of the adrenal cortex.

**afferent neuron** A nerve cell that carries impulses toward the central nervous system (CNS).

**agonist** A muscle that is opposed in action by another muscle, called the antagonist.

**aldosterone** A hormone produced by the adrenal cortex; it regulates sodium ion reabsorption in the kidney tubules and, thereby, influences the amount of water in the blood and the blood pressure.

**alkalosis** A state of abnormally high alkalinity in the extracellular fluid or blood.

**allantois** The embryonic membrane where the blood vessels of the umbilical cord develop.

**alpha cells** Pancreatic cells that secrete glucagon.

**alveolus** A microscopic air sac of the lung.

**amnion** The fetal membrane forming a fluid-filled sac around the embryo.

**amphiarthrosis** A slightly movable joint.

**ampulla** A saclike dilation of a tube.

**androgen** A hormone that controls male secondary sex characteristics.

**anemia** The reduced oxygen-carrying ability of blood, resulting from abnormal hemoglobin molecules or too few erythrocytes.

**angiotensin** A protein associated with blood pressure regulation.

**anorexia** The loss of desire for food.

**anoxia** A deficiency of oxygen.

**antagonist** A muscle that reverses or opposes the action of another muscle.

**anterior** The front of an organism or organ; the ventral surface.

**antidiuretic hormone (ADH)** The hormone produced by the hypothalamus and released from the posterior pituitary; it stimulates the kidneys to reabsorb water; it reduces the urine volume.

**aorta** The major artery of the systemic circulation arising from the left ventricle of the heart.

**apnea** Cessation of breathing.

**apocrine gland** A type of merocrine gland; it produces a secretion released through the cell membrane by exocytosis.

**aponeurosis** A fibrous or membranous sheet connecting a muscle and the part it moves.

**appendicular skeleton** The bones of the appendages and appendage girdles attached to the axial skeleton.

**aqueous humor** The watery fluid in both chambers of the eye.

**arachnoid** The weblike middle layer of the three meninges.

**arteries** Blood vessels transporting blood away from the heart.

**arteriole** A small artery.

**arthritis** Joint inflammation.

**articulation** A joint where two bones meet.

**atherosclerosis** A condition resulting from lipid deposits on the artery walls; the early stage of arteriosclerosis.

**atlas** The first cervical vertebra.

**atria** Paired chambers that receive blood returning to the heart.

**atrioventricular node (AV node)** The specialized mass of conducting cells located in the interatrial septum deep to the endocardium of the right atrium.

**auditory ossicles** The three tiny bones serving as transmitters of vibrations and located within the middle ear the malleus, incus, and stapes.

**auditory tube** The tube connecting the middle ear and pharynx; also called the Eustachian tube.

**autonomic nervous system** A division of the peripheral nervous system that involuntarily innervates glands and cardiac and smooth muscles.

**axial skeleton** The portion of the skeleton forming the central axis of the body; it includes the skull, vertebral column, and thorax.

**axilla** The area of the armpit.

**axis** The second cervical vertebra.

**axon** A neuron process that carries impulses away from the nerve cell body.

**B-lymphocytes** Cells functioning in antibody-mediated (humoral) immunity; also called B-cells.

**baroreceptor** A receptor stimulated by pressure changes.

**basal metabolic rate (BMR)** The rate at which energy is expended by the body per unit time under controlled conditions.

**basement membrane** The extracellular material between the epithelium and underlying connective tissue.

**basophil** The white blood cells whose granules stain deep blue with basic dyes.

**beta cells** Cells of the islets of Langerhans, the insulin-secreting cells of the pancreas.

**bile** The greenish-yellow or brownish fluid produced in the liver, stored in the gall bladder, and released into the small intestine.

**bilirubin** A bile pigment derived from hemoglobin.

**bipolar neuron** A neuron with an axon and a dendrite extending from opposite sides of the cell body.

**blastocyst** The stage of early embryonic development following the morula stage, which consists of a ball of cells.

**Bowman's capsule** A double-walled cup at the end of a renal tubule; it encloses a glomerulus.

**bradycardia** A heart rate below normal.

**brain stem** The brain region encompassing the midbrain, pons, and medulla.

**bronchus** One of the two large branches of the trachea leading to the lungs.

**bulbourethral gland** Reproductive gland inferior to the prostate gland; Cowper's gland.

**bursa** A fibrous sac lined with synovial membrane and containing synovial fluid; it is found between bones and muscle tendons, where it decreases friction.

**calcitonin** The hormone released by the thyroid to promote a decrease in calcium levels of the blood.

**calyx** A cuplike division of the pelvis; the inner open region of the kidney at its center.

**capillaries** The smallest blood vessels and the sites of gas and nutrient exchanges between the blood and tissue cells.

**carbonic anhydrase** The enzyme catalyzing the combination of carbon dioxide with water to form carbonic acid.

**carboxyhemoglobin** Carbon dioxide molecules bound to a hemoglobin molecule.

**cardiac cycle** The sequence of events encompassing one complete contraction and relaxation of the atria and ventricles of the heart.

**cardiac muscle** Specialized muscle of the heart consisting of intertwined filaments with intercalated disks.

**carotid body** A receptor in the common carotid artery sensitive to changing oxygen, carbon dioxide, and pH levels of the blood.

**cartilage** White, semiopaque connective tissue.

**castration** Removal of the testes.

**caudal** Referring to the tail portion of a structure.

**cecum** The blind-end pouch at the beginning of the large intestine.

**cell-mediated immunity** The immunity conferred by activated T-lymphocytes, which lyse microbial-infected cells or cells of foreign grafts and release chemicals to regulate the immune response.

**central nervous system (CNS)** The brain and the spinal cord.

**cerebellum** The brain region that coordinates skeletal muscle activity.

**cerebral aqueduct** The slender cavity of the midbrain connecting the third and fourth ventricles; also called the aqueduct of Sylvius.

**cerebral cortex** The outer gray matter region of the cerebral hemispheres.

**cerebrospinal fluid** The plasmalike fluid that fills the cavities of the central nervous system (CNS) and surrounds the CNS externally.

**cerebrum** The cerebral hemispheres and the structures of the diencephalon.

**cervix** The inferior necklike portion of the uterus.

**chemoreceptor** Receptors sensitive to various chemicals dissolved in a solution.

**cholecystokinin** An intestinal hormone that stimulates gall bladder contraction and pancreatic juice release.

**cholesterol** The steroid found in animal fats.

**cholinergic fibers** Nerve endings that release acetylcholine upon stimulation.

**chondroblast** A young cell that forms cartilage.

**chondrocyte** A mature cell that forms cartilage.

**chorion** The outermost fetal membrane; it helps form the placenta.

**chylomicrons** Accumulations of triglycerides and cholesterol that leave the intestinal lumen and enter the lacteal vessels, the milky-white vessels of the lymphatic system.

**chyme** The semifluid mass consisting of partially digested food and gastric juice.

**circumduction** Movement of a body part such as an arm or leg to outline a cone in space.

**cirrhosis** Liver disorder in which the functional cells are replaced by connective tissue.

**clitoris** The erectile organ of the female that is homologous to the male penis.

**coccyx** Fused vertebrae at the base of the vertebral column.

**cochlea** The snail-shaped chamber of the bony labyrinth housing the receptor for hearing.

**coenzyme** The nonprotein substance associated with an enzyme.

**colloidal osmotic pressure** The pressure exerted by blood plasma and body fluids; it results from the presence of proteins such as albumin in the blood; it encourages fluid to enter or leave the bloodstream.

**colon** A subdivision of the large intestine including ascending, transverse, descending, and sigmoid portions.

**colostrum** The first milk secreted from the mammary glands by a nursing mother.

**common bile duct** The tube formed from the hepatic and cystic ducts and leading to the duodenum.

**concentration gradient** The difference in the concentration of a particular substance between two different areas.

**condyle** A rounded projection at the end of a bone that articulates with another bone.

**cones** Photosensitive cells in the retina of the eye; they provide color vision.

**conjunctiva** The thin, mucous membrane lining the eyelids and covering the anterior surface of the eye.

**contralateral** Referring to a similar part on the opposite side of the body.

**cornea** The transparent anterior portion of the eyeball.

**corpus luteum** The ovarian structure resulting from the follicle after release of the ovum.

**cortex** The outer layer of an organ.

**corticosteroids** Steroid hormones released by the adrenal cortex.

**cortisol** A glucocorticoid produced by the adrenal cortex.

**cranial nerves** The 12 nerve pairs arising from the brain and extending to muscles and glands.

**creatine phosphate** A compound serving as an alternative energy source for muscle tissue.

**cutaneous** Pertaining to the skin.

**cyclic AMP** An intracellular second messenger that mediates hormonal effects in target cells.

**cystic duct** The duct transporting bile from the gall bladder to the common bile duct.

**cytokinesis** The division of cytoplasm occurring after the cell nucleus has divided.

**deamination** The removal of an amino group from an organic compound by hydrolysis, or oxidation.

**defecation** The elimination of the bowel contents.

**deglutition** The process of swallowing.

**dendrite** A branching neuron process that receives impulses and transmits them to the cell body.

**deoxyribonucleic acid (DNA)** A nucleic acid of living cells that carries the organism's hereditary information.

**depolarization** The loss of polarity in a nerve or muscle cell resulting in a wave of depolarization through the cell and the development of a nerve impulse.

**dermis** The layer of skin beneath the epidermis.

**diabetes insipidus** The disease characterized by passage of a large quantity of dilute urine plus intense thirst and dehydration; it is caused by the inadequate release of the antidiuretic hormone (ADH).

**diabetes mellitus** The disease caused by a deficiency of insulin, leading to the inability of the body cells to receive carbohydrates.

**diapedesis** The passage of cells through an intact vessel wall into the tissue.

**diaphragm (abdominal)** The muscle separating the thoracic cavity from the lower abdominopelvic cavity.

**diaphysis** The elongated shaft of a long bone.

**diarthrosis** A freely movable joint; also called a synovial joint.

**diastole** The period when the ventricles are relaxing.

**diencephalon** The part of the forebrain between the cerebral hemispheres and the midbrain including the thalamus, third ventricle, and hypothalamus.

**diffusion** The movement of molecules from a region of high concentration toward a region of low concentration.

**dilate** To expand or swell.

**distal** Away from the attached end of a limb or the origin of a structure.

**dorsal** Pertaining to the back; posterior.

**ductus deferens** *See* **vas deferens.**

**duodenum** The first 12 inches of the small intestine.

**dura mater** The outermost of the three meninges covering the brain and the spinal cord.

**ectoderm** The embryonic germ layer whose tissue forms the skin and nervous tissues.

**edema** An abnormal accumulation of fluid in extracellular body tissues.

**effector** An organ, gland, or muscle activated by nerve endings.

**efferent nerve fiber** A nerve fiber that carries impulses away from the central nervous system.

**elastin** A protein in the elastic fibers of connective tissue.

**electrocardiogram** A graphic record of the electrical activity of the heart.

**electrolyte** A chemical substance that exists as ions in water; it transmits an electrical charge.

**eleidin** A translucent skin substance related to the development of keratin, a protein that provides waterproofing and toughness to the skin tissues.

**embolism** Obstruction of a blood vessel by a floating clot in the blood.

**embryo** The developmental stage extending from gastrulation to the end of the eighth week after fertilization.

**endocarditis** Inflammation of the inner lining of the heart and heart values.

**endocrine glands** Ductless glands that empty their hormones directly into the blood.

**endocytosis** The chemical mechanism by which substances enter cells; examples are phagocytosis and pinocytosis.

**endoderm** The embryonic germ that later forms the lining of the digestive tube and its associated structures.

**endometrium** The mucous membrane lining the inner wall of the uterus.

**endosteum** The membrane lining the medullary cavity of the bone.

**endothelium** The single layer of simple squamous cells lining the walls of the heart, blood vessels, and lymphatic vessels.

**eosinophil** The white blood cell whose granules take up eosin (red) stain.

**epidermis** The superficial layer of the skin containing keratinized epithelium.

**epididymis** The portion of the male reproductive duct system that empties into the ductus (vas) deferens.

**epiglottis** The elastic cartilage at the back of the pharynx covering the glottis during swallowing.

**epinephrine** The hormone produced by the medulla of the adrenal gland; also called adrenaline.

**epiphyseal plate** The plate of hyaline cartilage at the junction of the diaphysis and the epiphysis.

**epiphysis** The end of a long bone; continuous with the shaft.

**epithelial** A primary tissue that covers the body surface, lines its internal cavities, and forms glands.

**erector pili** Tiny, smooth muscles attached to hair follicles; they cause the hair to stand upright on activation.

**erythrocytes** Red blood cells.

**erythropoietin** The kidney hormone that stimulates red blood cell production.

**estrogens** Female sex hormones.

**Eustachian tube** The auditory tube extending from the nasopharynx to the middle ear.

**exocrine glands** Glands having ducts for delivery of secretions to a target site.

**exocytosis** The mechanism by which substances move from the cell interior to the extracellular space.

**extension** A movement in which the angle increases between two body parts such as upper and lower arms.

**Fallopian tubes** The uterine tubes of the female reproductive system that receive the egg cell after its release by the ovary and lead it to the uterus; the site of fertilization of the egg cell by the sperm cell.

**fascia** Layers of fibrous tissue covering and separating muscles.

**fertilization** The union of sperm and egg cell in the uterine tube.

**fetus** The human developmental stage extending from the ninth week after fertilization to the time of birth.

**fibrin** Fibrous insoluble protein formed during blood clotting.

**fibrinogen** A blood protein converted to fibrin during blood clotting.

**fibroblast** The young, active cell that produces fibers of the connective tissue.

**fibrocyte** A mature fibroblast.

**filtration** The passage of dissolved substances through a membrane or filter; it occurs in the kidney nephron.

**fimbriae** Fingerlike projections at the openings of the uterine tubes.

**fissure** A deep inward fold of the brain.

**flexion** A bending of the joint in which the angle decreases between two bones, such as upper and lower arms.

**follicle** An ovarian structure containing a developing egg surrounded by supporting and nourishing cells.

**follicle-stimulating hormone (FSH)** The anterior pituitary hormone that stimulates ovarian follicle production in females and sperm production in males.

**fontanel** A membrane-covered area where bones come together before fusing.

**foramen** A hole or opening in a bone.

**forebrain** The anterior portion of the brain where many of the cranial nerves emerge and which is covered by the frontal bone.

**fossa** A shallow depression in a bone.

**fovea** The pit in the retina where nerve fibers meet.

**frontal (coronal) plane** The longitudinal plane dividing the body into anterior and posterior parts.

**fundus** The base of an organ such as the stomach or uterus; the part distant from the opening.

**gall bladder** The sac beneath the right lobe of the liver used for storing bile.

**gametogenesis** Gamete formation in the reproductive system.

**ganglion** A collection of cell bodies of neurons outside the central nervous system (CNS).

**gastrin** The gastric hormone that stimulates hydrochloric acid (HCl) release in the stomach; HCl is necessary for the digestion of protein.

**gastrulation** The developmental process resulting in three primary germ layers; gastrulation occurs by the invagination of a region of the blastocyst and occurs after implantation in the uterus has taken place.

**gestation** The period of human pregnancy; about 280 days or nine months.

**glomerulus** The tuft of capillaries forming part of the renal corpuscle.

**glottis** The opening to the larynx.

**glucagon** A hormone formed by alpha cells of the islets of Langerhans in the pancreas; it raises the glucose level of blood.

**glucocorticoids** Adrenal cortex hormones that increase the glucose levels of the blood.

**glycogen** The major polysaccharide stored in animal cells, especially liver cells.

**glycogenesis** The chemical synthesis of glycogen from glucose.

**glycogenolysis** The chemical breakdown of glycogen into glucose.

**glycolysis** The multistep enzyme-catalyzed chemical conversion of glucose to the three-carbon organic acid, pyruvic acid—during metabolism.

**gomphosis** A joint where a bone or tooth fits into a socket.

**gonad** A reproductive organ.

**gray matter** The outer area of the CNS consisting of cell bodies and unmyelinated nerve fibers.

**gustation** Taste.

**gyrus** An outward fold at the surface of the cerebral cortex.

**hair follicle** The tissue sheath in the epidermis where a hair is formed by epithelial cells.

**haustra** Pouches and outgrowths of the colon.

**haversian system** The system of interconnecting cells and canals in the microscopic structure of compact bone; the osteon.

**heme** The iron-containing pigment that transports oxygen in hemoglobin.

**hemocytoblast** The bone marrow stem cell that gives rise to all formed elements of the blood.

**hemoglobin** The oxygen-transporting red pigment of erythrocytes.

**hemolysis** The process in which erythrocytes rupture as a result of a chemical or physical change.

**hemorrhage** The loss of blood through ruptured walls of blood vessels.

**hepatic portal system** The transport system in which the hepatic portal vein carries dissolved nutrients toward the liver for processing.

**hernia** An abnormal protrusion of an organ through a containing membrane.

**histamine** The derivative of histidine causing smooth muscle contraction and increased vascular permeability during an allergic response.

**holocrine glands** Glands that accumulate their secretions within their cells and discharge them upon cell rupture.

**homeostasis** The maintenance of a stable internal environment in the body.

**hormones** Steroid or protein molecules released to the blood to act as chemical messengers on target cells.

**human growth hormone (HGH)** The anterior pituitary hormone that stimulates body growth; also called somatotropin.

**hyaluronic acid** A polysaccharide that binds cell in a tissue.

**hydrostatic pressure** Pressure exerted by a system's fluid; blood pressure.

**hymen** The thin mucous membrane at the vaginal orifice.

**hypertension** An alternate expression for high blood pressure.

**hypothalamus** The brain region at the floor of the third ventricle of the brain; it produces the hormones oxytocin and antidiuretic hormone, which are stored in the posterior pituitary gland before release.

**hypoxia** The condition in which inadequate oxygen is available to tissues.

**ileum** The last part of the small intestine.

**immunoglobulin** A protein molecule released by plasma cells to bring about immunity; an alternative expression for antibody.

**inferior** Pertaining to a position near the tail end of the body; caudal.

**insertion** The movable part or attachment of a muscle.

**inspiration** The act of drawing air into the lungs.

**insulin** The hormone produced by beta cells of the pancreas and facilitating passage of glucose molecules into cells; a deficiency results in diabetes mellitus.

**integumentary system** The skin and its derivatives.

**intercellular** Between body cells.

**interstitial fluid** Fluid between the cells; also called intercellular fluid.

**intracellular** Within a cell.

**ion** An atom having a positive or negative charge.

**ionic bond** The chemical bond formed by the attraction of ions.

**ipsilateral** Situated on the same side.

**ischemia** A decrease in blood supply.

**jejunum** The section of the small intestine between the duodenum and the ileum.

**juxtaglomerular apparatus** The region of cells located close to the glomerulus and Bowman's capsule; blood fluid flows from the bloodstream into the wall of Bowman's capsule here.

**keratin** The water-insoluble protein found in the epidermis, hair, and nails permitting those structures to be hard and water-repellent.

**keratohyalin** A preliminary substance in the production of keratin.

**ketosis** An abnormal condition during which an excess of ketone bodies (ketone molecules) is produced.

**Krebs cycle** The aerobic metabolic pathway in which metabolites are oxidized and $CO_2$ is liberated; high-energy coenzymes are also formed for electron transport; it occurs within the mitochondria.

**labia** Folds of tissue at the female genitalia.

**lacrimal glands** Tear glands.

**lactation** Production and secretion of milk by the mammary glands.

**lacteals** Lymphatic capillaries of the small intestine where lipids are transported.

**lacuna** A depression or space occupied by cells.

**lanugo** Fine hair covering the fetus.

**larynx** A cartilaginous organ between the trachea and pharynx; it contains the vocal cords; the voice box.

**lateral** The region away from the midline of the body.

**leukocytes** White blood cells.

**ligament** A band of fibrous tissue that connects bones.

**limbic system** The region of the brain cortex involved in emotional response.

**lumbar** The back portion between the thorax and the pelvis.

**lumen** A cavity inside a blood vessel or hollow organ.

**luteinizing hormone (LH)** The anterior pituitary hormone that assists maturation of cells in the ovary and triggers ovulation in females; it stimulates the interstitial cells to produce testosterone in males.

**lymph** The protein-rich fluid arising from the interstitial fluid and transported by lymphatic vessels back to circulation.

**lymph node** A lymphatic organ that filters lymph; the location of lymphocytes of the immune system.

**lymphatic system** The system consisting of lymphatic vessels, lymph nodes, and other lymphoid organs and tissues; it drains excess fluid from the extracellular space and is associated with the immune system.

**lymphocyte** An agranular white blood cell maturing and functioning in the lymphoid organs of the body; it functions in the immune system as a B-lymphocyte or T-lymphocyte.

**macrophage** The protective cell type found in connective tissue, lymphatic tissue, and certain body organs; it phagocytizes bacteria and other foreign debris.

**mammary glands** Milk-producing glands of the breasts.

**matrix** The specialized extracellular substance secreted by connective tissue cells.

**meatus** The external opening of a canal or system.

**medial** The region toward the midline of the body.

**mediastinum** The thoracic region between the lungs.

**medulla** The central portion of certain body organs, such as the kidney.

**medulla oblongata** The inferior part of the brain stem.

**melanin** The dark pigment formed by melanocytes; it imparts color to the skin and hair.

**meninges** The three protective coverings of the central nervous system.

**menopause** The end of menstrual cycles.

**menstruation** The periodic discharge of blood, secretions, and tissue from the female uterus in the absence of pregnancy.

**mesenteries** Double-layered extensions of the peritoneum supporting most abdominal organs.

**mesoderm** The central germ layer of the gastrula that forms the skeleton and muscles of the body; the other germ layers are the endoderm and the ectoderm.

**metabolism** The sum total of all the chemical reactions occurring in body cells; the combined reactions of anabolism and catabolism.

**micelles** Aggregates of bile salts and phospholipid molecules that dissolve fatty acids previous to transport.

**microglia** Neuroglial cells that engage in phagocytosis.

**microvilli** Microscopic projections on the free surface of some epithelial cells.

**micturition** The process of urination.

**midbrain** The region of the brain stem between the diencephalon and the pons.

**mineralocorticoids** Steroid hormones of the adrenal cortex that regulate mineral metabolism and fluid balance.

**mixed nerves** Nerves containing the processes of motor and sensory neurons.

**monocyte** A large agranulated white blood cell; it has a kidney-shaped nucleus.

**motor nerves** Nerves that carry impulses away from the brain and spinal cord.

**mucous membranes** Membranes forming the linings of body cavities open to the external environment.

**mucus** A thick fluid, secreted by mucous glands and mucous membranes to moisten the free surface of the membrane.

**multipolar neuron** A neuron having one long axon and numerous dendrites.

**muscle twitch** A single rapid contraction of a muscle.

**myelin sheath** The fatty insulating sheath that surrounds all but the smallest nerve fibers.

**myocardium** The major muscle layer of the heart wall composed of cardiac muscle.

**myofibril** A rodlike bundle of contractile filaments (myofilaments) found in muscle cells.

**myofilament** The filament found in muscle cells; two types of myofilaments are actin and myosin myofilaments.

**myoglobin** The oxygen-binding pigment found in muscle cells.

**myometrium** The thick musculature of the uterus.

**myosin** One of the principal proteins found in muscle.

**nephron** The structural and functional unit of the kidney; it consists of Bowman's capsule, distal and proximal tubules, and the loop of Henle.

**neuroglia** Cells of neural tissue used for support and insulation; they include astrocytes, microglia, and oligodendrocytes.

**neurohypophysis** The posterior pituitary gland.

**neuromuscular junction** The region where a motor neuron comes in close contact with a skeletal muscle cell.

**neuron** The cell of the nervous system specialized to generate and transmit nerve impulses.

**neurotransmitter** A chemical released by neurons to stimulate or inhibit receptors at the opposite side of the synapse.

**neutrophil** A type of granular white blood cell having a multilobed nucleus and specializing in phagocytosis.

**occipital** The area at the back of the head.

**oocyte** An immature egg cell.

**oogenesis** The process of ovum formation.

**optic chiasma** The partial crossover of fibers of the optic nerves.

**organ** A body part formed by two or more tissues and having a certain function.

**organ system** A group of organs working together to perform a certain body function.

**origin** The attachment site of a muscle remaining relatively fixed during contraction.

**osmoreceptor** A structure sensitive to osmotic pressure.

**osmosis** The diffusion of water through a semipermeable membrane from a region of higher concentration to a region of lower concentration.

**ossicles** The three bones of the middle ear: the malleus, incus, and stapes.

**osteoblast** An active bone-forming cell.

**osteoclast** A large cell that reabsorbs or breaks down bone matrix.

**osteocyte** A mature bone cell.

**osteogenesis** The process of bone formation.

**otolith** A calcium carbonate particle in the otolithic membrane of the semicircular ear canals that functions in the equilibrium process.

**ovary** The female sex organ in which ova are produced; the female gonad.

**ovulation** The ejection of an immature egg cell from a follicle in the ovary.

**ovum** The egg cell.

**oxyhemoglobin** The oxygen-bound form of hemoglobin.

**oxytocin** The hormone released by the posterior pituitary gland to stimulate uterine contractions during childbirth and the ejection of milk during nursing.

**pancreas** The gland located beneath the stomach and between the duodenum and spleen; it produces exocrine secretions to encourage the digestion of proteins, carbohydrates, and fats and endocrine secretions, such as insulin and glucagon.

**parasympathetic** The division of the autonomic nervous system that regulates internal functions.

**parathyroid glands** Small endocrine glands located on the posterior aspect of the thyroid gland.

**parathyroid hormone (PTH)** A hormone released by the parathyroid glands that regulates blood calcium levels.

**parietal cell** The cell of the gastric gland that produces hydrochloric acid.

**parietal pleura** The outer layer of the double membrane enclosing the lungs.

**pectoral** Pertaining to the chest region.

**pectoral girdle** The bones attaching the upper limbs to the axial skeleton.

**pelvic girdle** The bones attaching the lower limbs to the axial skeleton.

**penis** The male organ of copulation and urination.

**pericardium** The double-layered membrane enclosing the heart.

**perineum** The region of the body extending from the anus to the scrotum in males and to the vulva in females.

**periosteum** The membrane that covers and nourishes the bone.

**peripheral nervous system (PNS)** The subdivision of the nervous system consisting of nerves and ganglia lying outside the central nervous system (CNS).

**peristalsis** The wavelike contractions that move food materials through the gastrointestinal tract.

**peritoneum** The serous membrane lining the interior of the abdominal cavity and covering the surfaces of abdominal organs.

**pH** The measure of the relative acidity of alkalinity of a solution; it is inversely proportional to the hydrogen ion concentration of a solution.

**phagocytosis** The process of engulfing foreign particles by phagocytes.

**pharynx** The area extending from the nasal cavities to the esophagus; the throat area.

**photoreceptor** The specialized receptor cells that respond to light energy.

**pia mater** The innermost of the meninges.

**pineal gland** A tiny gland in the roof of the third ventricle of the brain; it is believed to secrete hormones that have to do with day-night cycles of activity.

**pinocytosis** The process by which phagocytes engulf small volumes of fluid.

**pituitary gland** The endocrine gland located at the base of the brain; it produces and/or stores numerous hormones in anterior and posterior portions.

**placenta** The temporary organ formed from both fetal and maternal tissues to provide nutrients and oxygen to the developing fetus, remove metabolic wastes, and produce hormones.

**plasma** The fluid component of blood containing the clotting agents.

**platelet** A formed element of blood involved in clotting; also called a thrombocyte.

**pleura** The two-layered serous membrane lining the thoracic cavity and covering the external surface of the lung.

**plexus** A network where nerve fibers converge.

**pons** The part of the brain stem connecting the medulla to the midbrain.

**posterior** The back region of an organism, organ, or part; the dorsal surface.

**postganglionic neuron** An autonomic motor neuron having its cell body in a peripheral ganglion and projecting its axon to an effector gland or muscle.

**preganglionic neuron** An autonomic motor neuron having its cell body in the central nervous system (CNS) and projecting its axon to a peripheral ganglion.

**progesterone** The hormone responsible for preparing the uterus for the fertilized ovum; it is produced by the corpus luteum and placenta.

**pronation** Inward rotation of the forearm causing the palms to face downward.

**proximal** Toward the end of a limb attached to the axial skeleton; the opposite of distal.

**pulmonary** Pertaining to the lungs.

**pulmonary circuit** The system of blood vessels allowing transport to and from the lungs.

**pulse** The rhythmic expansion and recoil of arteries resulting from heart contraction.

**pupil** The opening in the center of the iris through which light enters the eye.

**purkinje fibers** Modified cardiac muscle fibers of the conduction system of the heart.

**pyloric region** The distal portion of the stomach where it joins the duodenum.

**ramus** A branch of a nerve, artery, vein, or bone.

**receptor** A peripheral nerve ending specialized for response to particular types of stimuli.

**reflex** An automatic reaction to a stimulus.

**renal** Pertaining to the kidneys.

**renin** The substance released by the kidneys and involved with the substance angiotensin in the regulation of the blood pressure.

**rete** A network generally composed of nerve fibers or blood vessels.

**reticular formation** The system that regulates sensory input to the cerebral cortex.

**retina** The neural layer at the back of the eyeball containing its photoreceptors.

**rods** One of two types of photosensitive cells in the retina of the eye.

**rugae** Ridges in the mucosa of the stomach.

**saggital plane** A longitudinal plane dividing the body or parts into right and left portions.

**sarcomere** The contractile unit of skeletal muscle; it contains myofilaments composed of actin and myosin.

**sarcoplasmic reticulum** The specialized endoplasmic reticulum of muscle cells.

**sclera** The outer fibrous white layer of the eyeball.

**scrotum** The external sac in which the testes are found.

**sebaceous gland** An epidermal gland that produces sebum.

**sebum** The oily secretion of the sebaceous glands.

**semen** The fluid mixture of sperm and secretions produced by the male accessory reproductive glands.

**semilunar valves** Valves in the aorta and pulmonary artery that prevent blood from returning to the ventricles.

**seminiferous tubules** Coiled tubules within the testes where sperm cells are produced.

**sensory nerve** A nerve that carries nerve impulses toward the central nervous system (CNS).

**sensory neuron** A neuron that forms nerve impulses after stimulation and carries those impulses toward the central nervous system (CNS).

**serous fluid** The watery fluid secreted by cells of a serous membrane.

**serum** Straw-colored fluid of the blood containing no blood cells or clotting agents.

**sinoatrial (SA) node** Specialized myocardial cells in the wall of the right atrium; called the pacemaker.

**sinus** An air-filled cavity in certain cranial bones; also, a wide channel through which blood or lymph flows.

**smooth muscle** Muscle consisting of nonbanded muscle cells; it is found in visceral organs; it is not under voluntary control.

**sphincter** The muscle surrounding an opening to an organ and acting as a valve.

**spinal nerves** The 31 pairs of nerves arising from the spinal cord.

**squamous cells** Flat, thin cells forming the surface of certain epithelial tissues.

**steroids** A group of lipid substances having complex carbon rings and found in certain hormones.

**striated muscle** Muscle consisting of banded (striated) muscle fibers; it is found in cardiac and skeletal muscle.

**stroke volume** The amount of blood pumped out of a ventricle during a contraction.

**subcutaneous** Referring to beneath the skin.

**sudoriferous gland** An epidermal gland that produces sweat.

**sulcus** A shallow furrow on the brain.

**superficial** Referring to the region close to the body surface.

**superior** Referring to the head or upper body regions.

**supination** The outward rotation of the forearm causing the palms to face upward.

**suture** An immovable joint between skull bones.

**sympathetic nervous system** The division of the autonomic nervous system that activates body systems during time of stress.

**symphysis** A joint where the bones are connected by fibrocartilage.

**synapse** The junction between two neurons or between a neuron and a muscle cell.

**synarthrosis** An immovable joint.

**synovial fluid** The fluid secreted by the synovial membrane, which is the membrane that encapsulates a movable joint such as at the knee or elbow; it is used to lubricate the joint surfaces and nourish the articular cartilages.

**systemic** Referring to the entire body beneath the skin.

**systole** The period when either of the ventricles is contracting.

**T-lymphocytes** Lymphocytes involved in cell-mediated immunity; they include helper, killer, suppresser, and memory cells; they are also called T-cells.

**taste buds** Sensory receptors where gustatory cells are located in the tongue.

**tendon** The cord of dense fibrous tissue attaching muscle to bone.

**testis** The male sperm-producing sex organ.

**thalamus** A mass of gray matter in the floor of the brain.

**thermoreceptor** A body receptor sensitive to alterations in temperature.

**thoracic duct** The large duct that receives lymph from the lower body and the left side of the head and the thorax.

**thorax** The body cavity above the diaphragm.

**thrombin** The enzyme that converts fibrinogen to fibrin and, thereby, forms a clot.

**thrombocytes** An alternative expression for platelets; cell fragments participating in blood clotting and coagulation.

**thymus gland** The endocrine gland where T-lymphocytes are modified to their final form; it produces thymosins.

**thyroid gland** The endocrine gland in the neck tissues that produces thyroxin and other hormones.

**tissue** A group of similar cells specialized to perform a specific function.

**trachea** An alternative expression for the windpipe; a cartilage-reinforced tube carrying air from the larynx to the bronchi.

**transverse process** The pair of projections extending laterally from each neural arch of a vertebra.

**trochanter** A large blunt process on a bone where a muscle attaches.

**tuberosity** A broad process on a bone where a muscle attaches.

**twitch** A brief contraction of a muscle.

**umbilical cord** The structure connecting the placenta and the fetus; it carries nutrients, gases, and waste products between the mother and the fetus.

**unipolar neuron** A neuron having one process extending from the cell body.

**urea** The nitrogen-containing waste product of amino acid metabolism produced in the liver and excreted in urine.

**ureter** The tube carrying urine from the kidney to the bladder.

**urethra** The tube through which urine passes from the bladder to the exterior.

**uterine tube** The tube through which the ovum is transported to the uterus; it is also known as the Fallopian tube.

**uterus** The thick-walled hollow organ that receives and nourishes the fertilized egg; the organ where the embryo/fetus develops.

**uvula** The flap of tissue hanging from the soft palate into the pharynx.

**vagina** The muscular organ leading from the uterus to the vulva.

**vas deferens** The tube leading from the epididymis near the testis; also called the ductus deferens.

**vasoconstriction** The narrowing of blood vessels.

**vasodilation** The widening of blood vessels.

**vasopressin** The hormone that controls water reabsorption in the kidney tubules; it is also known as antidiuretic hormone (ADH).

**veins** Blood vessels returning blood toward the heart.

**ventral** Referring to the front region of the body; the anterior.

**ventricles** Paired heart chambers that pump blood to the pulmonary artery or aorta; also, fluid-filled chambers of the brain.

**venule** A small vein.

**villus** A fingerlike projection of the intestinal wall that increases its surface area for absorption.

**visceral** Referring to organs of the thoracic and abdominopelvic cavity.

**vital capacity** The largest volume of air that can be brought into the lungs by forcied inspiration.

**vitreous humor** The jellylike substance in the posterior cavity of the eye.

**vulva** The external genitalia of the female reproductive tract.

**white matter** The white substance of the central nervous system; it is found at the base of the brain and the exterior of the spinal cord and is composed of myelinated nerve fibers.

**xiphoid process** The lowest portion of the sternum.

**zygote** The fertilized egg.

# INDEX

521